T0231491

ELECTROCHEMICAL ENERGY CONVERSION AND STORAGE SYSTEMS FOR FUTURE SUSTAINABILITY

Technological Advancements

ELECTROCHEMICAL ENERGY CONVERSION AND STORAGE SYSTEMS FOR FUTURE SUSTAINABILITY

Technological Advancements

Edited by

Aneeya Kumar Samantara, PhD
Satyajit Ratha, PhD

APPLE
ACADEMIC
PRESS

First edition published [2021]

Apple Academic Press Inc.
1265 Goldenrod Circle, NE,
Palm Bay, FL 32905 USA

4164 Lakeshore Road, Burlington,
ON, L7L 1A4 Canada

CRC Press
6000 Broken Sound Parkway NW,
Suite 300, Boca Raton, FL 33487-2742 USA

2 Park Square, Milton Park,
Abingdon, Oxon, OX14 4RN UK

First issued in paperback 2021

© 2021 Apple Academic Press, Inc.

Apple Academic Press exclusively co-publishes with CRC Press, an imprint of Taylor & Francis Group, LLC

Reasonable efforts have been made to publish reliable data and information, but the authors, editors, and publisher cannot assume responsibility for the validity of all materials or the consequences of their use. The authors, editors, and publishers have attempted to trace the copyright holders of all material reproduced in this publication and apologize to copyright holders if permission to publish in this form has not been obtained. If any copyright material has not been acknowledged, please write and let us know so we may rectify in any future reprint.

Except as permitted under U.S. Copyright Law, no part of this book may be reprinted, reproduced, transmitted, or utilized in any form by any electronic, mechanical, or other means, now known or hereafter invented, including photocopying, microfilming, and recording, or in any information storage or retrieval system, without written permission from the publishers.

For permission to photocopy or use material electronically from this work, access www.copyright.com or contact the Copyright Clearance Center, Inc. (CCC), 222 Rosewood Drive, Danvers, MA 01923, 978-750-8400. For works that are not available on CCC please contact mpkbookspermissions@tandf.co.uk

Trademark notice: Product or corporate names may be trademarks or registered trademarks and are used only for identification and explanation without intent to infringe.

Library and Archives Canada Cataloguing in Publication

Title: Electrochemical energy conversion and storage systems for future sustainability: technological advancements / edited by Aneeya Kumar Samantara, PhD, Satyajit Ratha, PhD.

Names: Samantara, Aneeya Kumar, editor. | Ratha, Satyajit, editor.

Description: Includes bibliographical references and index.

Identifiers: Canadiana (print) 20200295772 | Canadiana (ebook) 20200295985 | ISBN 9781771888851 (hardcover) | ISBN 9781003009320 (ebook)

Subjects: LCSH: Electric power production from chemical action. | LCSH: Electrochemistry. | LCSH: Fuel cells. | LCSH: Energy storage.

Classification: LCC TK2901 .E44 2021 | DDC 621.31/242—dc23

Library of Congress Cataloging-in-Publication Data

....

CIP data on file with US Library of Congress

....

ISBN: 978-1-77188-885-1 (hbk)
ISBN: 978-1-77463-898-9 (pbk)
ISBN: 978-1-00300-932-0 (ebk)

About the Editors

Aneeya Kumar Samantara, PhD
School of Chemical Sciences, National Institute of Science Education and Research, Bhubaneswar, India

Aneeya Kumar Samantara, PhD, is presently working as a postdoctoral fellow in the School of Chemical Sciences, National Institute of Science Education and Research, Khordha, Odisha, India. Additionally, Dr. Samantara is working as an advisory board member for a number of publishing houses. Recently he joined as community board member of the *Materials Horizon*, one of the most leading journals in materials science of the Royal Society of Chemistry, London. He has authored 23 peer-reviewed international journal articles, 10 books (with Springer Nature, Nova Science, Arcler Press, Intech Open), and four book chapters. Several publications are in the press now and expected to be published soon. He pursued his PhD at the CSIR-Institute of Minerals and Materials Technology, Bhubaneswar, Odisha, India. Before joining the PhD program, he completed his master of philosophy (MPhil) degree in chemistry at Utkal University and his master in science degree in advanced organic chemistry at Ravenshaw University, Cuttack, Odisha, India. Dr. Samantara's research interests include the synthesis of metal oxide/chalcogenides and graphene composites for energy storage and conversion applications.

Satyajit Ratha
School of Basic Sciences, Indian Institute of Technology, Bhubaneswar, India

Satyajit Ratha, PhD, pursued his PhD at the Indian Institute of Technology Bhubaneswar, India. Recently he joined as community board member of *Nanoscale Horizon*, one of the most leading journals in materials science of the Royal Society of Chemistry, London. Prior to joining the Indian Institute of Technology Bhubaneswar, he received his bachelor of science degree, first class honors, from Utkal University, and his master of science degree from Ravenshaw University. Dr. Ratha's research interests include two-dimensional semiconductors, nanostructure synthesis and applications, energy storage devices, and supercapacitors. He has authored and coauthored over 21 peer-reviewed international journals and one book (Springer Nature).

Dedication

Dr. Aneeya K. Samantara would like to dedicate this work to Parents, Mr. Braja Bandhu Dash, Mrs. Swarna Chandrika Dash, and his beloved wife Elina.

Dr. Satyajit Ratha would like to dedicate this work to Parents, Mrs. Prabhati Ratha and Mr. Sanjaya Kumar Ratha.

Contents

9.   Methanol and Formic Acid Oxidation: Selective Fuel Cell Processes... 289

Tapan Kumar Behera, Pramod Kumar Satapathy, and Priyabrat Mohapatra

Contributors

Anil Arya
Department of Physical Sciences, Central University of Punjab, Bathinda – 151001, Punjab, India

Tapan Kumar Behera
North Orissa University, Baripada – 757003, Odisha, India

Surjendu Bhattacharyya
Institute of Atomic and Molecular Sciences (IAMS), Academia Sinica, P.O. Box 23-10617, Taipei – 10617, Taiwan

Avijit Biswal
College of Engineering and Technology, Ghatikia, Bhubaneswar – 751029, Odisha, India

Syed Mukulika Dinara
National Post-Doctoral Fellow, Indian Institute of Technology, School of Basic Sciences, Bhubaneswar, Odisha, E-mail: dr.smdinara@gmail.com

Anoop Kumar Kushwaha
School of Basic Sciences, Indian Institute of Technology, Bhubaneswar, Odisha – 752050, India

Avinna Mishra
College of Engineering and Technology, Ghatikia, Bhubaneswar – 751029, Odisha, India

Biswajit Mishra
Faculty of Chemical Sciences, Shri Ramswaroop Memorial University, Lucknow-Deva Road, Uttar Pradesh – 225003, India, E-mail: sahebm@gmail.com

Priyabrat Mohapatra
C.V Raman College of Engineering, Bhubaneswar – 752054, Odisha, India

Saswat Mohapatra
Department of Physics, IIT(ISM), Dhanbad, Jharkhand, India

Satyajit Ratha
School of Basic Sciences, Indian Institute of Technology, Bhubaneswar, Odisha – 752050, India, E-mail: satyajitratha89@gmail.com

Mihir Ranjan Sahoo
School of Basic Sciences, Indian Institute of Technology, Bhubaneswar, Odisha, India, E-mail: mrs10@iitbbs.ac.in

Pooja Sahoo
Department of Physics, IIT(ISM), Dhanbad, Jharkhand, India

Aneeya Kumar Samantara
Post-Doctorate Fellow, School of Chemical Sciences, National Institute of Science Education and Research, Bhubaneswar, Khordha, Odisha – 752050, India,
E-mails: cmrjitu@gmail.com, aneeya1986@gmail.com

Pramod Kumar Satapathy
North Orissa University, Baripada – 757003, Odisha, India

A. L. Sharma
Department of Physical Sciences, Central University of Punjab, Bathinda – 151001, Punjab, India,
E-mail: alsharma@cup.edu.in

Akash Sharma
Department of Physics, IIT(ISM), Dhanbad, Jharkhand, India, E-mail: akash.physics@gmail.com

Alfa Sharma
Discipline of Metallurgy Engineering and Materials Science, IIT, Indore, Madhya Pradesh, India,
E-mail: alfasharma89@gmail.com

Abbreviations

[NH$_2$C$_3$MIm][Br]	1-aminopropyl-3-methylimidazolium bromide
3DGN	three-dimensional graphene
ABPE	applied bias photon-to-current efficiency
AB-STH	applied bias solar to hydrogen conversion efficiency
AC	activated carbon
AEMFCs	anion-exchange membrane fuel cells
AFC	alkaline fuel cell
AFM	atomic force microscope
AgNPs	silver nanoparticles
ALD	atomic layer deposition
APCE	absorbed photon conversion efficiency
aq.	aqueous
AuNPs	gold nanoparticles
AZO	Al doped ZnO
BN	boron nitride
BZNs	branched ZnO nanofibers
C$_2$H$_4$	ethylene
C$_2$MIm	1-ethyl-3-methylimidazolium
C$_3$N$_4$	carbon nitride
CAES	compressed air energy storage
CB	conduction band
CBD	chemical bath deposition
CBM	conduction band minimum
CCNT	coupled carbon nanotube
CdS	cadmium sulfides
CE	counter electrode
CH$_3$OH	methanol
CH$_4$	methane
CHE	computational hydrogen electrode
CIS	complex impedance spectroscopy
CNTs	carbon nanotubes
CO	carbon monoxide
CO$_2$	carbon dioxide

COE	CO electrooxidation
CPE	composite polymer electrolyte
CPMEA	cross-linked poly(ethylene glycol) methyl ether acrylate
CRR	carbon dioxide reduction reactions
CTAB	cetyl trimethyl ammonium bromide
Cu_2O	copper oxide
$CuCl_2$	cupric chloride
CV	cyclic voltammogram
CVD	chemical vapor deposition
D	diffusion constant
DAFCs	direct alcohol fuel cells
dcbpy	4,4′-dicarboxylic acid-2,2′-bipyridine
DEC	diethyl carbonate
DEMS	differential electrochemical mass spectrometry
DENs	dendrimer-encapsulated nanoparticles
DFAFC	direct formic acid fuel cell
DFT	density functional theory
DFTFSI	[(difluoromethanesulfonyl)(trifluoromethanesulfonyl)imide]
DI	deionized
DMC	dimethyl carbonate
DMF	dimethyl formamide
DOS	density of states
DSC	differential scanning calorimetry
DST	department of science and technology
EC	electrochemical
EC	ethylene carbonate
ECR	electrochemical CO_2 reduction
ECs	electrochemical capacitors
ECSA	electrochemically active surface area
EDL	electrical double layer
EDLC	electric double layer capacitor
EES	electrical energy storage
EFF	efficiency for formate formation
EIS	electrochemical impedance spectroscopy
EMC	ethyl methyl carbonate
EQE	external quantum efficiency
ESD	electrostatic spray deposition

ESW	electrochemical stability window
ET	electron transfer
FA	formic acid
FAO	formic acid oxidation
Fe_2O_3	ferric oxide
FESEM	field emission scanning electron microscopy
FFI	fraction of free ions
FIP	fraction of ion pairs
FLC	fluorinated carbonates
FTIR	Fourier transform infrared
FTO	fluorine doped tin oxide
g	gram
GaP	gallium phosphide
GLAD	glancing-angle deposition
GO	graphene oxide
H	heyrovsky
H_2	hydrogen
H_2O	water
H_2O_2	hydrogen peroxide
H_2SO_4	sulfuric acid
H_3PO_4	phosphoric acid
$HAuCl_4$	chloroauric acid
HBE	hydrogen binding energy
$HClO_4$	perchloric acid
HCOOH	formic acid
HC-STH	half-cell solar to hydrogen conversion efficiency
HER	hydrogen evolution reaction
HMTA	hexamethylenetetramine
HNDR	hidden negative differential resistances
HOMO	highest occupied molecular orbital
HOR	hydrogen oxidation reaction
HPE	hybrid polymer electrolyte
HSE	hybrid solid electrolyte
IF	inorganic fullerene
IL	ionic liquid
$InCl_3$	indium chloride
InP	indium phosphide
IPCE	incident photon-to-current efficiency
IQE	internal quantum efficiency

IRENA	International Renewables Energy Agency
ISS	*in-situ* synthesis
ITO	indium tin oxide
$KHCO_3$	potassium bi-carbonate
LA	longitudinal acoustic
LATP	$Li_{1.3}Al_{0.3}Ti_{0.7}(PO_4)_3$
LBL	layer by layer
LCP	liquid crystal polymer
LED	light emitting diode
LIBs	lithium ion batteries
LIG	laser-induced graphene
LLZNO	$Li_7La_3Zr_{1.75}Nb_{0.25}O_{12}$
LLZO	$Li_7La_3Zr_2O_{12}$
LLZTO	$Li_{6.75}La_3Zr_{1.75}Ta_{0.25}O_{12}$
LSPR	localized surface plasmon resonance
LSV	linear sweep voltammetry
LUMO	lowest unoccupied molecular orbital
M	mole
MBE	molecular beam epitaxy
MCE	metal-complex electrocatalyst
MCFC	molten carbonate fuel cell
MEMS	micro-electro-mechanical systems
ML	monolayer
mM	mili molar
MMM	mechanical-mixing method
mmol	mili mole
MOR	methanol oxidation reaction
MSCs	micro-supercapacitors
mV	mili volt
mW	mili watt
MWCNT	multi-walled carbon nanotube
Na_2SO_3	sodium sulfite
Na_2SO_4	sodium sulfate
NaOH	sodium hydroxide
NCM	$LiNi_xCoyMn_zO_2$
NCNT	nitrogen-doped carbon nanotube
NDR	negative differential resistance
NFs	nanofibers
NHE	normal hydrogen electrode

NiAl-LDH	nickel-aluminum double layered hydroxide
Ni-Cd	nickel-cadmium
Ni-MH	nickel metal hydride
nM	nano molar
nm	nanometer
NMR	nuclear magnetic resonance
NO_2	nitrogen dioxide
NPs	nanoparticles
NRs	nanorods
NWs	nanowires
O_2	oxygen
OAD	oblique-angle deposition
OER	oxygen evolution reaction
OLA	oleylamine
OLJ	overlapping junction
ORR	oxygen reduction reaction
PAFC	phosphoric acid fuel cell
PANI	polyaniline
PC	propylene carbonate
PCE	photon-to-current conversion efficiency
PCM	polarizable continuum model
PCPSE	polymer/ceramic membrane/polymer sandwich electrolyte
PDMS	poly(dimethylsiloxane)
PdnDs	Pd dendrite shaped nanostructures
PE	polymer electrolyte
PEC	photoelectrochemical
PEDOT	poly(3,4-ethylenedioxythiophene)
PEGDA/[EMIM][TFSI]	poly(ethylene glycol) diacrylate/1-ethyl-3-methyl-imidazolium bis-(trifluoromethylsulfonyl)imide
PEM	polymer electrolyte membrane
PEM	proton exchange membrane
PEMFC	proton exchange membrane fuel cell
PEN	polyethylene naphthalate
PEO-LiTFSI	particles were dispersed in the polymer matrix
PET	polyethylene terephthalate
PGM	platinum group metals
PHES	pumped hydroelectric energy storage
PLD	pulsed layer deposition

PMMA	polymethyl methacrylate
PMMA/[EMIM][NTf2]	polymethyl methacrylate/1-ethyl-3-methylimid-azolium bis(trifluoromethylsulfonyl)imide
ppm	parts per million
Pt	platinum
PTFE	poly(tetrafluoroethylene)
PtNi	platinum-nickel
PtNPs	platinum nanoparticles
PVA	poly vinyl alcohol
PVP	polyvinylpyrrolidone
RE	reference electrode
RGO	reduced graphene oxide
RHE	reversible hydrogen electrode
S	symens
SC	semiconductor
SCE	saturated calomel electrode
SCMPU	self-charging micro-supercapacitor power unit
SCs	supercapacitors
SDS	sodium dodecyl sulfate
SEI	solid electrolyte interface
SEM	scanning electron microscope
SERB	science and engineering research board
SHE	standard hydrogen electrode
SiC	silicon carbide
SMD	solvation model on density
SMES	superconducting magnetic energy storage
SOFC	solid oxide fuel cell
SPA	sodium poly acrylate
SPE	solid polymer electrolyte
SPR	surface plasmon resonance
SS	stainless-steel
STH	solar to hydrogen
SWCNT	single walled carbon nanotube
T	Tafel
TA	transverse acoustic
TEM	transmission electron microscope
TENGs	triboelectric nanogenerators
TFSI	(trifluoromethanesulfonyl)imide
TGA	thermogravimetric analysis

TiO_2	titanium dioxide
TM	transition metal
TMCs	transition metal chalcogenides
TMDs	transition-metal dichalcogenides
TMOS/FA/EMITFSI	tetramethyl orthosilicate/formic acid/1-ethyl-3-methylimidazolium bis(trifluoromethylsulfonyl) imide
TMPs	transition metal phosphides
TOF	turnover frequency
TOP	trioctylphosphine
TOPO	trioctylphosphine oxide
TPa	tera pascal
UV	ultraviolet
UV-Vis	ultraviolet-visible
VB	valence band
VBM	valence band maximum
V_{OC}	open circuit voltage
VSe_2	vanadium selenide
VTF	Vogel-Tamman-Fulcher
W	watt
WC	tungsten carbide
WE	working electrode
WO_3	tungsten oxide
XPS	x-ray photoelectron spectroscopy
XRD	x-ray diffraction
ZnO NR	ZnO nanorod
ZnO NWs-OLJ	ZnO nanowires overlapping junction
ZnO	zinc oxide
ZnTe	zinc telluride

Symbols

μ	mobility
μmol	micro-molar
ΔG_f^0	standard Gibbs free energy of formation
λ	wavelength
CO_{ads}	adsorbed CO
e^-	electron
E^0_{redox}	standard reduction potential
E_{CBM}	energy difference between the conduction band minimum
eV	electro volt
F	farad
Hz	hertz
I^b	backward current
I^f	forward current
kJ	kilojoule
OH_{ads}	adsorbed hydroxyl groups
t_{cation}	cation transference number
$t_{electronic}$	electronic transference
T_g	glass transition temperature
t_{ion}	ionic transference
U_T	output cell potential
V	Volmer
V	volt

Forewarded

Foreword

The wide range of topics that have been put together in this book provides much significant information regarding the current state of energy generation and storage opportunities. The chapter-wise discussions on topics like photoelectrochemical catalysis by ZnO, hydrogen oxidation reaction (HOR) for fuel cell application, or realizing miniaturized energy storage devices in the form of micro-supercapacitors (MSCs) can invoke one's knowledge regarding the underlying mechanisms and acquire first-hand information on how to overcome some of the critical bottlenecks to achieve long-term and reliable energy solutions. Furthermore, the detailed synthesis process that has been tried and tested over time through rigorous attempts of a large number of researchers can help us in selecting the most effective and economical way to achieve maximum output and efficiency, without going through time-consuming and complex steps, whether it is about getting suitable electro-catalyst for fuel cell, hydrogen oxidation, and water splitting or fabrication of a suitable electrochemical energy storage devices that can achieve a balance between energy density and power density. Also, the theoretical analysis and computational results would further corroborate the experimental findings to help us get close to a better and more reliable energy solution. This book is, therefore, helpful in gathering basic understanding as well as some of the complex reaction mechanisms in a brief and vivid manner, which can be beneficial for a wide range of researchers and students working in areas closely related to energy harvesting and storage.

—**Aneeya Kumar Samantara, PhD**
Satyajit Ratha, PhD

Preface

Energy harvesting and storage are the two most important sectors that have gained significant attention from the scientific community. The dependency on fossil fuels limits the process of energy generation to a single domain, and therefore urgent actions are to be taken to realize the possibility of exploring several promising energy generation/harvesting options. In this context, renewable energy sources have gathered significant attention in recent decades, and revolutionary technologies have been implemented to enhance the overall efficiency in the generation of energy carriers, their transmission, and distribution among domestic consumers as well as industrial sectors.

However, the challenge remains due to the discreetness of renewable resources. Therefore, many laboratory-scale methods have been devised to produce synthetic fuels by consuming hazardous gases like CO, CO_2, etc. It is to be noted that, since these synthetic fuels do require a certain amount of threshold energy in order to get converted to meaningful energy, therefore, in most cases, catalysts are employed to overcome the same. There are state-of-art catalysts, which can effectively generate hydrogen and oxygen, or they can reduce carbon dioxide to produce organic fuels, which can be recycled in the system many times. However, these stete-of-art catalysts comprise precious metals; therefore, scalability is constrained. Storage of energy, on the other hand, is an important aspect to look for if the world is aiming for complete eradication of fossil fuel in the coming years. This is possible if reliable and stable energy storage mechanisms can be developed in parallel with energy harvesting. Considering the fact that most of the energy sources produce electricity as the final form of energy carrier, which is then distributed locally or transported to distant locations, electrical energy storage (EES) has been the most significant part of research since the past few years. Current electrical energy storage is dominated by battery, especially Li-ion battery, which is not going to provide a long term solution towards EES, because of the limited reserve of lithium in the earth's crust.

Supercapacitors [electrochemical capacitors (ECs)] are next-generation energy storage devices that can act as a power delivery system as well as an energy reservoir. Their lightweight design and environmentally friendly benign composition make them suitable for a wide range of applications starting from miniaturized electronic appliances to power-hungry heavy electric vehicles.

In this book, we will be briefly discussing some of the well-known energy harvesting techniques and will discuss electrical energy storage using super-capacitors including fuel cells. This will bring significant insight regarding the current progress in the said field and a check for the plausible options and areas that require a steadfast approach from the scientific community.

—Aneeya Kumar Samantara, PhD
Satyajit Ratha, PhD

Introduction

This book introduces a few important aspects of energy harvesting and storage that have a significant impact on the current environmental degradation and can provide much-needed insights (both experimental and theoretical) to address the future energy crisis.

The book has nine chapters addressing different areas. The first chapter discusses the photoelectrochemical performance of ZnO and further improvement by treating the same with a range of dopants. This is essential in realizing enhanced photochemistry in the case of ZnO, which is a cost-effective compound and can be easily synthesized by following a wide range of synthesis protocols.

The second chapter provides a detailed analysis of the progress in the development of hybrid polymer nanocomposites and their application in energy generation and storage. Polymers are highly stable electrochemically active materials that can impart significant improvement alone or as supportive frameworks for various other compounds or hybrids. Therefore, it is essential to gain insights into the formation and working principles of these hybrid polymer nanocomposites in order to address some of the issues like stability and reliability.

The third chapter is about the investigation of alkali metal-ion batteries and electrolytes through computational analyses. This can provide information regarding the feasibility of a system that will form a stable battery configuration and, therefore, is critical as far as electrical energy storage (EES) is concerned.

The fourth chapter contains a brief discussion of hydrogen oxidation reaction (HOR) and its importance in the context of polymer electrolyte (PE) membrane fuel cells (PEMFC). Furthermore, the authors discuss the current progress and state-of-art technologies being used and provide a detailed comparison as to how the process can be realized on a large scale, especially to meet industry standards.

Oxidizing CO and reducing CO_2 through photoelectrochemistry techniques have been detailed in the fifth chapter. Here, the authors explain the necessity of reducing and recycling greenhouse gases like CO_2 and toxic gases like CO through electrochemistry.

In the sixth chapter, a detailed study regarding two-dimensional transition metal chalcogenides (TMCs) has been carried out by the authors along with various synthesis protocols, promising properties and their application in energy storage.

The seventh chapter focuses on a theoretical approach to explain the enhanced catalytic activities of several nanostructured materials by getting insights about their electronic structures through density functional theory (DFT).

The eighth chapter emphasizes the need for miniaturization in the energy storage sector through the fabrication of lightweight and flexible micro-supercapacitors (MSCs) using a myriad of suitable electrode materials.

The ninth and last chapter discusses the fuel cell processes by the oxidation of methanol and formic acid (FA) to produce effective energy.

These chapters provide a clear understanding of the background and working principles of a few well-known energy harvesting and storage technologies. Also, new concepts such as micro-supercapacitors, CO oxidation, and CO_2 reduction can provide futuristic opportunities for further research and development in the said categories.

CHAPTER 1

Metal Chalcogenide-Based Electrochemical Capacitors

SATYAJIT RATHA[1] and ANEEYA KUMAR SAMANTARA[2]

[1]*School of Basic Sciences, Indian Institute of Technology, Bhubaneswar, Odisha – 752050, India*

[2]*School of Chemical Sciences, National Institute of Science Education and Research, Bhubaneswar, Khordha, Odisha – 752050, India.*
E-mails: cmrjitu@gmail.com, aneeya1986@gmail.com

ABSTRACT

Transition metal chalcogenides (TMCs) are wonder materials that can provide path-breaking results in various fields of application. The level of ongoing research on these materials is expected to yield several promising aspects of these excellent performers, whether it is catalysis or charge trapping through the faradic process, could pave the way for a myriad of possibilities in realizing the green energy revolution. The presence of edge plane and basal plane and resulting electrochemistry makes these TMCs highly versatile, and thus can be taken for virtually any kind of applications. They are highly electro-active, and are abundantly available in the earth's crust, which can address the issues that are currently faced by the Li-ion industries (owing to the limited lithium reserve). In the context of energy storage, especially supercapacitors, these materials provide a wide range of choices. They can be combined with EDLC materials to form an asymmetric arrangement of electrodes that could boost the working potential window and strike a balance between energy and power density. In this chapter, we have discussed on different synthetic methodologies for the preparation of TMCs. Further, the application of these materials towards electrochemical energy storage has been discussed with an updated literature survey. We presume that this chapter will be helpful to boost the knowledge of the energy researchers on these energy storage systems.

1.1   INTRODUCTION

With an increase in the energy demand due to population surge, the current reserve of fossil fuels is under severe stress. As the global policymakers are moving towards renewable energy resources to reduce the stress and to cut carbon footprint as well as check atmosphere degradation due to the combustion of fossil fuels, there have been increased interests toward electrical energy storage (EES) (Chen et al., 2009). Electricity, as an energy carrier, is essential for both domestic as well as industrial purposes. However, these renewables cannot produce a continuous supply of electricity because of their intermittent nature and dependency on several environmental factors (Chen et al., 2009). Thus, to form a continuous loop of electricity supply from these renewable resources, we need efficient and long term storage mechanisms, and this has to be done on an urgent basis to ward off an inevitable energy crisis, which is otherwise going to grapple us in no distant future. There has been a large pool of research dedicated to electric energy storage, a major portion of which has been focused on batteries. For almost 4 decades, battery technology has been dominant in the field of electric energy storage (Nazri and Pistoia, 2008). A further boost to the battery technology has been imparted by the Li-ion system, which was recognized in the early 60s. Lithium is lightweight and thus has high specific energy as compared to other metals in the periodic table. It, therefore, provides high energy densities and compactness in contrast to traditional rechargeable batteries, i.e., Ni-metal hydride (Ni-MH), Ni-Cd, Lead-acid batteries.

Batteries, despite their high energy densities, have several limitations, which hinder their market penetration when it comes to miniaturized electronic devices and heavy power consuming electric vehicles. First, batteries use bulk electrodes which prevents compactness and pure redox-based activities have a significant detrimental effect on the power density they have to offer. Also, the limited reserve of lithium in the earth's crust automatically puts the future of Li-ion technology on the brink (Egbue and Long, 2012). Recent advances in the field of EES lead to the discovery of new concepts such as electrochemical capacitors (ECs) which have high power densities, and are ideal for power-consuming electric vehicles and smart grids. These ECs (otherwise known as supercapacitors or ultracapacitors) have excellent discharge rates, which is at least 5–10 folds higher than that of a Li-ion battery. Supercapacitors basically store charge through surface adsorption, thus don't require bulk electrodes (as in the case of batteries), and can be constructed in a compact and lightweight configuration. The high power densities of supercapacitors can be combined with the high energy densities of Li-ion batteries to create battery-supercapacitor-hybrid structures/devices which can be advantageous for both miniaturized electronics and heavy electrical vehicles.

The development of the concept behind supercapacitance started off with electrodes based on carbon or carbon derived materials. These materials have high specific surface area values, e.g., graphene has a theoretical specific surface area of approximately 3000 m^2/g (Geim and Novoselov, 2007). This enormous value of an area can accumulate a large number of charged species, enhancing the capacitance of the supercapacitor device. The charge storage occurs through the formation of a charge double-layer at the electrode/electrolyte interface. The process of double-layer formation follows the principle of physisorption, hence it is fast and occurs only at the electrode surface (Contribution of surface/bulk diffusion is infinitesimal). Electrostatic double-layer formation imparts high power densities; ideal for systems where the start and stop activities are dominant. However, to store energy, a device must possess the desired value of energy density, which is limited in the case of supercapacitors operating on a double-layer mechanism. To improve the energy density of supercapacitors, as discussed earlier, electrode materials with surface redox-based properties are essential (Zhao et al., 2011). In these materials, surface diffusion of charge takes place, which is obviously slower than the physical adsorption process in EDLC materials. Fast and reversible charge transfer coupled with high redox peaks enhances the overall capacitance and energy density in these non-EDLC materials. Non-EDLC materials can be of pseudocapacitive or pure faradic nature depending upon the rate of surface/bulk diffusion, determined by the relationship between the potential sweep rate and the corresponding peak current response. While the electrochemical properties of a wide range of metal oxide compounds have been investigated in the recent past for their charge storage capabilities, they have one major limitation and that is their low conductivity. Metal chalcogenides, on the other hand, possess higher intrinsic conductivities than metal oxides. Furthermore, when chalcogen group elements form compounds with the metal cation belonging to the transition group, we can take advantage of both high conductivity of the chalcogen component and myriad of interesting properties shown by the transition group element. These transition metal-based chalcogen compounds are extremely popular for their wide range of applications in various areas, e.g., electronics, and optics, (Wang et al., 2015; Zibouche et al., 2014) bio-sensing, (Wen et al., 2018) energy harvesting and storage, polymer electrolyte (PE) membrane fuel cells, and reaction catalysts. This report will briefly discuss about the physicochemical properties, electrochemistry, and supercapacitor application of metal chalcogenide compounds and their futuristic role in clean and green energy drive.

1.1.1   COMPOUND FORMATION: METAL CHALCOGENIDES

After graphene took the scientific community by storm, revealing a wide range of exceptional and unprecedented physical and chemical properties, it has been extensively used in EES devices including batteries and supercapacitors. High specific surface area, high carrier mobility (due to electron hopping), quantum confinement, excellent thermal and mechanical stability, high mechanical strength, etc., are some of the interesting characteristics of graphene (Geim, 2009). However, the direct application of graphene in charge storage is limited for several reasons. First, taking out one atom thick graphite layer (graphene) requires sophisticated and cumbersome processes. Secondly, the surface adsorption process in the case of graphene is considerably poor due to its more or less flat surface. To improve the adsorption property and to minimize the cost, a close derivative known as the reduced graphene oxide (RGO) was synthesized and has largely been used in lieu of pure graphene. Nevertheless, the terms graphene and RGO are generally synonymous to the scientific community, at least on a laboratory scale. In order to enhance the charge storage of a typical capacitor, we can either increase the electrode surface area, or minimize the separation between the two charging electrodes (using an extremely thin separator) or both. Now, graphene or RGO provide rich specific surface area, and are extremely thin, therefore making the supercapacitor devices capable of storing large quantities of charge, while being exceptionally lightweight as compared to batteries. The huge popularity of graphene is also due to its two-dimensional structure, possible because of the weak van der Waals force between the layers. These layers can be separated from each other through various chemical and mechanical exfoliation techniques (Hernandez et al., 2008). Consequently, research on analogous two-dimensional materials rapidly multiplied in the following years, discovering a large pool of layered metal chalcogenides, which have well defined band gaps (both direct and indirect) and excellent reduction-oxidation properties (Xue et al., 2017). These two-dimensional metal chalcogenide compounds instantly gained popularity due to their close analogy with graphene and additional characteristics that are ideal for application in a wide range of research fields, i.e., energy storage and harvesting, (Xue et al., 2017) electronics, (Zhao et al., 2015) optoelectronics, (Kuc, 2014), resistive memory devices, and so forth. There are a large number of reports on metal chalcogenides till date. However, the majority of them have been devoted to the study of the oxide compounds, while the study on compounds having S, Se, and Te are rather limited, for various reasons. Therefore, in this chapter, we would be discussing the synthesis, properties, and energy applications of metal chalcogens having sulfur and its sub-group elements, i.e., Se, and Te as the active components. These metal chalcogenides (including metal chalcogenide nanostructures) can be synthesized in a variety of ways. Taking advantage of their layered structures, bulk metal chalcogenides can be structurally tuned by the use of techniques such as chemical or mechanical exfoliation. Similarly, in order to benefit from the outstanding properties of nanomaterials, their functionalization is essential, as any application in materials and devices is hindered by difficulties in processing and manipulation. Only the attachment of appropriate chemical functionalities on the nanoparticle surface allows the tailoring of their properties for the respective application. Tailoring of the surface chemical bonds might as well lead to an optimized interaction of the nanoparticles (NPs) with solvent molecules, polymer matrices, or biomolecules. Therefore, the nanoengineering of particle surfaces is key for the design and tailored construction of innovative nanomaterials. This chapter includes the discussion on the growth mechanism of metal chalcogenides synthesized using different methods and various protocols for their surface functionalization to improve the processability in technological applications.

1.1.2 SYNTHETIC PROTOCOLS FOR METAL CHALCOGENIDE COMPOUNDS

TMCs, as discussed previously, have drawn significant attention from the application viewpoint, both at the laboratory and industrial scale. This has led to the development of a large number of synthesis protocols over time. They can be broadly classified into two categories, i.e., Bottom-up approach and Top-down approach. The former technique includes simple methods such as hydrothermal synthesis to sophisticated chemical vapor deposition (CVD) method. Other methods such as oxide-to-sulfide conversion, solvothermal method, and wet chemical method are also included. The top-down approach includes chemical and mechanical exfoliation techniques to reduce three-dimensional bulk structures to achieve two-dimensional planar structures that are critical for energy applications. Figure 1.1 shows some of the known techniques that are currently being used for the synthesis of 2D TMCs and analogous materials at both laboratory and industrial scales.

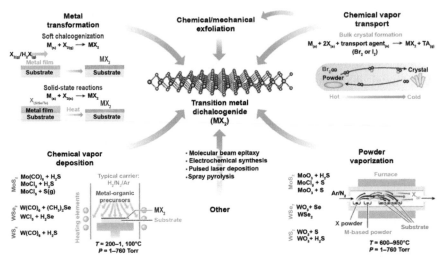

FIGURE 1.1 Summary of primary growth techniques for the formation of TMDC atomic layers. These methods include chemical vapor deposition, powder vaporization, metal transformation, chemical vapor transport, chemical exfoliation, pulsed laser deposition, molecular beam epitaxy, spray pyrolysis, and electrochemical synthesis.
Source: Reproduced with permission from Lin (2018). © Springer.

1.1.2.1 BOTTOM-UP APPROACH

One of the key steps to achieve materials with large areas and desired layered structures is to develop a number of effective synthesis protocols. For example, CVD has been phenomenal in producing high-quality few-or multi-layered graphene on the metal substrate (Suk et al., 2011). This method can be further extended for preparing other two-dimensional layered structures as well, especially materials like transition metal chalcogenides (TMCs). With this synthesis tool, one can control the growth, nucleation, and layer formation of TMCs by tuning a few reaction parameters. Reports on MoS_2, a widely studied member of the TMC group, suggests that it has been successfully synthesized through the CVD technique on several occasions with high yield and quality. This can be achieved through two approaches, i.e., either by depositing Mo on a suitable substrate/template first and subjecting it to the sulfurization process, or by letting Mo and S react with each other in gaseous states, inside the CVD chamber/tube. Thus, CVD can be implemented to prepare other members of the TMC group by taking suitable precursors and adjusting the reaction parameters.

1.1.2.1.1 Metal Chalcogenization

The two-step process in which MoS_2 is prepared involves the deposition of a nanometer-thick layer of Mo on to a SiO_2/Si substrate (through electron beam lithography), which is then exposed to vapor phase sulfurization process at a temperature of about 750°C (Zhan et al., 2012). Here, the two elements Mo and S directly react with each other to form MoS_2. The advantage of this technique is that the thickness of the deposited Mo layer determines the thickness of the finally prepared MoS_2. Thus, tuning the Mo layer during the lithography step, one can easily control the layered structure and quality of the MoS_2 prepared. However, molybdenum precursors with low melting points are to be used for the lithographic technique as pure Mo metal has a very high melting point which will limit the migration of Mo atoms, effectively suppressing the quality of the deposited layer. Furthermore, the ambient conditions for the sulfurization process could lead to the formation of both vertically aligned or in-plane grown MoS_2 and $MoSe_2$, depending on the sulfurization/selenization conditions in each case (Kong et al., 2013). Thus, the sulfur/selenium diffusion limits the growth process. The anisotropic structure of TMC layers facilitates the diffusion of sulfur/ selenium species along the van der Waals gaps that are induced by the vertical orientation of the TMDC layers. The as-formed TMCs are generally found with irregular edge growths indicating high surface energy values, which could be the reason behind their application as diverse catalysts. One of the most effective ways to achieve MoS_2 thin film is to employ MoO_3 as an alternative Mo precursor. This has an advantage due to the lower melting point of MoO_3, which is ~700°C. Through this technique, high-quality wafer-scale MoS_2 has been successfully grown on sapphire substrate by thermally evaporating MoO_3 prior to sulfurization and the final thickness is controlled by the amount of MoO_3 deposited in the first step of film growth (Pondick et al., 2018). In another report, crystalline micro-structures of MoO_2 were deposited on a SiO_2/Si substrate within an approximate temperature range of

650–850°C, under a weakly reducing environment of sulfur. Thereafter, the as-grown MoO_2 thin films were subject to the sulfurization process at 850°C to yield MoS_2 and the layers of the thin film can be controlled by varying the annealing temperature (Wang et al., 2013). Under certain conditions, heating MoO_2 thin films in vapor sulfur atmosphere can also result in the formation of MoS_2 nanoflowers with tens to hundreds petals self assembles within a single nanoflower (Li et al., 2003). Of special interest is screw-dislocation-driven growth of pyramidal structures, demonstrated for WSe_2 and MoS_2 (Chen et al., 2014; Yuan et al., 2015; Zhang et al., 2014). In such a case, the stacking sequence is different from the usual $2H$ stacking, which results in an increased second harmonic generation and photoluminescence intensities (Yuan et al., 2015; Zhang et al., 2014). It should be noted that while CVD growth of sulfides and selenides was developed rather easily, until very recently, $MoTe_2$ layer were only produced using exfoliation methods. There are several difficulties in growing stoichiometric $MoTe_2$ layers compared to other TMDCs. One of the reasons behind the failure of the vapor deposition technique to achieve $MoTe_2$ is that at high temperatures, instead of evaporating as a compound, $MoTe_2$ decomposes and loses Te as a vapor. These properties make it challenging to directly obtain atomically thin $MoTe_2$ films by physical vapor deposition, and there is very often a Te deficiency in as-prepared $MoTe_2$. Another distinguishing feature of $MoTe_2$ as compared to other TMCs is that the two polymorphs, i.e., 2H and 1T phases have very small energy differences, making it difficult to obtain any of the polymorphs in its pure crystal phase. Consequently, CVD growth of $MoTe_2$ was reported relatively late (Park et al., 2015; Zhou et al., 2015). It was found that the chemical composition of the Mo precursor was crucial for the CVD growth of $MoTe_2$. The resulting $MoTe_2$ phase and the efficiency of the tellurization were both strongly dependent on the oxidation state of the Mo precursor. Namely, it was found that MoO_3 reacts more easily with Te and forms $2H$-$MoTe_2$, while Mo and MoO_x ($x < 3$) precursors tend to form $1T$-$MoTe_2$ under the same conditions. It was further noted that Te vapor should be maintained during the growth of both the $2H$ and $1T$ phases to avoid Te deficiency in the as-grown film because $MoTe_2$ is unstable and sublimes at high temperatures (Zhou et al., 2015). An interesting observation was made for a Mo tellurization process (Park et al., 2015). It was found that initially the $1T$-$MoTe_2$ was formed and was converted gradually to $2H$-$MoTe_2$ over a prolonged growth time under an (excessive) Te atmosphere. The phase change could be reversed if the 2H phase is exposed to an annealing process at a low pressure of Teat the same temperature, followed by a rapid quenching process. It was further found that orientation of the $2H$-MoTe$_2$ phase was determined by the tellurization rate, namely, slow tellurization resulted in a highly oriented film over the entire area, while fast tellurization led to the formation of a $2H$-MoTe$_2$ film with a randomly oriented c-axis, the result being in agreement with an earlier observation (Kong et al., 2013).

1.1.2.1.2   Thermolysis Process

The thermolysis process is a slightly variant method of the metal chalcogenisation process. In this process, the substrate (on which MoS$_2$ is to be grown) is first put through a dip-coating step in a solvent containing ammonium thiomolybdates [(NH$_4$)$_2$MoS$_4$] (Liu et al., 2012). The next step is to anneal the dip-coated substrate at about 500°C in an Ar/H$_2$ environment to produce MoS$_2$ as well as remove the residual and unreacted solvents and molecules. The heat-treated substrate with MoS$_2$ deposited on it is again subject to the sulfurization process in the presence of Sulfur vapors at 1000°C. This process could yield high-quality MoS$_2$ thin films; however, the dip-coating process has few limitations which could result in a non-uniform ammonium thiomolybdate layer. Thus, producing ultra-thin and highly crystalline MoS$_2$ with large surface area through thermolysis process is still challenging. The reported results show that MX$_2$ synthesis via the sulfurization/decomposition of pre-deposited metal-based precursor layers is an effective means of preparing large-area MX$_2$ thin layers. This has again several limitations such as difficulty in controlling the thickness of the deposited sample (metal oxide or metallic thin film), and uniformity during large scale synthesis. An alternative method, atomic layer deposition (ALD), has been developed to control the initial precursor deposition, and to improve the growth of the TMC layers. Using this method, atomically thin TMDC nanosheets with good thickness controllability and uniformity could be achieved (Song et al., 2013).

1.1.2.1.3   Vapor Phase Reaction Synthesis

As has been discussed in the previous section, the direct reaction between the transition metal and chalcogenide precursors can be carried out in a CVD system. During the process, sulfur and MoO$_3$ powders were successively placed in the reaction chamber, with a SiO$_2$/Si substrate mounted face-down above the MoO$_3$ (Lee et al., 2012). Pre-treating this substrate by spin-casting graphene-like aromatic molecules was found to be preferable for inducing effective nucleation and layered growth of MoS_2. This method yielded near triangular MoS_2 sheets of 1–3 monolayers (ML). Planar aromatic molecules were also used as seeding promoters to grow MS_2 (M = Mo, W) on different substrates as shown in Figure 1.2 (Lee et al., 2013; Ling et al., 2014). It should be noted that the one-step CVD produced regularly triangular-shaped MoS_2 flakes with a typical domain size larger than 1 μm. It is suggested that the limiting factor for MoS_2 growth on bare SiO_2/Si was due to the nucleation process, (Najmaei et al., 2013) where the triangular domains preferably nucleated at the edges. The precursors for Mo and S source are crucial as different sources have different impacts on the morphology of the finally obtained MoS_2 films. The CVD growth of MoS_2 with two precursors MoO_3 and $MoCl_5$ has been compared, (Ganorkar et al., 2015), and results suggested that while the MoO_3 source gave a triangular-shaped MoS_2 ML, the use of $MoCl_5$ yielded uniform MoS_2 without triangles.

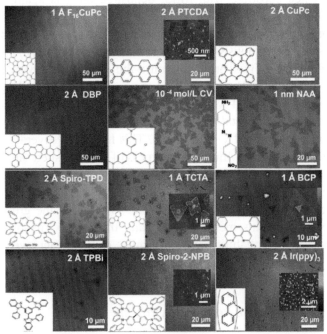

FIGURE 1.2 Typical optical images of the surface after the MoS_2 growth using different aromatic molecules as seeding promoters. The names and thicknesses of the seeding promoters are labeled on the images. The insets show the corresponding molecular structures or AFM images of the surface after MoS_2 growth. The color bars in the AFM images are 10 nm for PTCDA, 20 nm for TCTA and Spiro-2-NPB, 30 nm for BCP, and 50 nm for Ir(ppy)$_3$.
Source: Reprinted with permission from Ling et al. (2014). © American Chemical Society.

However, the mechanism of the absence of geometric shapes when using MoS$_2$ was not revealed in this work. Using ambient-pressure CVD, near-triangular MoS$_2$ flakes were fabricated (Figure 1.3) with the longest edge length so far of ~120 μm and in a setup in which a small quartz tube sealed at one end and used as the container for the precursors and the substrate (Figure 1.4) triangular WS$_2$ MLs with an edge length of up to 70 μm was realized (Cong et al., 2014a; van der Zande et al., 2013). MoS$_2$ ML flakes can also be produced on insulating substrates (such as a Si wafer and sapphire) by simple physical vapor transport of MoS$_2$ powder in an inert environment (Wu et al., 2013).

FIGURE 1.3 (a) Optical reflection image of a CVD growth of a typical large-grain MoS$_2$ on a SiO$_2$ (285 nm)/Si substrate. The image contrast has been increased for visibility; *magenta* is the bare substrate, and *violet* represents monolayer MoS$_2$. (b) Optical image of a monolayer MoS$_2$ triangle. The *triangle* is 123 μm from tip to tip.
Source: Reprinted with permission from Van Der Zande et al. (2013). © Springer Nature

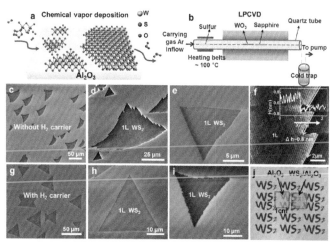

FIGURE 1.4 LPCVD synthesis of WS$_2$ on sapphire. (a) Schematic view of the related chemical reaction. (b) Experimental setup of the LPCVD system. (c, d, e) SEM images of jagged edge WS$_2$ flakes synthesized under pure Ar gas flow. (f) Corresponding AFM characterization of the monolayer nature. (g, h, i) SEM images of WS$_2$ flakes synthesized under mixed Ar and H$_2$ gas flow. (j) Photograph of bare sapphire and monolayer as-grown WS$_2$.
Source: Reprinted with permission from Zhang et al. (2013). © American Chemical Society.

This method allowed the authors to produce high optical quality MoS_2 MLs with near-unity valley polarization (Kolobov and Tominaga, 2016). Graphite was also proposed as an effective substrate. Thus, CVD-grown ML WS_2 on graphite gives rise to a single photoluminescence peak width a symmetric Lorentzian profile and very small peak width values of 21 meV at room temperature and 8 meV at 79 K (as opposed to 48 meV for the case of the SiO_2/Si substrate) (Kobayashi et al., 2015). It was argued that compared with WS_2 on sapphire and SiO_2/Si substrates, the WS_2 grown on graphite is less affected by charged impurities and structural defects. Both these routes can conceivably be extrapolated to obtain various MX_2 layers: WS_2, (Gutiérrez et al., 2013; Zhang et al., 2013) $MoSe_2$,(Chang et al., 2014; Kong et al., 2013; Lu et al., 2014; Shaw et al., 2014) and WSe_2 (Huang et al., 2014). Although SiO_2/Si is currently the preferred substrate for growing MX_2 thin layers, insulating single crystals such as sapphire, mica, and $SrTiO_3$ were considered by some as more suitable substrates for the CVD growth of MX_2 based on their ultra-flat surfaces, excellent thermal stability, and possible lattice registry with MX_2 adlayers. Chemical preparation of MoS_2 and $MoSe_2$ has also been demonstrated using hydrothermal synthesis (Matte et al., 2011; Peng et al., 2001). This method produces reasonably good-quality material with typical flake sizes of up to a few micrometers, but the fabrication of MLs was not conclusive. The composition and purity of CVD MoS_2 layers was analyzed by the X-ray photoemission spectra. Figure 1.5a shows the photoelectron spectrum of Mo $3d$. Peaks at 233.3 and 230.2 eV correspond to the binding energy of Mo $3d3/2$ and Mo $3d5/2$, respectively (Senthilkumar et al., 2014). The spin energy separation of Mo $3d$ doublet is 3.1 eV and in good agreement with the previous reports (Liu et al., 2012). The S 2p spectrum (Figure 1.5b) contains spin-orbit doublets of S $2p3/2$ centered at 162.9 eV, whereas the S $2p1/2$ is found at 164.1 eV with a spin energy separation of 1.2 eV for S (Liu et al., 2012). The obtained results are closely matched with the bulk MoS_2 crystal (Senthilkumar et al., 2014).

1.1.2.1.4   Alloy Formation

Earlier report on the synthesis of ternary $MoS_{2(1-x)}Se_{2x}$ nanosheets suggests that the composition of the as prepared sample can be tuned in an affordable path (Li et al., 2014). In this case, the chalcogen precursors (i.e., selenium and sulfur) were placed inside the quartz tube of the CVD system, and a temperature gradient along the tube was then maintained to enable control over the composition and spatial separation. Hence, simply placing the boat

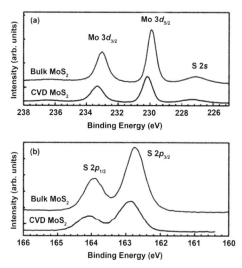

FIGURE 1.5 X-ray photoemission spectroscopy scans for (a) Mo and (b) S binding energies of the MoS$_2$ monolayer.
Source: Reprinted with permission from Senthilkumar et al. (2014). © Springer Nature.

filled with MoO$_3$ and the SiO$_2$/Si substrate above, at different positions, one can control the constituent stoichiometries of the resulting nanosheets, which can be varied from pure MoS$_2$ to pure MoSe$_2$. Selenization of CVD synthesized MoS$_2$ and sulfurization of CVD synthesized MoSe$_2$ (Figure 1.6) are alternative possibilities to produce TMDC alloys (Su et al., 2014a, b).

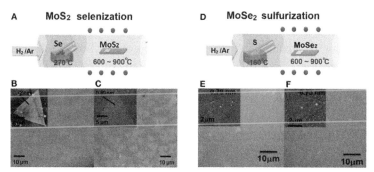

FIGURE 1.6 (A, D) Schematic illustration of the experimental set-up for the selenization/ sulfurization process, where the inlet gas carries the vaporized selenium (sulfur) to the heated MoS$_2$ (MoSe$_2$) flakes and optical micrographs for the (B, E) as-synthesized MoS$_2$/MoSe$_2$, and (C, F) selenized (sulfurized) MoS$_2$/MoSe$_2$ (at 800°C) on sapphire substrates. AFM images for the MoS$_2$/MoSe$_2$ flakes before and after selenization (sulfurization) (at 800°C) are shown as insets in OM images.
Source: Reprinted with permission from Su et al. (2014a).

1.1.2.1.5   Epitaxial Growth Technique

Homo- and hetero-epitaxy is one of the most important methods in thin-film technology. However, in order to obtain high-quality overlayers in hetero-epitaxy, one has to overcome the problem of lattice matching between the substrate and the grown layer. In tetrahedrally bonded semiconducting materials such as GaAs and Si, the condition of lattice-matching is highly essential (or we can say is mandatory). If this condition is not met, the numerous dangling bonds on the surface of the substrate will remain in an unsatisfied condition which may trigger severe instability and improper growth of the deposited material. Furthermore, as length and angle of covalently connected bonds are highly resistant to slightest of modifications, therefore perfect lattice matching is critical for successful and high-quality epitaxial growth processes. Koma et al. have reported that the lattice-matching requirement is relaxed drastically when the hetero-epitaxial growth proceeds with a van der Waals interaction in case of the growth of a layered material on a cleaved face of the other layered material having no dangling bonds (Koma et al., 1985). This type of epitaxy was termed van der Waals epitaxy. It was demonstrated that epitaxial growth by van der Waals epitaxy is possible even under the existence of lattice mismatch as large as 50% (Koma, 1999). Moreover, because of the non-existence of the dangling bonds, atomically abrupt interfaces can be fabricated (Koma, 1992; Koma and Yoshimura, 1986; Ohuchi et al., 1991). In the case of a three-dimensional substrate, the dangling bonds must be first terminated, e.g., by sulfur. The successful growth of epitaxial TMDCs on sulfur terminated GaAs (111) was reported back in 1990 (Ueno et al., 1990b). Interested readers may also check a review on epitaxial growth of 3D topological insulators (He et al., 2013). Since both TMDC and 3D topological insulators are van der Waals solids, the growth process has some common features. The interest in epitaxial growth of TMDC single and few layers using the van der Waals mode re-emerged in recent years using methods ranging from CVD to molecular beam epitaxy (MBE) and the substrates varying from graphene to SiO_2/Si (Cheng et al., 2013; Ji et al., 2013; Shi et al., 2012; Yue et al., 2015). High epitaxial quality of the grown films can be seen in Figure 1.7. Single layers of MoS_2 were also successfully grown on other substrates such as SnS_2 and mica, (Ji et al., 2013; Ueno et al., 1990a; Zhang et al., 2014) as is illustrated in Figure 1.8. ML $MoTe_2$ was grown by MBE on a bulk MoS_2 substrate. The morphology, structural transformations, compositional percentages, and oxidation states were investigated by scanning tunneling microscopy and X-ray photoelectron spectroscopy (XPS).

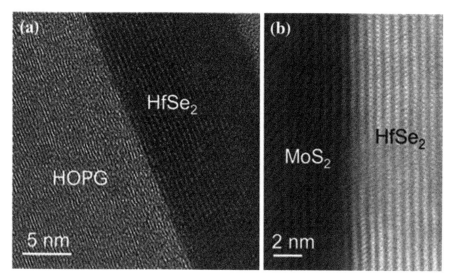

FIGURE 1.7 (a) TEM image of grown $HfSe_2$ on HOPG and (b) HADDF-STEM image of grown $HfSe_2$ on MoS_2 showing abrupt interfaces and layered crystalline films.
Source: Reprinted with permission from Yue et al. (2015). © American Chemical Society.

Of special interest is the growth by MBE of heterostructures between TMDC and topological insulators. The first such report has been published, where $MoSe_2/Bi_2Se_3$ heterostructures were successfully fabricated (Xenogiannopoulou et al., 2015). Theoretical analysis of single-layer MoS_2 on Cu (111) surface found that there was rather strong chemical interaction between the layer and the substrate, (Le et al., 2012) which is not surprising considering strong reactivity between copper and chalcogens. It is interesting to note that even such a technique as pulsed laser evaporation results in the formation of hexagonal MoS_2, which, in combination with the orienting effect of an appropriately chosen substrate, may open way to fabricate atomically thin layers by mass-production methods such as sputtering (Donley et al., 1988, 1989). Finally, it should be mentioned that by appropriately choosing the composition of the substrate material, self-organized van der Waals epitaxial growth of layered chalcogenides structures was successfully realized using industry-friendly RF-magnetron sputtering (Saito et al., 2015).

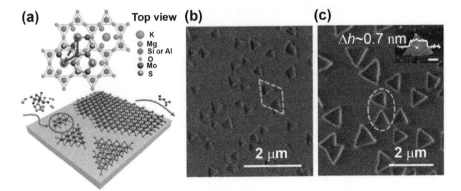

FIGURE 1.8   Low-pressure CVD synthesis of MoS$_2$ nano/microstructures on mica. (a) A schematic view illustrating the surface reaction during epitaxial growth of MoS$_2$. The upper panel gives probable occupation of MoS$_2$ on mica. (b–c) SEM images showing the initial growth of MoS$_2$ on mica. Inset in (c) is the AFM profile of the MoS$_2$ flake (scale bar 200 nm). *Source:* Reprinted with permission from Ji et al. (2013). © American Chemical Society.

1.1.2.1.6   Laser Ablation

Though known for its efficiency to produce carbon nanotubes (CNTs) and fullerenes of the highest quality, laser ablation technique is still far from reaching at par with methods like metal chalcogenisation, for the synthesis of TMCs having analogous structures. However, only recently MoS$_2$ nano-octahedral structures (known as the true inorganic fullerenes (IF)) have been synthesized through arc discharge, shock waves, solar ablation, and laser ablation techniques (Savva et al., 2017). These methods produced some excellent results regarding the synthesis of TMCs like MoS$_2$. In general, the reaction between Mo and S is triggered through the pulsed laser of suitable discharge power in the presence of a toxic/hazardous gaseous environment comprising H$_2$S and H$_2$. Nevertheless, Savva et al. have devised a unique technique with which metastable MX$_2$ phases can be achieved easily. Furthermore, they claim that being independent of the material that is being used for the synthesis, this laser ablation technique can be extended to synthesize other TMCs as well (Savva et al., 2017).

1.1.2.1.7   Arc Discharge

As already been discussed, arc discharge method is largely employed to produce CNTs and fullerenes. However, this technique was first implemented by Chhowalla et al. for the synthesis of IF like MoS_2 in the form of a thin film that exhibited an excellent lubricating behavior (Sano et al., 2003). Other fullerene-like particles, sometimes filled with different materials, e.g., CoS inside WS_2-IFs, were synthesized soon after (Alexandrou et al., 2003; Si et al., 2005). Filled IFs are of special interest because they can be advantageous as compared to hollow particles in areas where high-pressure conditions are deployed such as lubrication and cutting.

1.1.2.1.8 Microwave-Induced Plasma Treatment

This technique can be used to synthesize a wide range of compounds including TMCs. There have been reports on the synthesis of MX_2 (M=Mo, W; X=S, Se) nanostructures by microwave-treating $M(CO)_6$ in a H_2S or $SeCl_4$ rich environment (Vollath and Szabó, 2000). Nanoclusters of SnS_2 and ZrS_2 have also been prepared using a similar technique. Taking WO_3, ZrS_3, and HfS_3 as starting materials, Brooks et al. have obtained WS_2, HfS_2-IFs, and ZrS_2 nanotubes and nanorods (NRs), respectively by treating them with microwave-induced plasmas of H_2S and N_2/H_2 (Vollath and Szabó, 1998).

1.1.2.1.9 Irradiation Technique

Surface defects and folding of layered materials can be achieved by exposing them to an electron beam irradiation. IFs of MoS_2 have been synthesized utilizing this technique as per the reported literature (José-Yacamán et al., 1996). It has been found that the inner morphology of these onion-like structures have two different orientations depending on the radial growth during the synthesis process. The smaller radii show faceting, whereas the bending of layers occurs towards the larger radii. Other similar TMCs such as $NbSe_2$, $MoSe_2$, and silver-containing NbS_2 nanotubes could also be obtained by electron beam irradiation with adequate precursor materials (Galvan et al., 2000; Galván et al., 2001; Remskar et al., 2002).

1.1.2.1.10 Spray Pyrolysis Technique

To synthesize close structured nanomaterials comprising TMCs, the spray pyrolysis technique is used apart from the well-known arc discharge method. In spray pyrolysis, the precursors react with each other in a closed droplet environment which will provide control over the morphology of the finally obtained TMC nanostructures. IF structures and nanoboxes of MoS_2 and WS_2 have been obtained by dispersing ethanolic solutions of $(NH_4)_2MS_4$ (M = Mo, W) at a pyrolytic temperature of 900°C (Bastide et al., 2006).

1.1.2.1.11   Hydrothermal Synthesis

Hydrothermal techniques employ low-temperature synthesis procedures that are cost-effective, simplistic, and non-hazardous as compared to other methods. This has the advantage of avoiding the formation of high-crystallinity products usually obtained at higher temperatures. By taking precursors like MoO_3 and Na_2S of an appropriate amount, and treating them hydrothermally in an acidic ambiance (taking HCl) at a temperature of 260°C would yield MoS_2 samples with high specific area, or else taking ammonium molybdate and sulfur at 150–180°C using hydrazine as a reducing agent, (Li et al., 2003; Peng et al., 2001) one can also obtain good quality MoS_2. According to an earlier report, MoO_3 and KSCN, when treated hydrother-mally, yielded a mixture of MoS_2 NRs and nanotubes. In contrast to other nanotubes that comprise the parallel arrangement of wall layers, the walls of the tubes formed in the hydrothermal method were found to form agglomerates of small bent slabs (Tian et al., 2004).

1.1.2.1.12   Vapor-Liquid-Solid Growth

This synthesis technique has exclusively been used for the production of semiconductor nanowires (NWs) composed of metal chalcogenides. The nanowire structures are obtained through a metallic seed (gold) particle assisted catalytic growth process (Gudiksen and Lieber, 2000). The catalysts, i.e., gold nanoparticles (AuNPs) are provided a support in the form of a $SiO_2/$ Si substrate, and laser ablation technique is used to generate reactants from a bulk solid target. The growth process of the NWs trigger from underneath the vaporized metal droplet. This means the growth of NWs at non-catalyzed surfaces could be suppressed by a kinetically limiting process assisted by low-temperature conditions. Thus, we can achieve almost a unidirectional growth of the sample. The process proceeds via three stages: (1) formation of desired metal droplet by laser ablation, (2) Adsorption of the reactant in vapor phase which is gradually diffused into the metal droplet, (3) Oswald ripening (supersaturation) of the metal droplet and subsequent nucleation and growth at the solid-liquid interface triggers six consequent anisotropic crystal growth (Toyama, 1966). This synthesis method can be affected by a few factors such as surface tension and electronegativity, as they directly affect the interaction between catalyst and reactive elements.

1.1.2.1.13 Oxide-to-Sulfide Conversion

By exposing transition metal, oxides to a reducing environment, and in the presence of suitable gas reactant phase, high quality, and large area TMCs can be obtained. It is to be noted that this method has been leading all the synthesis protocols for the preparation of TMCs. In general, a metal oxide, which is taken as a precursor for the transition metal, is allowed to react with H_2S to trigger the sulfurization process. There have been numerous reports on the conversion of metal oxides to metal sulfides or selenides using this technique. According to one report, hexagonal WO_3 NRs were first obtained via a sol-gel process. The diameter of these NRs was 5–50 nm and the length varied within 150–250 nm. These NRs provide a perfect platform as precursors for the synthesis of multi-walled WS_2 nanotubes through a reduction process in an H_2S environment at 840°C for 30 min (Therese et al., 2005). Here, the wall thickness of the prepared nanotube can also be tuned by a controlled reduction of the precursor (metal oxide). According to Tenne and coworkers, WS_2 formation is initiated with the engulfing of WO_x particles. The gradual formation of WS_2 from within and outside of the WO_x particle as the latter is condensed from its vapor-phase (Rothschild et al., 1999, 2000). This mechanism can also applicable for the nanotube synthesis technique being reported here (in this section). The difference is, instead of using H_2/N_2 gas for the reduction process, Ar gas is used to treat the metal oxide samples. When treated with argon, these metal oxides form numerous intermediate phases with large numbers of defects and vacancies. However, the rod-like morphology remains intact. As the morphology of the precursor has a significant impact on the morphology of the final product, therefore, this method can be employed to produce many defects assisted TMC synthesis without altering the overall morphology (Murphy et al., 1976; Schneemeyer et al., 1980; Wiegers et al., 1974). The first report on the intercalation of organic molecules in between the layered structure of TMCs was on the amine-intercalated VS_2 nanotubes. Partially and fully sulfidized VS_2 nanotubes showed lattice fringes corresponding to layer distances of ~2.8 nm (partially sulfidized) and 1.6 nm (fully sulfidized). These values show large deviations from the layer separation calculated in bulk VS_2 (0.57 nm) (Rocker et al., 2004). Similar approach has been followed for the synthesis of Nb-W-S composite nanostructures. Here, Nb_2O_5 was first prepared via sol-gel process and then tungsten oxide was coated onto it through a solvothermal technique (Nanosheets et al., 2010). The tungsten coated niobium composites were then sulfidized by passing H_2S gas over under an Ar environment. The oxide-to-sulfide conversion follows exactly the same mechanism as described for pure WS_2 nanotubes. A closer look at the composite nanostructures reveals kinks, interruption of layers, orthogonal faults, and low contrast areas in the layered structures. These defects can be attributed to the large compressive lattice mismatch strain between NbS_2 and WS_2 during their growth. Increasing the percentage of tungsten in the niobium compound (at approximately 30%) forms stacked NWs to minimize surface energy and strain in structural geometry, due to internal strain, rather than to the formation of Nb-doped $Nb_xW_{1-x}S_2$ nanotubes.

1.1.2.1.14   *Hot-Injection Solution Method*

In the last few years, there has been intense interest in the controlled synthesis of metal chalcogenides at nano-level with better control over particle size, dispersity, and morphology. All these parameters are responsible for the fundamental properties related to nanotechnology like quantum confinement effects, especially when prepared with sizes less than the Bohr radius of the exciton. As discussed above, almost all traditional methods that are typically employed for the synthesis of chalcogenides materials use a high temperature and pressure. Therefore, most of these methods result in nanostructures with bare surfaces and tend to form aggregates, reducing surface areas, and precluding further functionalization or dispersion. Solution synthesis offers a plethora of variable parameters to design reactions, e.g., choice of reagents, solvents, concentrations, additives, temperature, pressure, heating, and cooling rates are some of the factors which can have a significant impact on the chemical composition of the product and its morphology. By varying the ratio between capping ligands and by choosing a suitable solvent, uniform, monodisperse, and ultra-small NPs can be synthesized. Conventional solution methods require separate sources of metal and chalcogen, reacted together in solvent, and the nanomaterials are finally collected as precipitates. For example, MoS_2 NPs of size <5 nm have been synthesized via the reaction of $Mo(CO)_6$ with elemental sulfur in the presence of trioctylphosphine oxide

(TOPO) and 1-octadecene at temperatures ranging from 270°C to 330°C. The MoS_2 NPs are discrete and dispersible in a variety of non-polar organic solvents, including toluene, chloroform, and pyridine. The size of the particles can be effectively tuned by varying the temperature, yielding nearly monodisperse samples (Yu et al., 2008). Paolo Ciambelli and co-workers have synthesized stable and free-standing nanosheets of MoS_2 and WS_2 by the decomposition of single-source precursors containing both metal and sulfur in oleylamine (OLA). This one-pot synthesis method resulted in high-quality 2D nanosheet crystals of MoS_2 and WS_2. The added benefit of this method is that materials obtained are covered by a dynamic protective coating of OLA that stabilizes the suspension, avoids aggregation, and prevents surface oxidation (Wu et al., 2011). In another example, when lead (II) acetate trihydrate ($Pb(Ac)_2.3H_2O$) and thioacetamide are reacted at room temperature using sodium dodecyl sulfate (SDS) as surfactant, it resulted in fern-like crystals of PbS. However, the addition of polyvinylpyrrolidone (PVP) into the reaction mixture, switched the anisotropic structures of the crystals to isotropic ones and produces nanocubes of PbS instead, (Dong et al., 2009) and reaction done at higher temperature produced nanocrystallites of very small dimensions. In a similar way, nanocrystalline Ni_3S_4 with a particle size of ~10 nm can be prepared from $NiCl_2$ dissolved in an OLA, n-trioctylphosphine (TOP) and sulfur solution (Ghezelbash and Korgel, 2005).

1.1.2.2   TOP-DOWN APPROACH

Top-down synthesis techniques include mechanical (scotch-tape method), liquid/chemical, and intercalation based exfoliation methods to produce high quality and large area TMCs. These techniques have largely been employed for the synthesis of graphene layered structures and recently being extended for layered TMCs. Here, a target bulk material is exfoliated through various techniques to obtain few-layered structures. These methods have been briefed in the following sections.

1.1.2.2.1   Mechanical Exfoliation Technique

The mechanical exfoliation technique is by far the most efficient technique to yield highly crystalline and extremely thin nanosheets of layered TMCs. In this technique, the bulk TMCs are the first subject to a mechanical peeling procedure in which loosely bound layers (connected to each other through van der Waals interactions) are gathered by an adhesive scotch tape (Figure 1.9, top panel). The scotch tape with cleaved samples are then rubbed onto the surface of a suitable substrate material, where the tape is repeatedly rubbed to cleave the sample further and upon removal of the tape, it would leave single- and/or multi-layered TMCs on the substrate surface (Figure 1.9, bottom panels a–d). Though the process of using a scotch tape produces high-quality single layers of TMDCs, it cannot be used for scalable synthesis of desired layered structures. Therefore, the technological application of scotch tape assisted mechanical exfoliation is rather limited. As a way to exfoliate large-area MLs, it was proposed to make use of chemically enhanced adhesion. The use of chemical affinity of sulfur atoms that can bind to a gold surface more strongly that to neighboring layers, single layers of various van der Waals bonded chalcogenides, such as MoS_2, WSe_2, and Bi_2Te_3 with lateral sizes of several hundreds of microns were successfully exfoliated (Dobrik

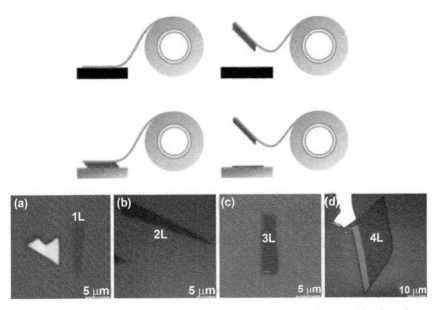

FIGURE 1.9 *Top*: Schematic of micromechanical cleavage technique (the Scotch tape method) for producing few-layer structures. *Top row:* adhesive tape is used to cleave the top few layers from a bulk crystal. *Bottom left* the tape with removed flakes is then pressed against the substrate of choice. *Bottom right:* some flakes stay on the substrate, even on the removal of the tape (*Source:* Reprinted with permission from Novoselov (2011)). *Bottom:* Mechanically exfoliated single and few-layer MoS_2 nanosheets on 300 nm SiO_2/Si. Optical microscopy (a–d) (*Source:* Reprinted with permission from Li, Wu, Yin, and Zhang (2014). © American Chemical Society.

et al., 2015). Nanomechanical cleavage of MoS_2 was studied *in-situ* using transmission electron microscopy and layers with thicknesses varying from a ML to 23 layers were successfully cleaved (Tang et al., 1AD). Thermal annealing and evaporative thinning using a focused laser spot were also used for layer-by-layer thinning of MoS_2 down to ML thickness by thermal ablation with micrometer-scale resolution (Castellanos-Gomez et al., 2012; Huang et al., 2014; Lu et al., 2013). Controlled MoS_2 layer etching using CF_4 plasma has been reported (Jeon et al., 2015). The damage and fluorine contamination of the etched MoS_2 layer could be effectively removed by exposure to H_2S.

1.1.2.2.2 Liquid Phase Exfoliation

Liquid exfoliation by direct ultrasonication, that was successfully used earlier to disperse graphene, was also employed to fabricate single-layer and multilayer nanosheets of a number of layered TMDCs, such as MoS_2, WS_2, $MoSe_2$, $NbSe_2$, $TaSe_2$, $MoTe_2$, $MoTe_2$, and others (Coleman et al., 2011; Nicolosi et al., 2013) where the commercially purchased samples (in powder form) were initially sonicated in a number of solvents to form uniform dispersions; the dispersion mixture was then centrifuged to collect the exfoliated samples. Solvents having high dispersion coefficient, polarity, and hydrogen bonding capability are generally preferred as they facilitate the exfoliation process during sonication. Two mostly used solvents are N-methylpyrrolidone and isopropanol. These solvents allow the separation of van der Waals layers and prevent the agglomeration or re-stacking of produced layers by forming solvation shell. Samples such as boron nitride (BN), MoS_2, and WS_2 have been prepared by this method using vacuum filtration or spraying techniques. Water has also been employed for this exfoliation technique on many occasions. In an aqueous medium, the samples are generally subject to vigorous sonication with the addition of a surfactant such as sodium cholate. The purpose of adding a surfactant is to coat the layered structures formed during sonication and protect them from getting re-stacked on each other (Smith et al., 2011). It may be worth noting that the dispersibility of exfoliated nanosheets varied only weakly between different TMDCs (Cunningham et al., 2012). The main challenge is to enhance the yield of the MLs and to maintain the lateral dimensions of the exfoliated sheets. Ion intercalation, such as lithium-intercalation or ultrasound-promoted hydration, is another approach, allowing fabrication of single-layer materials. The intercalation of TMDCs by ionic species allows the layers to be exfoliated in liquid (Dines, 1975; Joensen et al., 1986). In this exfoliation method, bulk TMCs, in powder form, is kept submerged in a lithium-based solvent such as n-butyllithium, for few days to facilitate the intercalation of lithium ions into the layers of TMCs. And then the dispersion mixture is exposed to water. The water reacts with the lithium between the layers; the process results in the formation of H_2 gas, which serves to separate the layers (Eda et al., 2012; Joensen et al., 1986). Such chemical exfoliation methods allow one to produce significant quantities of submicrometer-sized MLs (Tsai et al., 1997) but the resulting material differs structurally and electronically from the source bulk. For example, on exfoliation, the semi-metallic Mos_2 may change to metallic MoS_2, as the coordination sphere of Mo shifts from trigonal prismatic (2H) to octahedral (1T) phase. However, annealing the 1T phase at 300°C can reverse the phase to 2H. Lithium-based chemical exfoliation has been demonstrated for various TMDCs, in particular MoS_2, WS_2, $MoSe_2$, and SnS_2 (Gordon et al., 2002; Kirmayer et al., 2007). This method was also used to exfoliate topological insulators such as Bi_2S_3 and Bi_2Te_3 (Huang et al., 2009). An effective method for mass production of exfoliated TMD nanosheets is the ultrasound-promoted hydration of lithium-intercalated compounds. When the bulk TMCs are subject to exfoliation using organolithium compounds, intermediates are formed with the reduced phase, Li_xMX_n, and expanded lattice. If this intermediate phase is then treated in an aqueous medium (through ultrasound-assisted hydration process) (Dines, 1975; Frey et al., 2003; Tsai et al., 1997). Therefore, the important step in the formation of the reduced phase, i.e., Li_xMX_2, tuning which can provide vital control over the quality of the exfoliated layers.

TMDC nanosheets can also be exfoliated by thermal cyclings, such as rapid freezing (30 s in a liquid nitrogen bath) and heating (20 min in an oil bath at 60°C), of hydrated TMDC powder in water (Chakravarty and Late, 2015). The lithiation process can also be carried out in an alternative manner that uses an electrochemical cell with a lithium foil anode and TMDC-containing cathode (Figure 1.10) (Li et al., 2011; Zeng et al., 2012). As the intercalation occurs while a galvanic discharge is occurring in the electrochemical cell, the degree of lithiation can be monitored and controlled. The resulting Li-intercalated material is exfoliated by ultrasonication in water as before, yielding ML TMDC nanosheets.

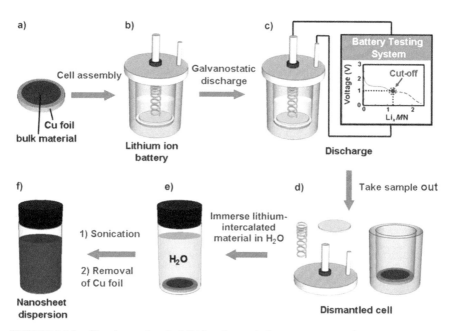

FIGURE 1.10 The electrochemical lithium intercalation process to produce 2D nanosheets from the layered bulk material (MN = BN, metal selenides, or metal tellurides in Li$_x$MN). *Source:* Reprinted with permission Zeng et al. (2012). © John Wiley and Sons.

This technique was first reported in the case of MoS$_2$, WS$_2$, TiS$_2$, TaS$_2$, ZrS$_2$ and graphene,(Li et al., 2011) and later extended for BN, NbSe$_2$, WSe$_2$, Sb$_2$Se$_3$, and Bi$_2$Te$_3$ (Zeng et al., 2012). This method is advantageous, considering the fact that it requires only a few hours to accomplish Li intercalation, as compared to longer time duration required in the case of *n*-butyl assisted method. During the whole experimental process, the lithium ions fulfill several important functions. First, the Li$^+$ ions are inserted into the interlayer space of the layered bulk material, which expands the interlayer distance and weakens the van der Waals interactions between the layers. Second, the inserted Li$^+$ ions are subsequently reduced to Li by accepting electrons during the discharge process. The metallic Li can react with water to form LiOH and produce H$_2$ gas (apparently, bubbles were observed during the experiments). The generated H$_2$ gas pushes the layers further apart. Under vigorous agitation by sonication, well-dispersed 2D nanosheets can be thus obtained (Zeng et al., 2012). For aqueous-based exfoliation procedures, lithium has been used extensively as compared to other alkali ions such as Na and K. Nevertheless, the atomic radii of Na and K are much larger than that of Li, and their reactivity towards water is more too. Thus, these ions can get intercalated into the layers of TMCs and expand them along the c-axis with more effectiveness. Taking this cue, naphthalenide adducts of Li, Na, and K was put to a comparison for their effectiveness in exfoliating MoS_2 layers (Zheng et al., 2014).

Figure 1.11 shows the schematic diagram of the processing steps involved in obtaining well-dispersed samples of LTMDs. In the atypical procedure, bulk MoS_2 powders were first treated with hydrazine to loosen the layers taking the help of hydrothermal method (panel a) Decomposition and gasification of intercalated N_2H_4 molecules expands the MoS_2 sheets by more than 100 times compared to its original volume. In a second step, the expanded MoS_2 crystal is intercalated by alkali naphthalenide solution (panel b).

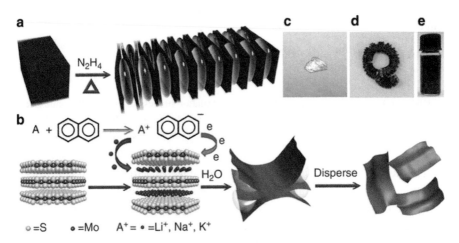

FIGURE 1.11   (a) Bulk MoS_2 is pre-exfoliated by the decomposition products of N_2H_4. (b) Pre-exfoliated MoS_2 reacts with $A^+C_{10}H_8$ to form an intercalation sample, and then exfoliates to single-layer sheets in water. (c) Photograph of bulk single-crystal MoS_2, d photograph of pre-exfoliated MoS_2, (e) photograph of Na-exfoliated single-layer MoS_2 dispersion in water. *Source:* Reprinted with permission from Zheng et al. (2014).

In the final step, the alkali-ion intercalated MoS_2 was exfoliated by dispersing in water under constant ultrasonication. This method was tested successfully on a wide range of TMDCs (Zheng et al., 2014). A tandem molecular intercalation was proposed for producing single-layer TMDCs from multi-layer colloidal TMDC nanostructures in solution phase, where short 'initiator' molecules first intercalate into TMDCs to open up the inter-layer gap, and the long 'primary' molecules then bring the gap to full width so that a random mixture of intercalates overcomes the interlayer force (Jeong et al., 2015). With the appropriate intercalates, single-layer nanostructures of group IV (TiS$_2$, ZrS$_2$), group V (NbS$_2$), and VI (WSe$_2$, MoS$_2$) TMDCs were successfully generated.

1.1.2.2.3 Electrochemical Exfoliation

A typical arrangement for the electrochemical exfoliation of bulk MoS$_2$ is shown in Figure 1.12a. A DC bias was applied between MoS$_2$ and the Pt wire for the electrochemical exfoliation, starting with a low positive bias to wet the bulk MoS$_2$ followed by a larger bias to exfoliate the crystal. As a result, many MoS$_2$ flakes dissociated from the bulk crystal and became suspended in the solution (panels b and c). The mechanism of electrochemical exfoliation of bulk MoS$_2$ crystals is described as follows (Figure 1.12e) (Liu et al., 2014). First, by applying a positive bias to the working electrode (WE), the

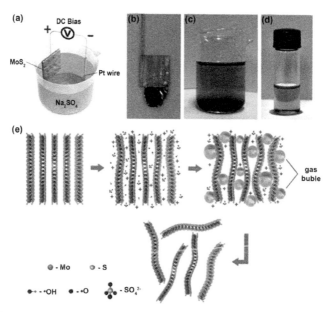

FIGURE 1.12 (a) Schematic illustration of the experimental setup for electrochemical exfoliation of bulk MoS$_2$ crystal. (b) Photograph of a bulk MoS$_2$ crystal held by a Pt clamp before exfoliation. (c) Exfoliated MoS$_2$ flakes suspended in Na$_2$SO$_4$ solution. (d) MoS$_2$ nanosheets dispersed in solution. (e) Schematic illustration for mechanism of electrochemical exfoliation of bulk MoS$_2$ crystal.
Source: Reprinted with permission from Liu et al. (2014).

oxidation of water produces -OH and -O radicals assembled around the bulk MoS_2 crystal. The radicals and/or SO_2^{-4} anions insert themselves between the MoS_2 layers and weaken the van der Waals interactions between the layers. Second, oxidation of the radicals and/or anions leads to a release of O_2 and/or SO_2, which causes the MoS_2 interlayers to greatly expand. Finally, MoS_2 flakes are detached from the bulk MoS_2 crystal by the erupting gas and are then suspended in the solution. A major problem here is that bulk MoS_2 should be oxidized during electrochemical exfoliation, which may affect the exfoliated MoS_2 nanosheets, unless the conditions are, optimized (Liu et al., 2014).

1.1.2.3   *LAYER TRANSFER TECHNIQUE AND ANALYSIS THEREOF*

When a thin TMDC layer is synthesized, it is important for fundamental and applied research for it to be transferred to an arbitrary substrate (Gurarslan et al., 2014; Lee et al., 2013). As the growth temperature of TMDC MLs are relatively high, temperature-sensitive substrates (such as polymer-based substrates) cannot be used in the synthetic process, while their use is essential for flexible electronics. It is thus essential to develop a transfer technique to implement large-area TMDC on different substrates. One such technique that maintains the quality of the as-grown ML has been reported (Lee et al., 2013). The as-grown MoS_2 sample was cut into three pieces and treated for 30 s with DI water, isopropyl alcohol, and acetone, respectively. The surface of the as-grown ML is hydrophobic, so that isopropyl alcohol and acetone were found to spread out on MoS_2, whereas water remained as a droplet. The treatment of MoS_2 MLs with DI within those 30 s resulted in the breakdown of floating cut-pieces into smaller ones, indicating that the as-grown MoS_2 ML can easily be detached from the substrate with the help of DI water. A surface-energy-assisted process has been reported that allowed the perfect transfer of centimeter-scale ML and few-layer TMDC films from original growth substrates onto arbitrary substrates with no observable wrinkles, cracks, or polymer residues. The unique strategies used in this process included leveraging the penetration of water between hydrophobic TMDC films and hydrophilic growth substrates to lift off the films and dry transferring the film after the lift-off. Scalable transfer of suspended TMDC layers on nanoscale patterned substrates (varying from polymers to Si to metals) using a capillary-force-free wet-contact printing method was demonstrated in (Li et al., 2015). As a proof-of-concept, a photodetector of suspended MoS_2 was fabricated using this method. As an advantage of this approach, the authors note the possibility of directly suspending a TMDC layer on nanoscale interdigitated electrodes.

1.1.2.4 COMPARISON OF EXFOLIATED AND CVD-GROWN TMDCS

While mono- and few-layer sheets can be obtained by both exfoliation and CVD growth, the samples are not perfectly identical. The issue of comparing exfoliated and CVD-grown samples has been discussed, (Plechinger et al., 2014) where MoS_2 single-layer samples obtained along these two pathways were compared using optical spectroscopy. In Figure 1.13a, individual spectra measured at different positions of the CVD sample are shown (the positions P1–P3 are separated by 100 μm). The vertical lines serve as a guide to the eye to mark maximum and minimum A exciton peak positions. The spectral shift indicates that the growth conditions, and the corresponding microscopic properties of the MoS_2 layer, vary in different parts of the layer, which may be associated with different strains (Plechinger et al., 2014). CVD-grown and exfoliated layers were also compared (Li et al., 2014). Figure 1.14b shows a comparison of the reflectance spectra for exfoliated and CVD-grown MoS_2 MLs. Shifts in the A and B excitonic peaks of ~40 meV are observed, although the overall dielectric function is very similar. For comparison, in Figure 1.14a, reflectance spectra for two different exfoliated samples of MoS_2 are shown. In addition to the shift of the A exciton peak position, it was also found that the FWHM of the peaks changed as a function of position, with values between 80 and 60 meV, while in an exfoliated single-layer MoS_2 flake has a significantly lower FWHM of about 37 meV, indicating a larger inhomogeneous broadening of the A exciton transition in the CVD-grown sample. In exfoliated MoS_2 flakes at low temperatures, a second, lower-energy PL peak was observed, which was previously associated with localized excitons bound to surface adsorbates (Plechinger et al., 2012). The difference was also observed in the temperature dependence of the PL emission (Figure 1.13b). While the maximum of the A exciton emission redshifts by 35 meV in the temperature range from 4 to 300 K in CVD samples, in exfoliated MoS_2 flakes, the authors observed a spectral redshift of the A exciton peak by 72 meV in the same temperature range.

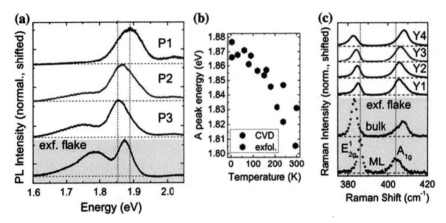

FIGURE 1.13   (a) Normalized PL spectra measured on different positions of a CVD film and on an exfoliated flake at liquid-helium temperature; (b) The A exciton peak position as a function of temperature for the CVD-grown sample (black dots) and an exfoliated MoS_2 flake (red hexagons) (c) Raman spectra measured on different positions of the CVD-grown film and on exfoliated monolayer (ML) and bulk-like flakes. All spectra are normalized to the amplitude of the A_{1g} mode. The vertical lines mark the positions of A_{1g} and E_{2g}^1 in an exfoliated monolayer and serve as a guide to the eye.
Source: Reprinted with permission from Plechinger et al. (2014).

According to an earlier report, shifting of the output spectra towards the red region suggests a temperature-induced reduction of the bandgap, resulted from the thermal expansion of the crystal lattice. Hence, the fact that the redshift is far less pronounced in the CVD-grown sample shows that the CVD-grown MoS_2 film strongly adheres to the SiO_2 substrate, which has a very small thermal expansion coefficient (Plechinger et al., 2014). Finally, Figure 1.13c shows four Raman spectra collected at different positions on the CVD film compared with the Raman spectrum from exfoliated flakes. All spectra were normalized to the A_{1g} mode amplitude. In the CVD-grown sample, the E_{2g}^1 amplitude was lower than the A_{1g}, while in the exfoliated flakes, the opposite was observed. Additionally, the line widths for both Raman modes in the CVD-grown film were larger than in the exfoliated flake, and the E_{2g}^1 mode was asymmetric. It was argued that these results suggest that the carrier density in the CVD-grown film may be significantly smaller than in the exfoliated flake (Plechinger et al., 2014). It should also be noted that stoichiometry variation during the CVD growth may have an effect on the optical and electrical properties of the grown layers. This issue has already been addressed for the case of MoS_2 (Kim et al., 2014).

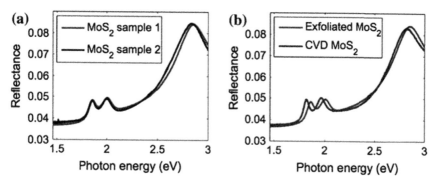

FIGURE 1.14 (a) Comparison of the reflectance spectra of two different exfoliated MoS$_2$ monolayers. (b) Comparison of the reflectance spectra of exfoliated (red) and CVD-grown (blue) MoS$_2$ monolayers.
Source: Reprinted with permission from Li et al. (2014). © American Chemical Society.

1.1.3 PHYSICO-CHEMICAL PROPERTIES

TMDCs have been the subject of study for over half a century. Some of the first examples of what we refer to as 2D crystals today were routinely produced and studied back in the 1960s (Frindt, 1966). Now the subject is revisited with a fresh perspective and inspiration from graphene.

1.1.3.1 MECHANICAL PROPERTIES

Elastic properties, stretching, and breaking of ultra-thin freely suspended MoS$_2$ were investigated in (Bertolazzi et al., 2011; van der Zant et al., 2012). In both studies, an AFM tip was placed on top of a film situated on top of a small (~1 μm) hole in a pre-patterned SiO$_2$ substrate and the MoS$_2$ membrane deflection under the applied force was measured. The same approach was used to determine the mechanical parameters of graphene (Lee et al., 2008). The experiment scheme and loading curves for 5–20 layers thick sample are shown in Figure 1.15. From a least-squares fit of the experimental curves, the pretension, σ_0^{2D} and the elastic modulus E^{2D} of the MoS$_2$ membrane were extracted. The numerical values somewhat vary between the two studies (Bertolazzi et al., 2011; van der Zant et al., 2012). Thus, for the 5–20 layer thick samples, the pre-tension was determined to be 0.05 ± 0.02 Nm^{-1} and Young's modulus, E_{Young}, 0.35 ± 0.02 TPa (van der Zant et al., 2012). At the same time, for ML thin membranes,(Bertolazzi et al., 2011) the average value of the elastic modulus E^{2D} of 180 ± 60 Nm^{-1} and pre-stress σ_0^{2D} in the 0.02–0.1 Nm^{-1} range were obtained. Assuming an effective ML thickness of 0.65 nm, Young's modulus $E_{\text{Young}} = 270 \pm 100$ GPa was obtained, close to Young's modulus of MoS$_2$ nanotubes (230 GPa,(Lee et al., 2008)), or steel (210 GPa) and only ~4 times smaller than that of graphene (1 TPa, (Lee et al., 2008)). The Young's modulus of bilayer MoS$_2$ was found to be 200 $\pm$ 60 GPa, a value little smaller than that for a ML, which was attributed as 'possibly due to defects or interlayer sliding' (Bertolazzi et al., 2011). Regarding the breaking of MoS$_2$, the average of maximum stress values were, respectively, 22 ± 4 GPa and 21 ± 6 GPa for a ML and a bilayer. On average, these values correspond to 8 and 10% of Young's modulus for ML and bilayer MoS$_2$. It was noted that these values are at the theoretical upper limit of a material's breaking strength and thus represent the intrinsic strength of interatomic bonds in MoS$_2$. It was further noted that the strength of ML MoS$_2$ is exceeded only by CNTs and graphene (Bertolazzi et al., 2011). Stretching and breaking of ML MoS$_2$ was also studied using atomistic simulations,(Lorenz et al., 2014b) with the obtained values of $E_{\text{Young}} = 262$ GPa and a breaking stress of 21 GPa being in good agreement with the experimental results. The mechanical properties of ML MoS$_2$ have been reported (Lorenz et al., 2014b).

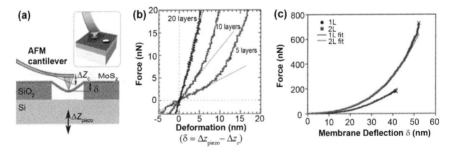

FIGURE 1.15 (a) Schematic diagram of the nanoscopic bending test experiment carried out on a freely suspended MoS$_2$ nanosheet. (b) Force versus deformation traces measured at the center of the suspended part of MoS$_2$ nanosheets with 5, 10, and 20 layers in thickness (*Source:* Reprinted with permission from van der Zant et al. (2012)) OA; (c) Loading curves for single and bilayer MoS$_2$ (*Source:* Reprinted with permission from Bertolazzi, Brivio, and Kis (2011)). © American Chemical Society.

Mechanical response of various TMDCs to large elastic deformation was studied using first-principles density functional calculations (Lorenz et al.,

2014b). In Figure 1.16a–c, the calculated stress-strain relations are shown under arm-chair (x), zig-zag (y), and biaxial tensions, respectively. For small strain, the stress for all MX_2 exhibits linear dependence on the applied strain for all loading directions. As MX_2 MLs are strained further ($\epsilon > 4\%$), the stress-strain response deviates from the linear behavior. For large strains, the hexagonal symmetry is broken, with the stress developed upon loading in the arm-chair direction (Figure 1.16a) much larger than in the zig-zag direction (Figure 1.16b). Upon straining further, the stress continues to increase until it reaches a maximum, termed the ultimate strength σ^*. It was found that in general, the chalcogens of W (WX_2) have larger moduli and tensile strength than those of Mo (MoX_2), while for the same transition metal, sulfides (MS_2) are the strongest, and tellurides (MTe_2) are the weakest. The anisotropy in stress response was found to be inversely correlated with the strength of the ML sheets: the MX_2 with lower Young's modulus and ultimate strength (for example, tellurides) are characterized by larger anisotropy factors. A direct correlation between the amount of charge transfer from the transition metal to the chalcogens and material's elastic properties were found. The results are shown in Figure 1.16d, e, that depict Young's modulus, E, and ultimate strength, σ^*, as a function of the (Bader) charge transfer, ΔQ and demonstrate that the mechanical properties of TMDCs exhibit a linearly increasing relation with the charge transfer from the transition metal atom to the chalcogen atoms (Li et al., 2013). Local deformations in free-standing MoS_2 sheets were theoretically studied and it was concluded that the electronic structure was robust with respect to local deformations (Lorenz et al., 2014a). Theoretical studies of the mechanical behavior of TiS_2 and MoS_2 arm-chair and zig-zag nanotubes were performed (Nanotubes et al., 2012). It was found that the Poisson ratio and Young's modulus depend on the tube diameter for smaller tubes (up to 20 Å in radius) and subsequently approach that of infinite MLs. The Young's moduli for armchair and zig-zag nanotubes were different for the $2H$ phases of both MoS_2 and TiS_2 (the former being larger in the case of TiS_2 and smaller for MoS_2) and essentially the same for $1T$-TiS_2. The influence of defects on the strain of rupture was also studied and it was found that the presence of various kinds of defects decreased the rupture stress from ~29 GPa for defect-free zig-zag MoS_2 nanotubes to ~19 GPa in nanotubes with defects. Cooper et al. (2013) have theoretically investigated the nonlinear elastic behavior of 2D MoS_2. Besides Young's modulus; the bending modulus is another fundamental mechanical property. The obtained elastic bending modulus of 9.61 eV in single-layer MoS_2 was significantly higher than the bending modulus of graphene (1.4 eV) (Jiang et al., 2013a).

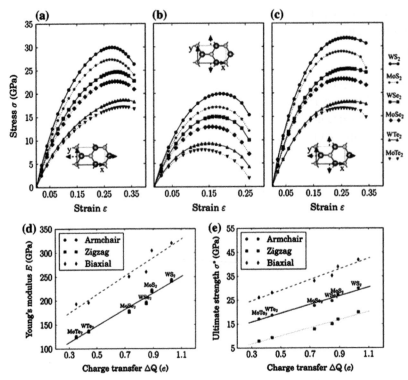

FIGURE 1.16 Tensile stress, σ, as a function of uniaxial strain, ε, along the (a) arm-chair and (b) zig-zag directions and (c) of biaxial strain, respectively, for monolayer MX_2 (M = Mo, W; X = S, Se, Te) TMDCs. Solid and dashed lines are used for WX_2 and MoX_2, respectively. Variation of (d) Young's modulus and (e) the ultimate strength of monolayer MX_2 (M = Mo, W; X = S, Se, Te) TMDCs with the charge transfer, ΔQ, from transition metal M to chalcogens, X. The Young's modulus and ultimate strength for biaxial strain are shifted rigidly upward for drawing purposes by 50 and 10 GPa, respectively.

Source: Reprinted with permission from Li, Medhekar, and Shenoy et al. (2013). © American Chemical Society.

1.1.3.2   THERMAL PROPERTIES

The thermal conductivity of ML MoS_2 was theoretically studied using the phonon Boltzmann transport equation combined with relaxation time approximation for transverse acoustic (*TA*), longitudinal acoustic (*LA*), and out-of-plane acoustic (*ZA*) phonons (Shen et al., 2014). At the same time, using first-principles simulations, the thermal conductivity was determined for samples of different sizes and it was concluded that for a typical sample size of 1 μm the thermal conductivity κ should be larger than 83 W/mK at room temperature, (Li et al., 2013) demonstrating a disagreement with the previous results. On the other hand, in a study of the anharmonic behavior of phonons and the intrinsic thermal conductivity associated with umklapp scattering in ML MoS_2, the room temperature thermal conductivity of ML MoS_2 was found to be around 23.2 W/mK (Cai et al., 2014). This value is contrasted by the DFT simulations which yielded thermal conductivity of only 1.35 W/mK in the case of a ML of MoS_2 (Wang and Tabarraei, 2016). Thermal conductivity was shown to be anisotropic. In particular, the thermal conductivity at room temperature for the armchair MoS_2 nanoribbon was found to be about 673.6 W/mK, while a value of 841.1 W/mK was obtained for the zig-zag nanoribbon (Jiang et al., 2013b). Experimentally determined values (from temperature-dependent Raman scattering) are in the range of 34–52 W/mK for MLs of both MoS_2 and WS_2 (Cong et al., 2014b; Sahoo et al., 2013; Yan et al., 2014). Thermal conductivity decreased with increasing temperature from 62.2 W/mK at 300 K to 7.45 W/mK at 450 K (Taube et al., 2015).

1.1.3.3   ELECTRICAL PROPERTIES

In 2D TMDC layers, transport, and scattering of the carriers are confined to the plane of the material. Carrier in-plane mobility is related to the momentum scattering time τ_D by $\mu = e\tau_D/m^*$, where m^* is the in-plane effective mass. The mobility of carriers is affected by (i) acoustic and optical phonon scattering; (ii) Coulomb scattering at charged impurities; (iii) surface interface phonon scattering; and (iv) roughness scattering. The degree to which these scattering mechanisms affect the carrier mobility is influenced by the layer thickness, carrier density, temperature, effective carrier mass, electronic band structure, and phonon band structure. At low temperatures (T <100 K), the acoustic component dominates, but at higher temperatures the optical component dominates. Coulomb scattering in 2D TMDCs is caused by random charged impurities located within the 2D TMDC layer or on its surfaces, and is the dominant scattering effect at low temperatures. The effect of surface phonon scattering and roughness scattering can be very important in extremely thin 2D materials. The phonon-limited room-temperature mobility was calculated for MoS_2 to be ~410 cm^2/Vs, (Kaasbjerg et al., 2012) and similar values are expected for other single-layer TMDCs. Using first-principle calculations combined with the Boltzmann transport equation, mobility of electrons and holes in different structural modifications of MoS_2 was calculated (Kan et al., 2014). In $2H$-MoS_2, the electron (hole) mobility was found to be 1.2 (3.8) $\times$ 10^2 cm^2/Vs and was isotropic. On the other hand, in ZT-MoS_2 the electron and hole mobility was 1–2 orders of magnitude larger and anisotropic (the electron (hole) mobility of ZT-MoS_2 was 4.1 (2.1) $\times10^3$ and 6.4 (5.7) $\times10^4$ cm^2/Vs along the x and y directions, respectively). The increase in the mobility was attributed to the reduction of electron (hole) effective mass from 0.49 (0.60) m_e to 0.12 (0.05) m_e when $2H$-MoS_2 was transformed to ZT-MoS_2. It was noted that the mobility of ZT-MoS_2 is higher than that of silicon and comparable to that of graphene nanoribbons. Experimentally the room temperature values of mobility were studied using device structures with the maximum value determined as up to 700 cm^2/Vs for multilayer MoS_2 at room temperature (Das et al., 2013; Kim et al., 2012; Pradhan et al., 2015). MoS_2 field-effect transistors on both SiO_2 and polymethyl methacrylate (PMMA) dielectrics have been reported and their charge carrier mobility was measured in a four-probe configuration (Das et al., 2013). For multilayer MoS_2 on SiO_2, the mobility was 30–60 cm^2/Vs, relatively independent of thickness (15–90 nm); most devices exhibited unipolar n-type behavior. In contrast, multilayer MoS_2 on PMMA showed increased mobility with thickness, up to 470 cm^2/Vs (electrons), and 480 cm^2/Vs (holes) at thickness ~50 nm (Bao et al., 2013). It is interesting to note that the obtained values depend on the measurement conditions. Thus, for a back-gated two-terminal configuration, mobility in MoS_2 of ~90 cm^2/Vs was observed, which is considerably smaller than 306 cm^2/Vs extracted from the same device when using a four-terminal configuration (Pradhan et al., 2015). This indicates the important limiting role of non-Ohmic contacts in some measurements. Using contact-less mobility determination based on THz spectroscopy, room temperature intrinsic mobility was determined to be ~250 cm^2/Vs at room temperature increasing up to 4200 cm^2/Vs at 30 K (Strait et al., 2014).

1.1.3.4   OPTICAL PROPERTIES

The dielectric function of ML TMDCs was determined by reflectance measurements of the samples at room temperature. The absolute reflectance spectra for the TMDC MLs on fused silica are presented in Figures 1.17a–d. For all four TMDC MLs, the two lowest energy peaks in the reflectance spectra correspond to the excitonic features associated with interband transitions at the $K(K')$ point in the Brillouin zone (Kastner, 1972). The two features, denoted by *A* and *B*, correspond to the splitting of the valence band (VB) by spin-orbit coupling. At higher photon energies spectrally broad response from higher-lying interband transitions were observed. Dielectric functions were obtained from the Kramers-Kronig constrained analysis. In Figures 1.17e–l, resultant real and imaginary parts are shown for $MoSe_2$, WSe_2, MoS_2, and WS_2 over the spectral range of 1.5–3.0 eV. The obtained dielectric functions for the ML TMDC crystals in comparison with the dielectric functions for the corresponding bulk materials are shown in Figure 1.18. While the two data sets show an overall similarity, differences in the spectral responses are readily seen, such as broadening of the resonance features in the bulk materials compared to the MLs, which was attributed to the additional optical transitions and carrier relaxation channels arising from interlayer coupling. Also, the resonance energies in the ML dielectric function are modestly shifted from the corresponding bulk material. Optical properties of ML transition metal dichalcogenides were investigated using spectroscopic ellipsometry (Hsu et al., 2014).

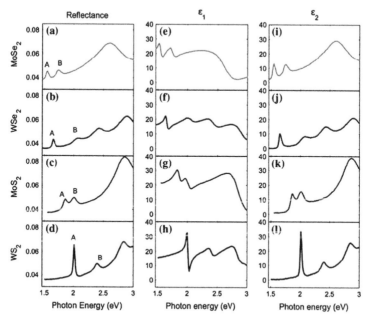

FIGURE 1.17 Optical response of monolayers of $MoSe_2$, WSe_2, MoS_2, and WS_2 exfoliated on fused silica: (a–d) Measured reflectance spectra. (e–h) Real part of the dielectric function, ϵ_1. (i–l) Imaginary part of the dielectric function, ϵ_1. The peaks labeled A and B in (a–d) correspond to excitons from the two spin-orbit split transitions at the K point of the Brillouin zone. *Source:* Reprinted with permission from Li et al. (2014).

Figure 19a–d shows the refractive index, n, and extinction coefficient, k, spectra of ML MoS_2, $MoSe_2$, WS_2, and WSe_2 thin films obtained from these measurements. In all cases, the refractive indices increase with increasing wavelength in the spectral range from 193–550 nm, and then approach the maxima, and decrease with wavelength until 1700 nm. It was noted that (i) the dispersive response in the refractive index exhibits several anomalous dispersion features below 800 nm and approaches a constant value of 3.5–4.0 in the near-infrared frequency range; and (ii) ML MoS_2 has the extraordinary large value of refractive index about 6.50 at 450 nm. Optical constants, n, and k, for ultrathin NbS_2 and MoS_2 crystals have been determined for the visible range (Agraït and Agraït, 2014). Non-linear optical properties of TMDC nanosheets were also studied (Dong et al., 2015).

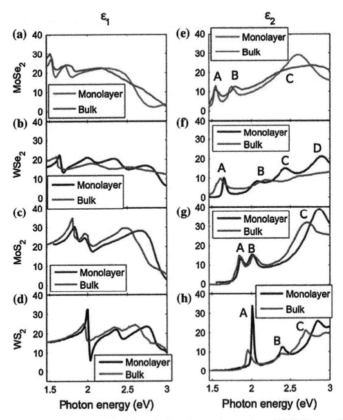

FIGURE 1.18 Comparison of the dielectric functions of monolayer TMDC crystals (colored lines) with those of the corresponding bulk materials (gray).
Source: Reprinted with permission from Li et al. (2014).

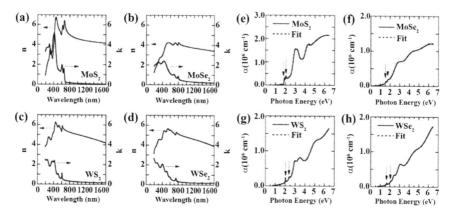

FIGURE 1.19 Refractive index n and extinction coefficient k of monolayer (a) MoS$_2$, (b) MoSe$_2$, (c) WS$_2$, and (d) WSe$_2$ thin films and optical absorption coefficient α of monolayer (e) MoS$_2$, (f) MoSe$_2$, (g) WS$_2$, and (h) WSe$_2$ thin films.
Source: Reprinted with permission from Hsu et al. (2014).

1.1.3.5   THERMOELECTRIC PROPERTIES

The thermoelectric performance of ML transition-metal dichalcogenides (TMDs), MoS$_2$, MoSe$_2$, WS$_2$, and WSe$_2$ was studied using a two-dimensional ballistic transport approach based on the full electronic band structures, (Guo et al., 2014; Huang et al., 2013) and phonon energy dispersion relations obtained from first-principles calculations with different crystal orientations and temperatures for *n*-type and *p*-type materials. It was found that figures of merit of these materials are generally low. The thermoelectric properties of bulk and ML MoSe$_2$ and WSe$_2$ were also studied by first-principles calculations and semi-classical Boltzmann transport theory (Kumar and Schwingenschlögl, 2015). WSe$_2$ was found to be superior to MoSe$_2$ for thermoelectric applications. Analytical results for thermoelectric transport in ML MoS$_2$ and related group-VI dichalcogenides in the presence of off-resonant light were derived (Tahir and Schwingenschlögl, 2014). It was shown that an increased intensity of light reduces the direct bandgap and results in a strong spin splitting in the conduction band (CB) and, therefore, in a dramatic enhancement of thermoelectric transport.

1.2   ENERGY APPLICATIONS OF METAL CHALCOGENIDE COMPOUNDS

Owing to their excellent redox behavior and electrochemical stability, TMCs have been investigated extensively for their direct application in energy harvesting and storage. Loosely bound layer structures, highly defect oriented and large number of active sites not only enables TMCs to show high catalytic behavior towards water splitting reactions, but also their large surface area and 2D analogy with graphene allows for the fabrication of highly efficient, low-cost, and stable energy storage devices such as battery and ECs (supercapacitors). The following sections will emphasize the plausibility of these excellent analogs of graphene in energy sectors, especially renewable energy storage along with their extended use towards various other important applications.

1.2.1   ENERGY STORAGE: SUPERCAPACITOR

The plethora of physicochemical properties shown by TMCs showcases their excellent versatility when it comes to a myriad of applications. Along with carbon-based electric double-layer capacitors (EDLC), these redox rich materials are revolutionizing the supercapacitor research with their high values of energy densities, retaining at the same time their fast and reversible charge transfer properties inside suitable electrolytic solvents. TMCs have been used both in batteries and supercapacitors, due to their atomically layered structure, high surface area, and excellent electrochemical properties. The layered structures facilitate intercalation or trapping of charged species without significant degradation over thousands of repeated charging-discharging cycles. The layered structures and numerous surface irregularities (due to the presence of functional groups and dangling bonds) provide ample opportunity for the ad-atoms to get adsorbed on the surface through both physisorption and chemisorption process. Another advantage with these layered materials is their strong analogy with graphene. This opens up numerous hybridization options, where TMCs are hybridized with certain percentages of graphene to enhance their cyclic stability and provide robust electrochemical performance which would otherwise not possible in the case of TMCs in their pristine forms. There are several reports on the excellent supercapacitive behavior of TMCs and their graphene hybrids (MarriThese Authors Contributed Equally et al., 2017; Ratha et al., 2015, 2016; Ratha and Rout, 2013).

Among TMCs, only MnO_2 and RuO_2 have been distinguished as purely pseudocapacitive materials, i.e., their characteristic current vs. potential response more or less mimic those obtained from pure EDLC materials (Ratha and Samantara, 2018). The underlying mechanism has been detailed by many researchers. One of the most extensively followed concepts is the rapid and highly reversible redox couples of these two TMC materials inside electrolytic solvents. The redox couples consist of underpotential, normal, and overpotential redox peaks that are fast and extremely close to each other, making it hard to distinguish them. The combination of series of such fast and reversible redox peaks tend to add up effecting a peak broadening process and imparting a cyclic voltammetry curve just like that of an EDLC material.

However, the concept of redox peak overlapping and peak broadening can't be conceptually correct, considering the fact that the distribution of faradic activity over a potential range of even 0.5 V (working potential window) as has been claimed previously would result in erratic and unrealistic standard potential distribution value (Costentin et al., 2017). During the electrochemical reaction, when the oxidation potential is overcome, the material turns into a band-conducting material and accumulates charge on the electrode surface rather than following a chemisorption process (as has been believed to be so far). However, other binary and/or ternary metal oxides also have been investigated for their charge storage application in supercapacitor devices (Qiu et al., 2015; Ratha et al., 2017; Ratha and Rout, 2015; Sahoo et al., 2015; Samantara et al., 2018).

However, in comparison to metal oxide-based compounds, transition metal sulfides, selenides have recently shown better electrochemical activities. This might be attributed to the high intrinsic conductivity of sulfide and selenide based compounds. In addition, numerous unique and exclusive synthesis processes have also been developed to afford these TMCs, so that they can be implemented on a large scale basis, to meet the huge demand for the manufacture of EES devices. The most prominent member of TMC group is MoS_2, which has been studied for its wide range of applicability. Similar layered metal sulfides have also been reported to have excellent charge storage prospects. Sulfides of transition metals, e.g., MoS_2, TiS_2, VS_2, WS_2, CoS, and NiS, have been studied for their possible supercapacitor applications (Hu et al., 2016; Wang et al., 2016). Their high intrinsic conductivity and fast surface diffusion controlled redox activities are advantageous for both battery and supercapacitor electrodes. These materials have shown better electrochemical activities as compared to their oxide counterparts. However, the long cyclic stability and electrical conductivity are still far from that of graphene-based EDLCs. Thus, in most cases, these sulfides, or selenide based compounds are hybridized with RGO to get benefitted from the high electrochemical stability provided by graphene and excellent faradic charge storage capabilities of the TMCs.

Qu et al. have reported to have synthesized β-cobalt sulfide ($CoS_{1.097}$) NPs decorated on conductive graphene nano-composite, displaying a superior specific capacitance of 1535 F/g at a normalized current of 2.0 A/g, and a high capacitance of 725 F/g at even extremely high normalized current of 40 A/g (Qu et al., 2012). In another report, Wang et al. have afforded uniformly decorated NiS NPs on the GO film, which yielded a specific capacitance of 800 F/g at 1.0 A/g (Wang et al., 2013). This clearly suggests that hybridizing these TMC layer structures with graphene imparts high stability and charge storage capability on the hybrids. However, the electrical conductivity of RGO is restricted by the deficiency of conjugate electron on RGO panel. To address this issue, Yan et al. have successfully synthesized α-NiS dispersed on the RGO surface and hybridized them with single-walled carbon nano-tubes (SWCNTs) to obtain nanohybrids (Yan et al., 2015). Upon investigating the resultant electrical conductivity of the nanohybrids, it was found that the electrochemical activity of the ternary hybrid showed enhanced charge storage as compared to α-NiS/RGO hybrid. This can attributed to the higher electrical conductivity of SWCNTs and its role in inhibiting NiS agglom-eration. The rate of diffusion of electrolytic components (ion channeling) is crucial to take advantage of the electrochemical accessible surface area, and plays a significant role in the charge storage process. In this context, Lin et al. have developed a glucose-assisted hydrothermal method coupled with CVD for preparing Co_9S_8/3DG nanocomposites (Lin et al., 2015). Owing to the uniform deposition of Co_9S_8 NPs on conductive 3DG, and high electrical conductivity of 3DG combined with the open-pore channels for electrolyte penetration, a high specific capacitance of 1721 F/g could be achieved at a normalized current of 16 A/g. Apart from the electron/ion transfer, the morphology and structure of sulfides can also influence the performance of hybrids. Recently, Abdel Hamid et al. developed novel graphene-wrapped NiS nanoprisms for application in Li-ion batteries and supercapacitors by controlling the morphology and structure of NiS (Abdel et al., 2016). As a supercapacitor electrode, graphene-wrapped nickel sulfide nanoprisms demonstrated a high specific capacitance that exceeded 1000 F/g at a normal-ized current of 5.0 A/g. Other graphene-based metal sulfide nanocomposites, such as CuS/RGO, (Huang et al., 2015b) MoS_2/N-doped graphene, (Xie et al., 2016) WS_2/rGO, (Ratha and Rout, 2013) have also been investigated as electrode materials. Mixed metal sulfides, especially ternary nickel-cobalt sulfides, exhibited great potential to be implemented as high-performance supercapacitor electrodes owing to their richer redox activity as compared to metal sulfides having single core metallic cation (Yu and Lou, 2018). Peng et al. have reported ultrathin $NiCo_2S_4$ nanosheets anchored on RGO sheets as electrode materials of supercapacitors, exhibiting higher specific capacitance, better rate performance, and superior cycling life than bare $NiCo_2S_4$ (Peng et al., 2013). Simply preparing physical mixtures by mixing RGO and mixed metal sulfides at a graphene percentage as low as 5 wt% could still yield enhanced capacitance in the case of $CoNi_2S_4$/graphene hybrids (Du et al., 2014). Recently, Yang et al. ingeniously developed edge site-enriched nickel-cobalt sulfide (Ni-Co-S) NPs loaded on graphene frameworks via an in-situ anion-exchange process (Figure 1.20) (Yang et al., 2016). They considered that the etching-like behavior resulted from the S^{2-} ions was the

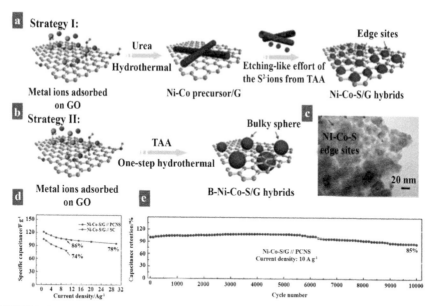

FIGURE 1.20 Schematic illustrations of (a) the in situ integration process of edge site-enriched Ni-Co-S nanoparticles on graphene substrate (Strategy I) and (b) the directly growth of bulk Ni-Co-S particles on graphene via the one-step hydrothermal method (Strategy II). (c) TEM image of the integrated edge site-enriched Ni-Co-S/graphene hybrids. (d) The specific capacitance at different current densities and (e) cycling performance at a constant current density of 10 A/g of an asymmetric supercapacitor made of the Ni-Co-S/graphene hybrid (positive electrode)//porous carbon nanosheets (negative electrode).
Source: Reprinted with permission from Yang et al. (2016). © Royal Society of Chemistry.

major contribution towards sufficient edge active sites on Ni-Co-S NPs; moreover, the edge sites were certified to afford strong interaction with OH$^-$. Therefore, the synergistic effect of edge sites and graphene dramatically facilitated the electrochemical reaction kinetics, delivering a high specific capacitance of 1492 F/g at a normalized current of 1.0 A/g and an ultrahigh retention of 96% at 50 A/g.

The hybrids of metal selenides and graphene were also applied in supercapacitors. There are reports on excellent electrochemical energy storage performances of VSe$_2$ and its RGO hybrids, (MarriThese Authors Contributed Equally et al., 2017) MoSe$_2$/RGO, (Balasingam et al., 2016) CoSe$_2$, etc. Huang et al. proposed a MoSe$_2$-graphene grown on the Ni foam substrate through a facile hydrothermal method (Huang et al., 2015a). The MoSe$_2$/graphene with an optimum proportion of 7:1 yielded a superior specific capacitance of 1422 F/g; more importantly, the loss of capacity was hardly observed even after 1500 cycles. CoSe NPs in-situ grown on the graphene sheets were also evaluated as nanohybrid electrodes, offering a high energy density of 45.5 Wh/kg and capacitance retention of 81% after 5000 cycles (Kirubasankar et al., 2017).

1.3 CONCLUSION AND FUTURE PROSPECTS

TMCs are wonder materials that can provide path-breaking results in various fields of application. The level of ongoing research on these materials are expected to yield several promising aspects of these excellent performers, whether it is catalysis or charge trapping through faradic process, could pave the way for a myriad of possibilities in realizing the green energy revolution. The presence of edge plane and basal plane and resulting electrochemistry makes these TMCs highly versatile, and thus can be taken for virtually any kind of applications. They are highly electro-active, and are abundantly available in the earth's crust, which can address the issues that are currently faced by the Li-ion industries (owing to the limited lithium reserve). In the context of energy storage, especially supercapacitors, these materials provide wide range of choices. They can be combined with EDLC materials to form an asymmetric arrangement of electrodes that could boost the working potential window and strike a balance between energy and power density. Furthermore, TMCs having purely faradic behavior can be coupled with EDLC materials to create hybrid systems taking advantage of the fast charging of EDLC electrode and slow discharging of the faradic/battery electrode.

These designs based on TMCs could solve several bottlenecks that are still posing challenges towards realizing the complete global implementation of renewables and eradication of fossil fuel assisted energy generation.

KEYWORDS

- **chemical vapor deposition**
- **electrical energy storage**
- **metal chalcogenides**
- **molecular beam epitaxy**
- **physicochemical properties**
- **supercapacitor**

REFERENCES

Abdel, H. A. A., Yang, X., Yang, J., Chen, X., & Ying, J. Y., (2016). Graphene-wrapped nickel sulfide nanoprisms with improved performance for Li-ion battery anodes and super capacitors. *Nanoenergy, 26*, 425–437.

Agraït, N., & Agraït, N., (2014). Optical identification of atomically thin dichalcogenide crystals optical identification of atomically thin dichalcogenide crystals. *Appl. Phys. Lett., 213116*, 55–58.

Alexandrou, I., Sano, N., Burrows, A., Meyer, R. R., Wang, H., Kirkland, A. I., Kiely, C. J., et al., (2003). Structural investigation of MoS$_2$ core-shell nanoparticles formed by an arc discharge in water. *Nanotechnology, 14*, 913–917.

Balasingam, S. K., Lee, J. S., & Jun, Y., (2016). Molybdenum diselenide/reduced graphene oxide based hybrid nanosheets for super capacitor applications. *Dalt. Trans., 45*, 9646–9653.

Bao, W., Cai, X., Kim, D., Sridhara, K., & Fuhrer, M. S., (2013). High mobility ambipolar MoS$_2$ field-effect transistors: Substrate and dielectric effects. *Appl. Phys. Lett., 102*, 42104.

Bastide, S., Duphil, D., Borra, J. P., & Lévy-Clément, C., (2006). WS$_2$ closed nanoboxes synthesized by spray pyrolysis. *Adv. Mater., 18*, 106–109.

Bertolazzi, S., Brivio, J., & Kis, A., (2011). Supplement: Stretching and breaking of ultrathin MoS$_2$. *ACS Nano, 5*, 9703–9709.

Cai, Y., Lan, J., Zhang, G., & Zhang, Y. W., (2014). Lattice vibrational modes and phonon thermal conductivity of monolayer MoS$_2$. *Phys. Rev. B-Condens. Matter Mater. Phys., 89*, 35438.

Castellanos-Gomez, A., Barkelid, M., Goossens, A. M., Calado, V. E., Van, D. Z. H. S. J., & Steele, G. A., (2012). Laser-thinning of MoS$_2$: On demand generation of a single-layer semiconductor. *Nano Lett., 12*, 3187–3192.

Chakravarty, D., & Late, D. J., (2015). Exfoliation of bulk inorganic layered materials into nanosheets by the rapid quenching method and their electrochemical performance. *Eur. J. Inorg. Chem.*, 1973–1980.

Chang, Y. H., Zhang, W., Zhu, Y., Han, Y., Pu, J., Chang, J. K., et al., (2014). Monolayer $MoSe_2$ grown by chemical vapor deposition for fast photo detection. *ACS Nano, 8*, 8582–8590.

Chen, H., Cong, T. N., Yang, W., Tan, C., Li, Y., & Ding, Y., (2009). Progress in electrical energy storage system: A critical review. *Prog. Nat. Sci.*

Chen, L., Liu, B., Abbas, A. N., Ma, Y., Fang, X., Liu, Y., & Zhou, C., (2014). Screw-dislocation-driven growth of two-dimensional few-layer and pyramid-like WSe_2 by sulfur-assisted chemical vapor deposition. *ACS Nano, 8*, 11543–11551.

Cheng, Y., Yao, K., Yang, Y., Li, L., Yao, Y., Wang, Q., Zhang, X., Han, Y., & Schwingenschlögl, U., (2013). Van Der Waals epitaxial growth of MoS_2 on SiO_2/Si by chemical vapor deposition. *RSC Adv., 3*, 17287–17293.

Coleman, J. N., Lotya, M., O'Neill, A., Bergin, S. D., King, P. J., Khan, U., Young, K., et al., (2011). Two-dimensional nanosheets produced by liquid exfoliation of layered materials. *Science, 80, 331*, 568–571.

Cong, C., Shang, J., Wu, X., Cao, B., Peimyoo, N., Qiu, C., Sun, L., et al., (2014a). Synthesis and optical properties of large-area single-crystalline 2D semiconductor WS_2 monolayer from chemical vapor deposition. *Adv. Opt. Mater., 2*, 131–136.

Cong, C., Wang, Y., Shang, J., Peimyoo, N., Yu, T., & Yang, W., (2014b). Thermal conductivity determination of suspended mono- and bilayer WS_2 by Raman spectroscopy. *Nano Res., 8*, 1210–1221.

Cooper, R. C., Lee, C., Marianetti, C. A., Wei, X., Hone, J., & Kysar, J. W., (2013). Nonlinear elastic behavior of two-dimensional molybdenum disulfide. *Phys. Rev. B-Condens. Matter Mater. Phys., 87*, 35423.

Costentin, C., Porter, T. R., & Savéant, J. M., (2017). How do pseudocapacitors store energy? theoretical analysis and experimental illustration. *ACS Appl. Mater. Interfaces, 9*, 8649–8658.

Cunningham, G., Lotya, M., Cucinotta, C. S., Sanvito, S., Bergin, S. D., Menzel, R., Shaffer, M. S. P., et al., (2012). Solvent exfoliation of transition metal dichalcogenides: Dispersibility of exfoliated nanosheets varies only weakly between compounds. *ACS Nano, 6*, 3468–3480.

Das, S., Chen, H. Y., Penumatcha, A. V., & Appenzeller, J., (2013). High performance multilayer MoS_2 transistors with scandium contacts. *Nano Lett., 13*, 100–105.

Dines, M. B., (1975). Lithium intercalation via view the math ML source-butyllithium of the layered transition metal dichalcogenides. *Mater. Res. Bull., 10*, 287–291.

Dobrik, G., Magda, G. Z., Tapasztó, L., Biró, L. P., Hwang, C., & Pető, J., (2015). Exfoliation of large-area transition metal chalcogenide single layers. *Sci. Rep., 5*, 14714.

Dong, L., Chu, Y., Zhuo, Y., & Zhang, W., (2009). Two-minute synthesis of PbS nanocubes with high yield and good dispersibility at room temperature. *Nanotechnology, 20*, 125301.

Dong, N., Li, Y., Feng, Y., Zhang, S., Zhang, X., Chang, C., Fan, J., et al., (2015). Optical limiting of layered transition metal dichalcogenide semiconductors. *Sci. Rep., 5*, 14646.

Donley, M. S., McDevitt, N. T., Haas, T. W., Murray, P. T., & Grant, J. T., (1989). Deposition of stoichiometric MoS_2 thin films by pulsed laser evaporation. *Thin Solid Films, 168*, 335–344.

Donley, M. S., Murray, P. T., Barber, S. A., & Haas, T. W., (1988). Deposition and properties of MoS_2 thin films grown by pulsed laser evaporation. *Surf. Coatings Technol., 36*, 329–340.

Du, W., Wang, Z., Zhu, Z., Hu, S., Zhu, X., Shi, Y., Pang, H., & Qian, X., (2014). Facile synthesis and superior electrochemical performances of $CoNi_2S_4$/graphene nanocomposite suitable for supercapacitor electrodes. *J. Mater. Chem. A, 2*, 9613–9619.

Eda, G., Yamaguchi, H., Voiry, D., Fujita, T., Chen, M., & Chhowalla, M., (2012). Photoluminescence from chemically exfoliated MoS$_2$ (Vol. 11, pp. 5111, 2011). *Nano Lett., 12*, 526.

Egbue, O., & Long, S., (2012). Critical issues in the supply chain of lithium for electric vehicle batteries. *EMJ-Eng. Manag. J., 24*, 52–62.

Frey, G. L., Reynolds, K. J., Friend, R. H., Cohen, H., & Feldman, Y., (2003). Solution-processed anodes from layer-structure materials for high-efficiency polymer light-emitting diodes. *J. Am. Chem. Soc., 125*, 5998–6007.

Frindt, R. F., (1966). Single crystals of MoS$_2$ several molecular layers thick. *J. Appl. Phys., 37*, 1928–1929.

Galván, D. H., Kim, J. H., Maple, M. B., & Adem, E., (2001). Effect of electronic irradiation in the production of NbSe$_2$ nanotubes. *Fuller. Sci. Technol., 9*, 225–232.

Galvan, D. H., Kim, J. H., Maple, M. B., Avalos-Borja, M., & Adem, E., (2000). Formation of NbSe$_2$ nanotubes by electron irradiation. *Fuller. Sci. Technol., 8*, 143–151.

Ganorkar, S., Kim, J., Kim, Y. H., & Kim, S. Il., (2015). Effect of precursor on growth and morphology of MoS$_2$ monolayer and multilayer. *J. Phys. Chem. Solids, 87*, 32–37.

Geim, A. K., & Novoselov, K. S., (2007). The rise of graphene. *Nat. Mater., 6*, 183–191.

Geim, A. K., (2009). Graphene: Status and prospects. *Science, 324*, 1530–1534.

Ghezelbash, A., & Korgel, B. A., (2005). Nickel sulfide and copper sulfide nanocrystal synthesis and polymorphism. *Langmuir, 21*, 9451–9456.

Gordon, R. A., Yang, D., Crozier, E. D., Jiang, D. T., & Frindt, R. F., (2002). Structures of exfoliated single layers of WS$_2$, MoS$_2$, and MoSe$_2$ in aqueous suspension. *Phys. Rev. B, 65*, 125407.

Gudiksen, M. S., & Lieber, C. M., (2000). Diameter-selective synthesis of semiconductor nanowires [13]. *J. Am. Chem. Soc., 122*, 8801–8802.

Guo, H. H., Yang, T., Tao, P., & Zhang, Z. D., (2014). Theoretical study of thermoelectric properties of MoS$_2$. *Chinese Phys. B, 23*, 10866–10874.

Gurarslan, A., Yu, Y., Su, L., Yu, Y., Suarez, F., Yao, S., Zhu, Y., et al., (2014). Surface-energy-assisted perfect transfer of centimeter-scale monolayer and few-layer MoS$_2$ films onto arbitrary substrates. *ACS Nano, 8*, 11522–11528.

Gutiérrez, H. R., Perea-López, N., Elías, A. L., Berkdemir, A., Wang, B., Lv, R., López-Urías, F., et al., (2013). Extraordinary room-temperature photoluminescence in triangular WS$_2$ monolayers. *Nano Lett., 13*, 3447–3454.

He, L., Kou, X., & Wang, K. L., (2013). Review of 3D topological insulator thin-film growth by molecular beam epitaxy and potential applications. *Phys. Status Solidi-Rapid Res. Lett., 7*, 50–63.

Hernandez, Y., Nicolosi, V., Lotya, M., Blighe, F. M., Sun, Z., De, S., McGovern, I. T., et al., (2008). High-yield production of graphene by liquid-phase exfoliation of graphite. *Nature Nanotechnol., 3*, 563–568.

Hsu, C. L., Li, L. J., Su, S. H., Li, M. Y., Liu, H. L., & Shen, C. C., (2014). Optical properties of monolayer transition metal dichalcogenides probed by spectroscopic ellipsometry. *Appl. Phys. Lett., 105*, 201905.

Hu, H., Guan, B. Y., Lou, X. W., & David, (2016). Construction of complex CoS hollow structures with enhanced electrochemical properties for hybrid super capacitors. *Chem., 1*, 102–113.

Huang, J. K., Pu, J., Hsu, C. L., Chiu, M. H., Juang, Z. Y., Chang, Y. H., Chang, W. H., Iwasa, Y., Takenobu, T., & Li, L. J., (2014). Large-area synthesis of highly crystalline WSe$_2$ monolayer's and device applications. *ACS Nano, 8*, 923–930.

Huang, K. J., Zhang, J. Z., & Cai, J. L., (2015a). Preparation of porous layered molybdenum selenide-graphene composites on Ni foam for high-performance super capacitor and electrochemical sensing. *Electrochim. Acta, 180*, 770–777.

Huang, K. J., Zhang, J. Z., Liu, Y., & Liu, Y. M., (2015b). Synthesis of reduced graphene oxide wrapped-copper sulfide hollow spheres as electrode material for super capacitor. *Int. J. Hydrogen Energy, 40*, 10158–10167.

Huang, S. C., Kaner, R. B., Chen, T. H., Ding, Z., Bux, S. K., Chang, F. L., & King, D. J., (2009). Lithium intercalation and exfoliation of layered bismuth selenide and bismuth telluride. *J. Mater. Chem., 19*, 2588.

Huang, W., Da, H., & Liang, G., (2013). Thermoelectric performance of MX$_2$ (M Mo, W; X S, Se) monolayers. *J. Appl. Phys., 113*, 104304–1043047.

Huang, Y. K., Cain, J. D., Peng, L., Hao, S., Chasapis, T., Kanatzidis, M. G., Wolverton, C., et al., (2014). Evaporative thinning: A facile synthesis method for high quality ultrathin layers of 2D crystals. *ACS Nano, 8*, 10851–10857.

Jeon, M. H., Ahn, C., Kim, H. U., Kim, K. N., Lin, T. Z., Qin, H., Kim, Y., et al., (2015). Controlled MoS$_2$ layer etching using CF$_4$ plasma. *Nanotechnology, 26*, 355706.

Jeong, S., Yoo, D., Ahn, M., Miro, P., Heine, T., & Cheon, J., (2015). Tandem intercalation strategy for single-layer nanosheets as an effective alternative to conventional exfoliation processes. *Nat. Commun., 6*, 5763.

Ji, Q., Zhang, Y., Gao, T., Zhang, Y., Ma, D., Liu, M., Chen, Y., et al., (2013). Epitaxial monolayer MoS$_2$ on mica with novel photoluminescence. *Nano Lett., 13*, 3870–3877.

Jiang, J. W., Qi, Z., Park, H. S., & Rabczuk, T., (2013a). Elastic bending modulus of single-layer molybdenum disulfide (MoS$_2$): Finite thickness effect. *Nanotechnology, 24*, 435705.

Jiang, J. W., Zhuang, X., & Rabczuk, T., (2013b). Orientation dependent thermal conductance in single-layer MoS$_2$. *Sci. Rep., 3*, 2–5.

Joensen, P., Frindt, R. F., & Morrison, S. R., (1986). Single-layer MoS$_2$. *Mater. Res. Bull., 21*, 457–461.

José-Yacamán, M., López, H., Santiago, P., Galván, D. H., Garzón, I. L., & Reyes, A., (1996). Studies of MoS$_2$ structures produced by electron irradiation. *Appl. Phys. Lett., 69*, 1065–1067.

Kaasbjerg, K., Thygesen, K. S., & Jacobsen, K. W., (2012). Phonon-limited mobility in MoS$_2$ from first principles. *Phys. Rev. B, 85*, 115317.

Kan, M., Wang, J. Y., Li, X. W., Zhang, S. H., Li, Y. W., Kawazoe, Y., Sun, Q., et al., (2014). Structures and phase transition of a MoS$_2$ monolayer. *J. Phys. Chem. C, 118*, 1515–1522.

Kastner, M., (1972). Bonding bands, lone-pair bands, and impurity states in chalcogenide semiconductors. *Phys. Rev. Lett., 28*, 355–357.

Kim, I. S., Sangwan, V. K., Jariwala, D., Wood, J. D., Park, S., Chen, K. S., Shi, F., et al., (2014). Influence of stoichiometry on the optical and electrical properties of chemical vapor deposition derived MoS$_2$. *ACS Nano, 8*, 10551–10558.

Kim, S., Konar, A., Hwang, W. S., Lee, J. H., Lee, J., Yang, J., Jung, C., et al., (2012). High-mobility and low-power thin-film transistors based on multilayer MoS$_2$ crystals. *Nat. Commun., 3*, 1011.

Kirmayer, S., Aharon, E., Dovgolevsky, E., Kalina, M., & Frey, G. L., (2007). Self-assembled lamellar MoS_2, SnS_2, and SiO_2 semiconducting polymer nanocomposites. *Philos. Trans. R. Soc. A Math. Phys. Eng. Sci., 365*, 1489–1508.

Kirubasankar, B., Murugadoss, V., & Angaiah, S., (2017). Hydrothermal assisted: In situ growth of CoSe onto graphene nanosheets as a nanohybrid positive electrode for asymmetric supercapacitors. *RSC Adv., 7*, 5853–5862.

Kobayashi, Y., Sasaki, S., Mori, S., Hibino, H., Liu, Z., Watanabe, K., Taniguchi, T., et al., (2015). Growth and optical properties of high-quality monolayer WS2 on graphite. *ACS Nano, 9*, 4056–4063.

Kolobov, A. V., & Tominaga, J., (2016). *Two-Dimensional Transition-Metal Dichalcogenides.* Springer.

Koma, A., & Yoshimura, K., (1986). Ultra sharp interfaces grown with Van Der Waals epitaxy. *Surf. Sci., 174*, 556–560.

Koma, A., (1992). Van Der Waals epitaxy: A new epitaxial growth method for a highly lattice-mismatched system. *Thin Solid Films, 216*, 72–76.

Koma, A., (1999). Van Der Waals epitaxy for highly lattice-mismatched systems. *J. Cryst. Growth, 201, 202*, 236–241.

Koma, A., Sunouchi, K., & Miyajima, T., (1985). Summary abstract: Fabrication of ultrathin heterostructures with Van Der Waals epitaxy. *J. Vac. Sci. Technol. B Microelectron. Process Phenom., 3*, 724.

Kong, D., Wang, H., Cha, J. J., Pasta, M., Koski, K. J., Yao, J., & Cui, Y., (2013). Synthesis of MoS_2 and $MoSe_2$ films with vertically aligned layers. *Nano Lett., 13*, 1341–1347.

Kuc, A., (2014). Low-dimensional transition-metal dichalcogenides. *Chem. Model., 11*, 1–29.

Kumar, S., & Schwingenschlögl, U., (2015). Thermoelectric response of bulk and monolayer $MoSe_2$ and WSe_2. *Chem. Mater., 27*, 1278–1284.

Le, D., Sun, D., Lu, W., Bartels, L., & Rahman, T. S., (2012). Single layer MoS_2 on the Cu(111) surface: First-principles electronic structure calculations. *Phys. Rev. B, 85*, 75429.

Lee, C., Kysar, J. W., & Hone, J., (2008). Measurement of the elastic properties and intrinsic strength of monolayer graphene energy transfer from quantum dots to graphene view project. *Science, 80, 321*, 385–389.

Lee, Y. H., Yu, L., Wang, H., Fang, W., Ling, X., Shi, Y., et al., (2013). Synthesis and transfer of single-layer transition metal disulfides on diverse surfaces. *Nano Lett., 13*, 1852–1857.

Lee, Y. H., Zhang, X. Q., Zhang, W., Chang, M. T., Lin, C. T., Chang, K. D., Yu, Y. C., et al., (2012). Synthesis of large-area mos_2 atomic layers with chemical vapor deposition. *Adv. Mater., 24*, 2320–2325.

Li, B., He, Y., Lei, S., Najmaei, S., Gong, Y., Wang, X., Zhang, J., et al., (2015). Scalable transfer of suspended two-dimensional single crystals. *Nano Lett., 15*, 5089–5097.

Li, H., Boey, F., Huang, X., Lu, G., Yin, Z., Zhang, H., Zeng, Z., et al., (2011). Single-layer semiconducting nanosheets: High-yield preparation and device fabrication. *Angew. Chemie. Int. Ed., 50*, 11093–11097.

Li, H., Duan, X., Wu, X., Zhuang, X., Zhou, H., Zhang, Q., Zhu, X., et al., (2014). Growth of alloy $MoS_2xSe_2(1-x)$ nanosheets with fully tunable chemical compositions and optical properties. *J. Am. Chem. Soc., 136*, 3756–3759.

Li, H., Wu, J., Yin, Z., & Zhang, H., (2014). Preparation and applications of mechanically exfoliated single-layer and multilayer MoS_2 and WSe_2 nano sheets. *Acc. Chem. Res., 47*, 1067.

Li, J., Medhekar, N. V., & Shenoy, V. B., (2013). Bonding charge density and ultimate strength of monolayer transition metal dichalcogenides. *J. Phys. Chem. C, 117*, 15842–15848.

Li, W. J., Shi, E. W., Ko, J. M., Chen, Z. Z., Ogino, H., & Fukuda, T., (2003). Hydrothermal synthesis of MoS$_2$ nanowires. *J. Cryst. Growth, 250*, 418–422.

Li, Y. B., Bando, Y., & Golberg, D., (2003). MoS$_2$ nanoflowers and their field-emission properties. *Appl. Phys. Lett., 82*, 1962–1964.

Li, Y., Chernikov, A., Zhang, X., Rigosi, A., Hill, H. M., Van, D. Z. A. M., Chenet, D. A., et al., (2014). Measurement of the optical dielectric function of monolayer transition-metal dichalcogenides: MoS$_2$, MoSe$_2$, WS$_2$, and WSe$_2$. *Phys. Rev. B, 90*, 205422.

Lin, T. W., Dai, C. S., Tasi, T. T., Chou, S. W., Lin, J. Y., & Shen, H. H., (2015). High-performance asymmetric supercapacitor based on Co$_9$S$_8$/3D graphene composite and graphene hydrogel. *Chem. Eng. J., 279*, 241–249.

Lin, Y. C., (2018). In: Lin, Y. C., (ed.), *Synthesis and Properties of 2D Semiconductors BT-Properties of Synthetic Two-Dimensional Materials and Heterostructures* (pp. 21–43). Springer International Publishing, Cham.

Ling, X., Lee, Y. H., Lin, Y., Fang, W., Yu, L., Dresselhaus, M. S., & Kong, J., (2014). Role of the seeding promoter in MoS$_2$ growth by chemical vapor deposition. *Nano Lett., 14*, 464–472.

Liu, K. K., Zhang, W., Lee, Y. H., Lin, Y. C., Chang, M. T., Su, C. Y., Chang, C. S., Li, H., Shi, Y., & Zhang, H., (2012). Growth of large-area and highly crystalline MoS$_2$ thin layers on insulating substrates. *Nano Lett., 12*, 1538–1544.

Liu, N., Kim, P., Kim, J. H., Ye, J. H., Kim, S., & Lee, C. J., (2014). Large-area atomically thin MoS$_2$ nanosheets prepared using electrochemical exfoliation. *ACS Nano, 8*, 6902–6910.

Lorenz, T., Ghorbani-Asl, M., Joswig, J. O., Heine, T., & Seifert, G., (2014a). Is MoS$_2$ a robust material for 2D electronics? *Nanotechnology, 25*, 445201.

Lorenz, T., Joswig, J. O., & Seifert, G., (2014b). Stretching and breaking of monolayer MoS$_2$: An atomistic simulation. *2D Mater, 1*, 11007.

Lu, X., Utama, M. I. B., Lin, J., Gong, X., Zhang, J., Zhao, Y., Pantelides, S. T., et al., (2014). Large-area synthesis of monolayer and few-layer MoSe$_2$ films on SiO$_2$ substrates. *Nano Lett., 14*, 2419–2425.

Lu, X., Utama, M. I. B., Zhang, J., Zhao, Y., & Xiong, Q., (2013). Layer-by-layer thinning of MoS$_2$ by thermal annealing. *Nanoscale, 5*, 8904–8908.

Marri These Authors Contributed Equally, S. R., Ratha, S., Rout, C. S., & Behera, J. N., (2017). 3D cuboidal vanadium diselenide embedded reduced graphene oxide hybrid structures with enhanced supercapacitor properties. *Chem. Commun., 53*, 228–231.

Matte, H. S. S. R., Plowman, B., Datta, R., & Rao, C. N. R., (2011). Graphene analogs of layered metal selenides. *Dalt. Trans., 40*, 10322–10325.

Murphy, D. W., Di Hull, F. J., Hull, G. W., & Waszczak, J. V., (1976). Convenient preparation and physical properties of lithium intercalation compounds of group 4B and 5B layered transition metal dichalcogenides. *Inorg. Chem., 15*, 17–21.

Najmaei, S., Liu, Z., Zhou, W., Zou, X., Shi, G., Lei, S., Yakobson, B. I., et al., (2013). Structure of molybdenum disulfide atomic layers. *Nat. Mater., 12*, 754–759.

Nanosheets, S., Yella, A., Mugnaioli, E., Panthöfer, M., Kolb, U., & Tremel, W., (2010). Mismatch strain versus dangling bonds : Formation of coin-roll. *Angew. Chemie Int. Ed., 49*, 3301–3305.

Nanotubes, M., Lorenz, T., Teich, D., Joswig, J., & Seifert, G., (2012). Theoretical study of the mechanical behavior of individual TiS$_2$ and. *J. Phys. Chem. C, 116*, 11714–11721.

Nazri, G. A., & Pistoia, G., (2008). *Lithium Batteries: Science and Technology*. Springer Science & Business Media.

Nicolosi, V., Chhowalla, M., Kanatzidis, M. G., Strano, M. S., & Coleman, J. N., (2013). Liquid exfoliation of layered materials. *Science, 80, 340*, 1226419.

Novoselov, K. S., (2011). Nobel lecture: Graphene: Materials in the flatland. *Rev. Mod. Phys., 83*, 837.

Ohuchi, F. S., Shimada, T., Parkinson, B. A., Ueno, K., & Koma, A., (1991). Growth of $MoSe_2$ thin films with Van Der Waals epitaxy. *J. Cryst. Growth, 111*, 1033–1037.

Park, J. C., Yun, S. J., Kim, H., Park, J. H., Chae, S. H., An, S. J., Kim, J. G., et al., (2015). Phase-engineered synthesis of centimeter-scale 1T'- and 2H-molybdenum ditelluride thin films. *ACS Nano, 9*, 6548–6554.

Peng, S., Li, L., Li, C., Tan, H., Cai, R., Yu, H., Mhaisalkar, S., et al., (2013). In situ growth of $NiCo_2S_4$ nanosheets on graphene for high-performance supercapacitors. *Chem. Commun., 49*, 10178–10180.

Peng, Y., Meng, Z., Zhong, C., Lu, J., Yu, W., Jia, Y., & Qian, Y., (2013). Hydrothermal synthesis and characterization of single-molecular-layer MoS_2 and $MoSe_2$. *Chem. Lett., 30*, 772–773.

Plechinger, G., Mann, J., Preciado, E., Barroso, D., Nguyen, A., Eroms, J., Schüller, C., Bartels, L., & Korn, T., (2014). A direct comparison of CVD-grown and exfoliated MoS_2 using optical spectroscopy. *Semicond. Sci. Technol., 29*, 64008.

Plechinger, G., Schrettenbrunner, F., Eroms, J., Weiss, D., Schueller, C., & Korn, T., (2012). Low-temperature photoluminescence of oxide-covered single-layer MoS_2. *Phys. Status Solidi (RRL)-Rapid Res. Lett., 6*, 126–128.

Pondick, J. V., Woods, J. M., Xing, J., Zhou, Y., & Cha, J. J., (2018). Stepwise sulfurization from MoO_3 to MoS_2 via chemical vapor deposition. *ACS Appl. Nano Mater., 1*, 5655–5661.

Pradhan, N. R., Rhodes, D., Zhang, Q., Talapatra, S., Terrones, M., Ajayan, P. M., Balicas, L., et al., (2015). Intrinsic carrier mobility of multi-layered MoS_2 field-effect transistors on SiO_2 Intrinsic carrier mobility of multi-layered MoS_2 field-effect transistors on SiO_2. *Appl. Phys. Lett., 123105*, 10–14.

Qiu, K., Lu, Y., Zhang, D., Cheng, J., Yan, H., Xu, J., Liu, X., et al., (2015). Mesoporous, hierarchical core/shell structured $ZnCo_2O_4/MnO_2$ nanocone forests for high-performance supercapacitors. *Nano Energy, 11*, 687–696.

Qu, B., Chen, Y., Zhang, M., Hu, L., Lei, D., Lu, B., Li, Q., et al., (2012). β-cobalt sulfide nanoparticles decorated graphene composite electrodes for high capacity and power supercapacitors. *Nanoscale, 4*, 7810–7816.

Ratha, S., & Rout, C. S., (2013). Supercapacitor electrodes based on layered tungsten disulfide-reduced graphene oxide hybrids synthesized by a facile hydrothermal method. *ACS Appl. Mater. Interfaces, 5*, 11427–11433.

Ratha, S., & Rout, C. S., (2015). Self-assembled flower-like $ZnCo_2O_4$ hierarchical superstructures for high capacity supercapacitors. *RSC Adv., 5*, 86551–86557.

Ratha, S., & Samantara, A. K., (2018). *Supercapacitor: Instrumentation, Measurement, and Performance Evaluation Techniques*. Springer.

Ratha, S., Marri, S. R., Behera, J. N., & Rout, C. S., (2016). High-energy-density supercapacitors based on patronite/single-walled carbon nanotubes/reduced graphene oxide hybrids. *Eur. J. Inorg. Chem.*, 259–265.

Ratha, S., Marri, S. R., Lanzillo, N. A., Moshkalev, S., Nayak, S. K., Behera, J. N., & Rout, C. S., (2015). Supercapacitors based on patronite-reduced graphene oxide hybrids: Experimental and theoretical insights. *J. Mater. Chem. A, 3*, 18874–18881.

Ratha, S., Samantara, A. K., Singha, K. K., Gangan, A. S., Chakraborty, B., Jena, B. K., & Rout, C. S., (2017). Urea-assisted room temperature stabilized metastable β-NiMoO₄: Experimental and theoretical insights into its unique bifunctional activity toward oxygen evolution and supercapacitor. *ACS Appl. Mater. Interfaces, 9*, 9640–9653.

Remskar, M., Mrzel, A., Jesih, A., & Lévy, F., (2002). Metal-alloyed NbS₂ nanotubes synthesized by the self-assembly of nanoparticles. *Adv. Mater., 14*, 680–684.

Rocker, F., Tremel, W., Glasser, G., Therese, H. A., Kolb, U., Reiber, A., Li, J., & Stepputat, M., (2004). VS₂ nanotubes containing organic-amine templates from the NT-VOx precursors and reversible copper intercalation in NT-VS₂. *Angew. Chemie Int. Ed., 44*, 262–265.

Rothschild, A., Frey, G. L., Homyonfer, M., Tenne, R., & Rappaport, M., (1999). Synthesis of bulk WS₂ nanotube phases. *Mater. Res. Innov., 3*, 145–149.

Rothschild, A., Sloan, J., & Tenne, R., (2000). Growth of WS₂ nanotubes phases. *J. Am. Chem. Soc., 122*, 5169–5179.

Sahoo, S., Gaur, A. P. S., Ahmadi, M., Guinel, M. J. F., & Katiyar, R. S., (2013). Temperature-dependent Raman studies and thermal conductivity of few-layer MoS₂. *J. Phys. Chem. C, 117*, 9042–9047.

Sahoo, S., Ratha, S., & Rout, C. S., (2015). Spinel NiCo₂O₄; nanorods for super capacitor applications. *Am. J. Eng. Appl. Sci., 8*, 371–379.

Saito, Y., Fons, P., Kolobov, A. V., & Tominaga, J., (2015). Self-organized Van Der Waals epitaxy of layered chalcogenide structures. *Phys. Status Solidi, 252*, 2151–2158.

Samantara, A. K., Kamila, S., Ghosh, A., & Jena, B. K., (2018). Highly ordered 1D NiCo 2 O 4 nanorods on graphene: An efficient dual-functional hybrid materials for electrochemical energy conversion and storage applications. *Electrochim. Acta, 263*, 147–157.

Sano, N., Wang, H., Chhowalla, M., Alexandrou, I., Amaratunga, G. A. J., Naito, M., & Kanki, T., (2003). Fabrication of inorganic molybdenum disulfide fullerenes by arc in water. *Chem. Phys. Lett., 368*, 331–337.

Savva, K., Višić, B., Popovitz-Biro, R., Stratakis, E., & Tenne, R., (2017). Short pulse laser synthesis of transition-metal dichalcogenide nanostructures under ambient conditions. *ACS Omega, 2*, 2649–2656.

Schneemeyer, L. F., Stacy, A., & Sienko, M. J., (1980). Effect of nonstoichiometry on the periodic lattice distortion in vanadium diselenide. *Inorg. Chem., 19*, 2659–2662.

Senthilkumar, V., Tam, L. C., Kim, Y. S., Sim, Y., Seong, M. J., & Jang, J. I., (2014). Direct vapor phase growth process and robust photoluminescence properties of large area MoS₂ layers. *Nano Res., 7*, 1759–1768.

Shaw, J. C., Zhou, H., Chen, Y., Weiss, N. O., Liu, Y., Huang, Y., & Duan, X., (2014). Chemical vapor deposition growth of monolayer MoSe₂ nanosheets. *Nano Res., 7*, 511–517.

Shen, Y., Xiao, H., Zhang, G., Wei, X., Zhong, J., Xie, G., & Wang, Y., (2014). Phonon thermal conductivity of monolayer MoS₂ : A comparison with single layer graphene. *Appl. Phys. Lett., 105*, 103902.

Shi, Y., Zhou, W., Lu, A. Y., Fang, W., Lee, Y. H., Hsu, A. L., Kim, S. M., Kim, K. K., Yang, H. Y., Li, L. J., Idrobo, J. C., & Kong, J., (2012). Van Der Waals epitaxy of MoS₂ layers using graphene as growth templates. *Nano Lett., 12*, 2784–2791.

Si, P. Z., Zhang, M., Zhang, Z. D., Zhao, X. G., Ma, X. L., & Geng, D. Y., (2005). Synthesis and structure of multi-layered $WS_2(CoS)$, $MoS_2(Mo)$ nanocapsules and single-layered $WS_2(W)$ nanoparticles. *J. Mater. Sci., 40*, 4287–4291.

Smith, R. J., King, P. J., Lotya, M., Wirtz, C., Khan, U., De, S., O'Neill, A., et al., (2011). Large-scale exfoliation of inorganic layered compounds in aqueous surfactant solutions. *Adv. Mater., 23*, 3944–3948.

Song, J. G., Park, J., Lee, W., Choi, T., Jung, H., Lee, C. W., Hwang, S. H., et al., (2013). Layer-controlled, wafer-scale, and conformal synthesis of tungsten disulfide nanosheets using atomic layer deposition. *ACS Nano, 7*, 11333–11340.

Strait, J. H., Nene, P., & Rana, F., (2014). High intrinsic mobility and ultrafast carrier dynamics in multilayer metal-dichalcogenide MoS_2. *Phys. Rev. B-Condens. Matter Mater. Phys., 90*, 245402.

Su, S. H., Hsu, W. T., Hsu, C. L., Chen, C. H., Chiu, M. H., Lin, Y. C., Chang, W. H., et al., (2014a). Controllable synthesis of band-gap-tunable and monolayer transition-metal dichalcogenide alloys. *Front. Energy Res.* https://doi.org/10.3389/fenrg.2014.00027 (accessed on 16 May 2020).

Su, S. H., Hsu, Y. T., Chang, Y. H., Chiu, M. H., Hsu, C. L., Hsu, W. T., Chang, W. H., et al., (2014). Band gap-tunable molybdenum sulfide selenide monolayer alloy. *Small, 10*, 2589–2594.

Suk, J. W., Kitt, A., Magnuson, C. W., Hao, Y., Ahmed, S., An, J., Swan, A. K., et al., (2011). Transfer of CVD-grown monolayer graphene onto arbitrary substrates. *ACS Nano, 5*, 6916–6924.

Tahir, M., & Schwingenschlögl, U., (2014). Tunable thermoelectricity in monolayers of MoS_2 and other group-VI dichalcogenides. *New J. Phys., 16*, 115003.

Tang, D. M., Kvashnin, D. G., Najmaei, S., Bando, Y., Kimoto, K., Koskinen, P., Ajayan, P. M., et al. (2014). 1AD: Nanomechanical cleavage of molybdenum disulfide atomic layers. *Nat. Commun., 5*, 1–8.

Taube, A., Judek, J., Łapińska, A., & Zdrojek, M., (2015). Temperature-dependent thermal properties of supported MoS_2 monolayers. *ACS Appl. Mater. Interfaces, 7*, 5061–5065.

Therese, H. A., Li, J., Kolb, U., & Tremel, W., (2005). Facile large scale synthesis of WS_2 nanotubes from WO_3 nanorods prepared by a hydrothermal route. *Solid State Sci., 7*, 67–72.

Tian, Y., He, Y., & Zhu, Y., (2004). Low temperature synthesis and characterization of molybdenum disulfide nanotubes and nanorods. *Mater. Chem. Phys., 87*, 87–90.

Toyama, M., (1966). Kinetics of the vapor growth of II-VI compounds crystals. *Jpn. J. Appl. Phys., 5*, 1204.

Tsai, H. L., Heising, J., Schindler, J. L., Kannewurf, C. R., & Kanatzidis, M. G., (1997). Exfoliated-restacked phase of WS_2. *Chem. Mater., 9*, 879–882. https://doi.org/10.1021/cm960579t (accessed on 16 May 2020).

Ueno, K., Saiki, K., Shimada, T., & Koma, A., (1990a). Epitaxial growth of transition metal dichalcogenides on cleaved faces of mica. *J. Vac. Sci. Technol. A., 8*, 68–72.

Ueno, K., Shimada, T., Saiki, K., & Koma, A., (1990b). Heteroepitaxial growth of layered transition metal dichalcogenides on sulfur-terminated GaAs{111} surfaces. *Appl. Phys. Lett., 56*, 327–329.

Van, D. Z. A. M., Huang, P. Y., Chenet, D. A., Berkelbach, T. C., You, Y., Lee, G. H., Heinz, T. F., et al., (2013). Supplementary information for grains and grain boundaries in highly crystalline monolayer molybdenum disulfide. *Nat. Mater., 12*, 554–561.

Van, D. Z. H. S. J., Poot, M., Castellanos-Gomez, A., Rubio-Bollinger, G., Steele, G. A., & Agrait, N., (2012). Mechanical properties of freely suspended semiconducting graphene-like layers based on MoS_2. *Nanoscale Res. Lett., 7,* 233.

Vollath, D., & Szabó, D. V. V., (1998). Synthesis of nanocrystalline MoS_2 and WS_2 in a microwave plasma. *Mater. Lett., 35,* 236–244.

Vollath, D., & Szabó, D. V., (2000). Nanoparticles from compounds with layered structures. *Acta Mater., 48,* 953–967.

Wang, A., Wang, H., Zhang, S., Mao, C., Song, J., Niu, H., Jin, B., & Tian, Y., (2013). Controlled synthesis of nickel sulfide/graphene oxide nanocomposite for high-performance supercapacitor. *Appl. Surf. Sci., 282,* 704–708.

Wang, F., Wang, Z., Wang, Q., Wang, F., Yin, L., Xu, K., Huang, Y., & He, J., (2015). Synthesis, properties and applications of 2D non-graphene materials. *Nanotechnology, 26,* 292001.

Wang, R., Luo, Y., Chen, Z., Zhang, M., & Wang, T., (2016). The effect of loading density of nickel-cobalt sulfide arrays on their cyclic stability and rate performance for supercapacitors. *Sci. China Mater, 59,* 629–638.

Wang, X., & Tabarraei, A., (2016). Phonon thermal conductivity of monolayer MoS_2. *Appl. Phys. Lett., 108,* 133113.

Wang, X., Feng, H., Wu, Y., & Jiao, L., (2013). Controlled synthesis of highly crystalline MoS_2 flakes by chemical vapor deposition. *J. Am. Chem. Soc., 135,* 5304–5307.

Wen, W., Song, Y., Yan, X., Zhu, C., Du, D., Wang, S., Asiri, A. M., & Lin, Y., (2018). Recent advances in emerging 2D nanomaterials for biosensing and bioimaging applications. *Mater. Today, 21,* 164–177.

Wiegers, G. A., Van, D. M. R., Van, H. H., Kloosterboer, H. J., & Alberink, A. J. A., (1974). The sodium intercalates of vanadium disulfide and their hydrolysis products. *Mater. Res. Bull., 9,* 1261–1265.

Wu, J., Li, B., & Guo, J., (2011). A novel wet chemistry approach for the synthesis of hybrid 2D free-floating single or multilayer nanosheets of MS_2@oleylamine (MdMo, W). *Chem. Mater., 23,* 3879–3885.

Wu, S., Huang, C., Aivazian, G., Ross, J. S., Cobden, D. H., & Xu, X., (2013). Vapor-solid growth of high optical quality MoS_2 monolayers with near-unity valley polarization. *ACS Nano, 7,* 2768–2772.

Xenogiannopoulou, E., Tsipas, P., Aretouli, K. E., Tsoutsou, D., Giamini, S. A., Bazioti, C., Dimitrakopulos, G. P., et al., (2015). High-quality, large-area $MoSe_2$ and $MoSe_2/Bi_2Se_3$ heterostructures on AlN(0001)/Si(111) substrates by molecular beam epitaxy. *Nanoscale, 7,* 7896–7905.

Xie, B., Chen, Y., Yu, M., Sun, T., Lu, L., Xie, T., Zhang, Y., & Wu, Y., (2016). Hydrothermal synthesis of layered molybdenum sulfide/N-doped graphene hybrid with enhanced supercapacitor performance. *Carbon N.Y., 99,* 35–42.

Xue, Y., Zhang, Q., Wang, W., Cao, H., Yang, Q., & Fu, L., (2017). Opening two-dimensional materials for energy conversion and storage: A concept. *Adv. Energy Mater, 7,* 1602684.

Yan, J., Lui, G., Tjandra, R., Wang, X., Rasenthiram, L., & Yu, A., (2015). α-NiS grown on reduced graphene oxide and single-wall carbon nanotubes as electrode materials for high-power supercapacitors. *RSC Adv., 5,* 27940–27945.

Yan, R., Simpson, R. J., Bertolazzi, S., Brivio, J., Watson, M., Wu, X., Kis, A., et al., (2014). Supporting information for: "Thermal conductivity of monolayer molybdenum disulfide obtained from temperature dependent raman spectroscopy." *ACS Nano, 8,* 986–993.

Yang, J., Yu, C., Fan, X., Liang, S., Li, S., Huang, H., Ling, Z., et al., (2016). Electroactive edge site-enriched nickel-cobalt sulfide into graphene frameworks for high-performance asymmetric supercapacitors. *Energy Environ. Sci., 9,* 1299–1307.

Yu, H., Liu, Y., & Brock, S. L., (2008). Synthesis of discrete and dispersible MoS_2 nanocrystals. *Inorg. Chem., 47,* 1428–1434.

Yu, X. Y., David, & Lou, X. W., (2018). Mixed metal sulfides for electrochemical energy storage and conversion. *Adv. Energy Mater, 8,* 1701592.

Yuan, C., Cao, Y., Luo, X., Yu, T., Huang, Z., Xu, B., Yang, Y., et al., (2015). Monolayer-by-monolayer stacked pyramid-like MoS_2 nanodots on monolayered MoS_2 flakes with enhanced photoluminescence. *Nanoscale, 7,* 17468–17472.

Yue, R., Barton, A. T., Zhu, H., Azcatl, A., Pena, L. F., Wang, J., Peng, X., et al., (2015). $HfSe_2$ thin films: 2D Transition metal dichalcogenides grown by molecular beam epitaxy. *ACS Nano, 9,* 474–480.

Zeng, Z., Sun, T., Zhu, J., Huang, X., Yin, Z., Lu, G., Fan, Z., Yan, Q., Hng, H. H., & Zhang, H., (2012). An effective method for the fabrication of few-layer-thick inorganic nanosheets. *Angew. Chemie-Int. Ed., 51,* 9052–9056.

Zhan, Y., Liu, Z., Najmaei, S., Ajayan, P. M., & Lou, J., (2012). Large-area vapor-phase growth and characterization of MoS_2 atomic layers on a SiO_2 substrate. *Small, 8,* 966–971.

Zhang, L., Liu, K., Wong, A. B., Kim, J., Hong, X., Liu, C., Cao, T., Louie, S. G., Wang, F., & Yang, P., (2014). Three-dimensional spirals of atomic layered MoS_2. *Nano Lett., 14,* 6418–6423.

Zhang, X., Meng, F., Christianson, J. R., Arroyo-Torres, C., Lukowski, M. A., Liang, D., Schmidt, J. R., & Jin, S., (2014). Vertical heterostructures of layered metal chalcogenides by Van Der Waals epitaxy. *Nano Lett., 14,* 3047–3054.

Zhang, Y., Zhang, Y., Ji, Q., Ju, J., Yuan, H., Shi, J., et al., (2013). Controlled growth of high-quality monolayer WS_2 layers on sapphire and imaging its grain boundary. *ACS Nano, 7,* 8963–8971.

Zhao, W., Ribeiro, R. M., & Eda, G., (2015). Electronic structure and optical signatures of semiconducting transition metal dichalcogenide nanosheets. *Acc. Chem. Res., 48,* 91–99.

Zhao, X., Sánchez, B. M., Dobson, P. J., & Grant, P. S., (2011). The role of nanomaterials in redox-based supercapacitors for next generation energy storage devices. *Nanoscale, 3,* 839–855.

Zheng, J., Zhang, H., Dong, S., Liu, Y., Tai, N. C., Suk, S. H., Young, J. H., et al., (2014). High yield exfoliation of two-dimensional chalcogenides using sodium naphthalenide. *Nat. Commun., 5,* 2995.

Zhou, L., Xu, K., Zubair, A., Liao, A. D., Fang, W., Ouyang, F., Lee, Y. H., et al., (2015). Large-area synthesis of high-quality uniform few-layer $MoTe_2$. *J. Am. Chem. Soc., 137,* 11892–11895.

Zibouche, N., Kuc, A., Musfeldt, J., & Heine, T., (2014). Transition-metal dichalcogenides for spintronic applications. *Ann. Phys., 526,* 395–401.

CHAPTER 2

Photoelectrochemical Reduction of CO_2 and Electrochemical Oxidation of CO

SURJENDU BHATTACHARYYA[1] and BISWAJIT MISHRA[2]

[1]*Institute of Atomic and Molecular Sciences (IAMS), Academia Sinica, P.O. Box 23-10617, Taipei – 10617, Taiwan*

[2]*Faculty of Chemical Sciences, Shri Ramswaroop Memorial University, Lucknow-Deva Road, Uttar Pradesh – 225003, India, E-mail: sahebm@gmail.com*

ABSTRACT

The disparity between energy demand and fossil fuel sources is increasing very fast and the air quality is deteriorating recklessly with the exponential growth of the world population. These force us not only to think about the greener alternate of fossil fuel as the energy source as soon as possible but also immediate removal of the pollutants which are already accumulated in the air in a large excess. CO and CO_2 are the two among the major pollutants present in air, both of which mainly originate due to the partial and full combustion of the fossil fuels, respectively. CO has immediate hazards to our health, whereas CO_2 acts as a greenhouse gas and consequently the major contributor to the global warming. Therefore, photooxidation of CO and photoreduction of CO_2 have become inevitable to make the air free of CO and CO_2. In addition, the photoreduction of CO_2 to high energy density fuels can potentially be a route to shun the heavy use of fossil fuel. This chapter includes the important outcomes so far, major challenges associated and future prospects as well. This will be broadly divided in two sections of photochemical oxidation of CO and photoelectrochemical reduction of CO_2.

2.1   INTRODUCTION

Global energy demand is rising very steeply with the exponential growth of the world population as well as their improving basic amenities. On the contrary, the major source for fulfilling the demand for energy has been fossil fuels whose stocks are not only depleting very fast but a large amount of polluting gases is released during the course of the energy conversion. As a result, a large disparity between the increasing energy demand and the depleting fossil fuel sources is quite obvious and the air quality is also deteriorating uncontrollably day by day. Therefore, it already became inevitable to think about the greener alternate of fossil fuel as the energy source as soon as possible and immediate removal of the pollutants which have already been accumulated in the air in large excess.

During the conversion of energy from the fossil fuels through the combustion of coal, natural gases, or oils in the presence of oxygen, CO_2 is among the major pollutants released in the air. In addition, a significant amount of CO also comes out due to the partial combustion of the fossil fuels. CO_2 is known as the major contributor to greenhouse gases and consequently to the global warming. Conversion of CO_2 to high-density fuels like methane, methanol, etc., and utilizing them as the energy source, whenever it is needed, can potentially shun the over-reliance of the fossil fuels and reduce the abnormal CO_2 level in the air. On the contrary, the immediate health hazards of CO are well known. In addition, CO is the major intermediate formed during the energy extraction from the fuels through their catalytic oxidation in fuel cells. Platinum and gold are conventionally used as the catalysts in fuel cells. These catalysts show a high affinity towards CO and so, their catalytic activity diminishes drastically due to the CO poisons. Therefore, the study of the electrochemical oxidation of CO is also important. This chapter sums up the recent progresses in photoelectrochemical reduction of CO_2 and the current understanding of the mechanisms involved and the development in designing state-of-the-art electrocatalysts for the electrochemical oxidation of CO.

2.2   PHOTOELECTROCHEMICAL REDUCTION OF CO_2

2.2.1   BACKGROUND

Sequestering carbon dioxide (CO_2) by plants and convert light energy into chemical energy in carbohydrate molecules, such as sugars, is a natural reduction process of CO_2 which is vital for sustaining the life. CO_2 exists in Earth's atmosphere in comparably small concentrations (0.041% by volume of the atmosphere which is equal to 410 ppm) also known as a greenhouse gas that absorbs and emits thermal radiation, creating the 'greenhouse effect.' The CO_2 emission has increased at an alarming rate due to the energy-driven consumption of fossil fuels since the industrial revolution. This has disrupted the global carbon cycle resulting to the warming of the whole planet. Global warming has an adverse effect on the weather (such as floods, droughts, storms, and heatwaves); rise of sea-level; crop growth, etc. In this perspective, to maintain the uninterrupted carbon cycle, catalytic reduction of CO_2 to usable energy-rich fuels in the form of hydrocarbons, alcohols, and aldehydes become none other than a mandate. Once CO_2 transformed into fuels, it is easy to handle and maneuvered within the existing infrastructure of natural gas and liquid fuels transportation. This is an additional advantage on contrary to H_2 generation from solar water splitting. Since hydrogen is the lightest element, therefore, the storage and transportation of H_2 become an additional difficulty. The first report on photoreduction of CO_2 to formic acid (FA), formaldehyde, methyl alcohol, and methane by Inoue et al. used powders of several semiconductors such as TiO_2, ZnO, CdS, GaP, SiC, and WO_3 suspended in water as catalysts (Inoue et al., 1979). However, the challenges for making efficient catalysts of CO_2 reduction are manifold and so are the rewards. However, numerous challenges are possessing obstacles in upbringing this domain of research from its infancy. Till date, the highest reported rates of product formation are only in the range of few tens of μmoles per hour of illumination using 1 g of photocatalyst. Therefore, a major breakthrough in making efficient CO_2 reduction catalyst might be one of the largest innovation in realizing the sustainable developments. Unlike single electron water splitting, the CO_2 reduction mechanism involved multiple electrons, therefore, far more complex and yet to be understood clearly. It is very important to understand the stepwise mechanism to further improve the photocatalytic activity of CO_2 reduction which is a major energy-related concern. The two major challenges hindering the progress are selectivity and energetics which are explained in the succeeding sections.

2.2.2 CHALLENGES IN CO_2 PHOTOREDUCTION ENERGETICS AND SELECTIVITY

One can think of CO_2 reduction process in two steps: activation and reduction. CO_2 is an extremely stable molecule with −394.4 kJ/mole of standard

Gibbs free energy of formation (ΔG_f^0). The carbon atom in CO_2 possesses the highest of all of its probable oxidation states (-4 to $+4$). The simplest thermal dissociation of CO_2 also requires high energy (749 kJ/mole of C=O bond) to form CO which can be chemically transformed into liquid fuels like gasoline applying Fischer-Tröpsch synthesis (Khodakov et al., 2007). However, this barrier is too high to bear with. An enormous amount of effort from numerous researchers have been devoted to improve the reactivity by changing the chemical environment. One such example is using catalyst which offers specific binding site to activate CO_2 for its chemical transformation following a lower energy barrier. The next step is the reduction of CO_2 into its anion, CO_2^-. Here is confronted to the additional barrier. The reduction potential (E^0_{redox}) of CO_2 is -1.9 V which is again having too high driving force. However, the proton assisted reduction involving multiple electrons (Eqs. 2.2 to 2.6) reduces the reduction potential of the CO_2 reduction. Figure 2.1 indicates the redox potentials required to drive important reactions during the course of CO_2 photoreduction and also represents the positions of E_c and E_v of various semiconductor photocatalysts with respect to the potentials of the redox reactions mentioned below (Habisreutinger et al., 2013). But the CO_2 reduction is often accompanying by the undesired hydrogen evolution reaction (HER) (Eqs. 2.7 and 2.8) due to the presence of water (vapor or liquid phase) (Chang et al., 2015; Habisreutinger et al., 2013). Therefore, it is evident that two primary challenges for the CO_2 reduction process are the energetics and product selectivity. Photoelectrochemical reduction of CO_2 still in the developing phase was introduced to overcome these drawbacks. In this chapter, we will discuss about the working principle of the photoelectrochemical cell and the recent progress in CO_2 photoelectrochemical reduction.

$$CO_2 + e^- = CO_2^- \quad E^0 = -1.9V \; vs \; NHE \; at \; pH \; 7 \tag{2.1}$$

$$CO_2 + 2H^+ + 2e^- = HCOOH \quad E^0 = -0.61V \; vs \; NHE \; at \; pH \; 7 \tag{2.2}$$

$$CO_2 + 2H^+ + 2e^- = CO + H_2O \quad E^0 = -0.51V \; vs \; NHE \; at \; pH \; 7 \tag{2.3}$$

$$CO_2 + 4H^+ + 4e^- = HCHO + H_2O \quad E^0 = -0.48V \; vs \; NHE \; at \; pH \; 7 \tag{2.4}$$

$$CO_2 + 6H^+ + 6e^- = CH_3OH + H_2O \quad E^0 = -0.38V \text{ vs NHE at pH 7} \quad (2.5)$$

$$CO_2 + 8H^+ + 8e^- = CH_4 + H_2O \quad E^0 = -0.24V \text{ vs NHE at pH 7} \quad (2.6)$$

$$2H^+ + 2e^- = H_2 \quad E^0 = -0.41V \text{ vs NHE at pH 7} \quad (2.7)$$

$$2H_2O = O_2 + 4H^+ + 4e^- \quad E^0 = 0.82V \text{ vs NHE at pH 7} \quad (2.8)$$

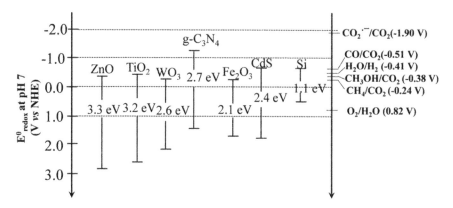

FIGURE 2.1 Band gaps, positions of E_cs, and E_vs of most widely used semiconductor photocatalysts and the potentials of the redox couples participating in fuel generation reactions (Eqs. 2.1–2.8).
Source: Reprinted with permission from Mishra and Chaudhary (2017). © CRC Press.

2.2.3 WORKING PRINCIPLE OF PHOTOELECTROCHEMICAL ELECTROCHEMICAL CELL

The photoelectrochemical (PEC) cells are still in the developing stage; however, they have displayed promising potential in the utilization of a wider range of solar spectrum efficiently. The simplest form of a PEC cell comprises of the working electrode (WE) and counter electrode (CE) consist of a semiconductor thin film (a photoactive material, photoelectrode) and a platinum wire/mesh, respectively. In some cases, both the electrodes used are photoactive. To observe the half-cell reactions, another electrode, called a reference electrode (RE) is also used. The whole system is immersed into an appropriate aqueous electrolyte solution. The reactor consists of an optical window, which allows the illuminated light to reach the electrode surface. An external bias is also applied at the WE (Jiang et al., 2017; Yang et al., 2017) in addition to the light irradiation.

An electron-hole pair is generated when the photoactive WE absorb a photon of energy that is equal or higher than the bandgap. If the WE is n-type semiconductor, then the excited electrons are transported to the CE through an external circuit. Consequently, the leaving behind the holes in WE, in this case, a photoanode are utilized in oxidation and the electrons in the CE, cathode, are utilized in reduction. Conversely, if the WE is p-type semiconductor, then the photogenerated electrons are consumed in reduction, acting as a photocathode. The holes are transported through an external circuit to the CE where they take part in the oxidation (anode) process. In summary, n-type semiconductors produce an anodic photocurrent by transferring holes into electrolyte and p-type semiconductors produce a cathodic photocurrent by transferring electrons into electrolyte (Jiang et al., 2017; Yang et al., 2017).

When a semiconductor is in contact with the electrolyte, a depletion region is formed on the surface of the semiconductor due the transfer of electrons (for n-type) or holes (for p-type) to the electrolyte. Therefore, the band edges of the semiconductor bend to attain the equilibrium with the redox potential of the electrolyte redox couple. The direction of the band bending of the semiconductor at the electrode-electrolyte interface depends upon the type of semiconductor (n/p-type) is being used. Fermi energy level lies just below the conduction band (CB) in n-type and at a more negative potential than that of the electrolyte redox couple. Being the n-type semiconductor electron-rich, they flow from the semiconductor to the electrolyte through the interface. This causes a lowering of the Fermi level of semiconductor until it reaches the equilibrium potential with electrolyte redox couple. Therefore, the band bends upward at the interface (Figure 2.2a) (Bott, 1998). Now a depletion region is formed at the semiconductor electrolyte interface as the electrons have flown out of that region to electrolyte, and become devoid of any charge carriers. Unlike the n-type semiconductors, the Fermi level lies just above the valence band (VB) in p-type semiconductor and at a more positive potential than that of the electrolyte redox couple. Therefore, the majority of charge carriers and holes flow from the semiconductor to the electrolyte through the interface. In other sense, the electrons move from electrolytes to the semiconductor electrode. As a result, the Fermi level of the semiconductor increases until it becomes equal in potential with that of the electrolyte redox couple. In this case, the band bends downward at the interface (Bott, 1998). Owing to the lack of majority charge carriers at the interface, a depletion region (Figure 2.2b) is formed like in the case of n-type semiconductor.

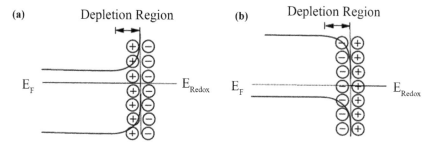

FIGURE 2.2   Band bending for an n-type semiconductor (a) and a p-type semiconductor (b) in equilibrium with an electrolyte.
Source: Reprinted with permission from Bott (1998).

Under this equilibrium condition, when a negative potential (forward bias) is applied to the n-type semiconductor WE, the electrons flow from this external bias source, the negative end of the battery, into the semiconductor. As a result, the Fermi level starts to rise with an increment in the applied negative potential. Consequently, the band edges start to bend downward. The situation arises such that at a certain forward bias potential, the band edges become flat and there is no net flow of electrons between the semi-conductor and the electrolyte. This potential is called *flat band potential* (Bott, 1998) (Figure 2.3b). In this condition of flat band potential, there will be no net flow of current. Owing to the absence of any electric field at the semiconductor electrolyte interface even if the light irradiation falls on the semiconductor electrodes, the photogenerated charge carriers are not possible to be separated (Region II, Figure 2.5). Now, if forward bias potential is further increased above the flat band potential, then the Fermi level of the semiconductor will be pushed at a potential higher than that of the electrolyte redox couple. The band edges will bend more downward due to the accumulation of excess electrons at the semiconductor-electrolyte interface. This accumulation region (Figure 2.3a) allows a continuous flow of electrons to the electrolyte through the interface. As a result, the n-type semiconductor performs like a dark cathode. With further increase in the

forward bias potential, the electrons flow like in a metallic conductor. Therefore, the light illumination on the semiconductor does not show any effect
on the flow of electrons, i.e., current density (Figure 2.3c and Region I in
Figure 2.5).

When the WE containing n-type semiconductor is connected to the
reverse bias, i.e., to a positive potential, then the electrons are extracted from
the semiconductor. As a result, the Fermi level starts to decrease. In such
reverse bias condition, the depletion region increases with the increment
in the applied bias (positive) which in turn does not allow the flow of any
current in absence of light illumination even if the potential more positive
than the flat band potential. It occurs because of the absence of majority
charge carriers (here electrons) in the depletion region to flow across the
interface (Figure 2.3a and Curve a, Region III in Figure 2.5). However, the
situation changes in the presence of light irradiation. In the presence of light,

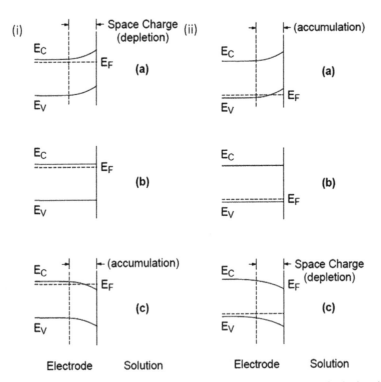

FIGURE 2.3 Effect of varying the applied potential (E) on the band edges in the interior of
a (i) n-type; and (ii) p-type semiconductor at (a) $E > E_{fb}$, (b) $E = E_{fb}$, and (c) $E < E_{fb}$.
Source: Reprinted with permission from Bott (1998).

electrons, and holes are generated. Owing to the presence of an electric field at the depletion region, these photogenerated electrons and holes are separated from each other. Thus, the electrons move from the electrolyte to the holes present in the depletion region. Therefore, we obtain an anodic current due to the flow of these electrons from the electrolyte to the semiconductor through the interface (Curve b, Region III in Figure 2.5). Hence, the n-type semiconductor acts as a photoanode under reverse bias condition. Following the converse analogy, the p-type semiconductor at a potential more positive than the flat band potential allows a continuous flow of holes to the electrolyte through the interface, i.e., acts as a dark anode and a photocathode under reverse bias condition as the holes move from the electrolyte to the semiconductor through the interface. Since we need to reduce CO_2 under light irradiation using the photoelectrochemical method, therefore, the required electrode has to be photocathode. This boils down to the only choice of using the p-type semiconductor under the reverse bias (Figure 2.4).

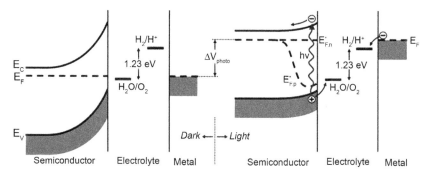

FIGURE 2.4   Band diagram for a PEC cell based on an n-type semiconducting photoanode in equilibrium in the dark (left) and under illumination (right).
Source: Reprinted with permission from van de Krol and Gratzel (2012). © Springer Nature.

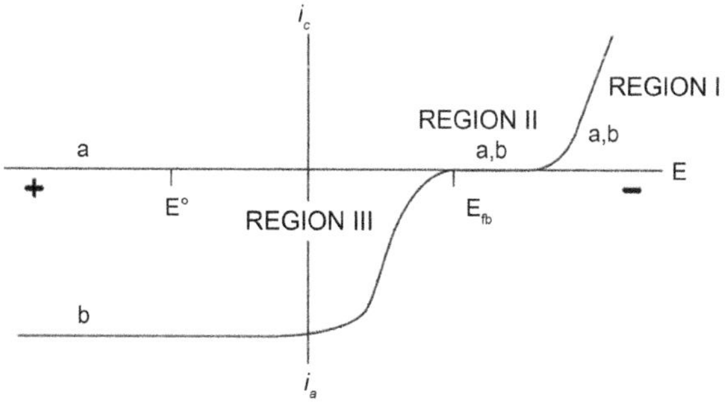

FIGURE 2.5   Ideal behavior for an n-type semiconductor in the dark (a) and under irradiation.
Source: Reprinted with permission from Bott (1998).

2.2.4   STATE OF THE ART PHOTOELECTROCATALYSTS

Electrochemical reactions always happen at a potential higher than the thermodynamic potential. The potential, which is required in excess of the calculated thermodynamic potential, is called over-potential. Often electrocatalysts, which make the reaction kinetics faster by reducing the associated activation energy, hence, are used to reduce the over-potential as much as possible. Now, for a material to be photoelectrocatalyst, it needs to be an efficient electrocatalysts along with a light absorber with an appropriate bandgap and band edges position as discussed in section 2.2.2. The involvement of light reduces the over-potential further. However, it is a difficult task to design an efficient photoelectrocatalyst having a single material which plays the dual roles of an electrocatalyst and a light absorber. Rather, the photoelectrocatalysts often consist of hybrid materials. The electrocatalysts are generally dispersed on the surface of a semiconductor support. The semiconductor support actually helps to sensitize the reaction through light absorption. This type of hybrid systems shows multiple benefits as compared to the single component photoelectrocatalysts. In a hybrid system, electrocatalyst provides active sites for the reaction to happen. In addition, it helps in better separation of the photogenerated charge carriers. It often acts as the sink for the photogenerated electron in the CB of semiconductor support. Till date, there are only a few reports on the photoelectrochemical reduction of CO_2 utilizing the hybrid semiconductor/metal complex. Important contribution in this area was done by Sato et al. (2011). Their idea was to combine photoactive semiconductors with metal complex catalysts having ability to reduce CO_2. Therefore, the electron transfer (ET) from the CB of a photoexcited semiconductor to the metal-complex catalyst is very critical. It has been observed that a hybrid photocatalyst prepared by modifying a semiconductor (SC) surface with a metal-complex electrocatalyst (MCE), SC/[MCE], can reduce CO_2 to $HCOO^-$ (formate) in H_2O with an electrical bias (Arai et al., 2010). This SC/[MCE] hybrid catalyst consists of a p-type SC, zinc-doped indium phosphide (InP), and a ruthenium complex polymer electrocatalyst, $[Ru\{4,4'\text{-di}(1H\text{-pyrrolyl-3-propylcarbonate})\text{-2,2'-bipyridine}\}(CO)_2]_n$, as the MCE (Takeda et al., 2008). The driving force to promote the ET from the SC to the MCE was attributed to the energy difference between the conduction band minimum (E_{CBM}) of the SC and the CO_2 reduction potential of the MCE (Sato et al., 2010; Yamanaka et al., 2011). However, they wanted to extend this idea in such a way that H_2O can be used as both an electron donor and a proton source concomitantly. Therefore, the SC/[MCE] needs to be combined with a photocatalyst capable of H_2O oxidation in aqueous media in renowned Z-scheme (or two-step photoexcitation) system. Such a reaction can lead to artificial photosynthesis for combining H_2O and CO_2 to produce carbohydrates and oxygen. In this regard, we must remember that CO_2 reduction in aqueous solution suffers from low quantum efficiencies due to preferential H_2 production and low selectivity for the carbon species produced. For this purpose, the authors have introduced another Ru complex containing anchor ligand into MCE which significantly improved the two-electron reduction of CO_2 on SC/[MCE] by enhancing the ET from SC to MCE. The synthesized Ru complex polymer electrocatalysts were: [Ru{4,4'-di(1H-pyrrolyl-3-propyl   carbonate)-2,2'-bipyridine}$(CO)_2Cl_2$] (MCE1), [Ru(4,4'-diphosphate ethyl-2,2'-bipyridine)$(CO)_2Cl_2$] (MCE2-A), [Ru(4,4'-dicarboxylic acid-2,2'-bipyridine)$(CO)_2Cl_2$] (MCE3-A), and [Ru{4,4'-di(1H-pyrrolyl-3-propyl carbonate)-2,2'-bipyridine}(CO)(MeCN)Cl_2] (MCE4). The anchor ligands 4,4'-diphosphate ethyl-2,2'-bipyridine, and 4,4'-dicarboxylic acid-2,2'-bipyridine (dcbpy) in MCE2-A, and MCE3-A, respectively, were expected to promote the ET between the SC and MCE-A. The decreasing order of photocatalytic reaction rate for $HCOO^-$ was found to be InP/[MCE2-A+MCE4] >InP/[MCE4] >InP/[MCE1] under a xenon light ($\lambda > 400$ nm) and at an applied potential of -0.4 V (vs. Ag/AgCl). One of the possible reasons why the mixture of MCE2-A and MCE4 showed higher reaction rate than MCE4 is that the 4,4'-diphosphate ethyl-2,2'-bipyridine anchor ligand in MCE2-A can link tightly with the surface of InP, therefore, accelerates ET from the photoexcited InP to [MCE2-A+MCE4]. In the subsequent step, platinum-loaded anatase titanium dioxide on conducting glass (TiO_2/Pt), photocatalyst for H_2O oxidation, was combined with InP/[MCE2-A+MCE4] to reduce CO_2 in H_2O with no external electrical bias. Figure 2.6 is displaying the schematic of the catalytic reaction. The estimated potential difference of CB minimum of TiO_2 toward valence band maximum (VBM) of InP was found to be -0.5 V which allowed successful ET between two photocatalysts with no external electrical bias (Ida et al., 2010). Here, the Pt performed as a cocatalyst in O_2 production from H_2O_2 which was originated from H_2O (Eqs. 2.9 and 2.10). The selectivity in the formation $HCOO^-$ was above 70%with 0.03–0.04% conversion efficiency of solar energy to chemical energy and this value is one-fifth of that in switchgrass (0.2%). Thus, InP/[MCE2-A+MCE4]-TiO_2/Pt system mimics photosynthesis in plants.

$$2H_2O = H_2O_2 + 2H^+ + 2e^-$$

(2.9)

$$2H_2O_2 = 2H_2O + O_2 \qquad (2.10)$$

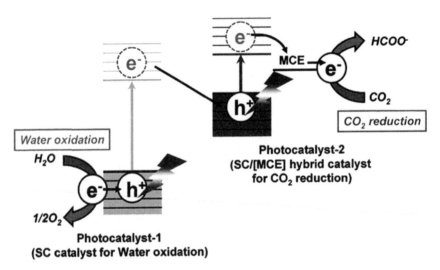

FIGURE 2.6 Total reaction of the Z-scheme system for CO_2 reduction using hybrid semiconductor/metal complex.
Source: Reprinted with permission from Sato et al. (2011). © American Chemical Society.

Zhao et al. have discussed the most recent progress in hybrid catalyst systems for photoelectrochemical reduction of CO_2 (2014). The light-harvesting semiconductors used in the photoelectrochemical reduction of CO_2 in most recent days included N-Ta_2O_5 (Sato et al., 2010), C_3N_4 (Maeda et al., 2013), p-Si (Kumar et al., 2010, 2012; Alenezi et al., 2013), p-GaAs (Arai et al., 2010), p-InP (Sato et al., 2011; Arai et al., 2010), p-GaP (Arai et al., 2010), etc. The primarily used complex co-catalysts were rhenium (Kumar et al., 2010, 2012), ruthenium (Sato et al., 2011; Arai et al., 2010, 2011) and iron-based complexes (Alenezi et al., 2013). In the simplest approach for CO_2 reduction (Kumar et al., 2010), hydrogen-terminated p-type silicon was used as the photocathode light harvester and Re(bipy-But)(CO)$_3$Cl (bipy-But=4,4′-di-tert-butyl-2,2′-bipyridine) metal complex was dissolved into acetonitrile. The major product was found to be CO with Faradaic efficiency of 97± 3%. Similar outcomes were observed in the hybrid catalytic system consists of meso-tetraphenylporphyri Fe^{III} chloride and p-type silicon using CF_3CH_2OH as a proton source (Alenezi et al., 2013).

The ET from the semiconductor to the metal complex can be improved by linking them together or grafting metal complex co-catalysts onto semiconductors as it inhibits complexes moving from one electrode to another. Arai et al. modified a p-type InP-Zn photocathode (zinc doped InP) with an electropolymerized ruthenium complex $[Ru(L-L)(CO)_2Cl_2]$ (2010). The efficiency for formate (HCOO$^-$ from CO_2) formation (EFF) in water was found to be 34.3%. A two-step polymerization strategy improved the EFF to 62.3%. The EEF was furthermore improved to 78% by introducing an (4,4'-diphosphate ethyl-2,2'-bipyridine) anchor complex (Sato et al., 2011). Faradaic efficiency for the photoelectrochemical reduction of CO_2 to HCOO$^-$ was achieved about 80% by modification of Cu_2ZnSnS_4 with a polymerized Ru complex (Arai et al., 2011).

So far, we have discussed the hybrid catalytic systems for the reduction of CO_2 to CO and HCOO$^-$ that involves the addition of only two electrons. Recently, Yuan and Hao showed interesting results of methanol formation which involves the addition of six electrons (2013). They have used chalcopyritep-$CuInS_2$ thin film as a photocathode for solar-driven photoelectrochemical reduction of CO_2 to methanol. The $CuInS_2$ thin film was fabricated in two steps. At first, the Cu-In alloy layers were prepared by electrodeposition using an electrodeposition bath of 5 mmol/L $CuCl_2$, 5 mmol/L $InCl_3$, 0.2 mol/L triethanolamine and 0.015 mol/L sodium citrate at pH 4.0. The ITO (indium tin oxide) conductive glass, platinum foil, and saturated calomel electrode (SCE) were used as the working, counter, and REs, respectively for the electrodeposition process at 1 V (vs. SCE) for 1800 s at 30°C. In the next step, Cu-In alloy layers were annealed in a tube furnace with sulfur powder for 0.5 h in N_2 atmosphere at 400°C for sulfurization. The resulting $CuInS_2$ thin film showed a very strong peak at 27.9 degree indicating the preferential orientation growth is along (112) direction. When the cathode current density of $CuInS_2$ thin film was measured, it was observed that the current density is larger in magnitude under illumination than in dark. Therefore, it confirmed that $CuInS_2$ thin film was a p-type semiconductor. The bandgap estimated from the UV-Vis spectrum of the $CuInS_2$ thin film was found to be 1.53 eV and the flat band potential was measured to be –0.11 V in a buffer solution of 0.1 M acetate containing 10 mM pyridine by the open-circuit method. Since the flat band potential for p-type semiconductors is almost equal to its VB potential, therefore the positions of CB and VB edges were calculated at –1.64 and –0.11 eV, respectively. At pH 5.2, the potential required for reduction of CO_2 to methanol has been reported to be –0.27 V (vs. NHE) which is in between the conduction and VB edges of $CuInS_2$ thin film (Barton et al., 2008). Therefore, $CuInS_2$ thin films can drive thephotocatalyticreductionofCO_2 to methanol. Figure 2.7 describes the schematic setup of photoelectrochemical reduction of CO_2 into methanol using three electrodes. The reaction was performed in a quartz glass beaker containing 0.1 M acetate buffer solution and 10 mM pyridine. The pyridinium ion in worked as a co-catalyst. It is important to mention here that the graphite sheet minimized the re-oxidation of the reduced CO_2 species. The photo-illumination intensity in the visible region emitted from a Xenon lamp (AULTT, Beijing, PR China) on the WE was 100 mW/cm². The faradaic efficiency of this $CuInS_2$ thin film for reduction of CO_2 to methanol was 97% at 20 mV over-potential and shows a stable performance for 11 h.

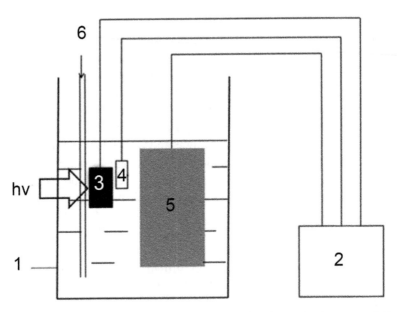

FIGURE 2.7 Schematic diagram of the photoelectrochemical reduction setup. 1-Quartz glass beaker, 2-electrochemistry workstation, 3-$CuInS_2$ thin film electrode, 4-SCE, 5-graphite sheet counter electrode, 6-CO_2 inlet.
Source: Reprinted with permission from Yuan and Hao (2013). © Elsevier.

Lu et al. utilized a photoelectrochemical system to capture and convert CO_2 into FA by a fairly new concept (Lu et al., 2017). An aqueous solution of ionic liquid (IL), 1-aminopropyl-3-methylimidazolium bromide ([NH_2C_3MIm][Br]), functioned as an absorbent and electrolyte at ambient temperature and pressure. It has been reported that ILs with an amide-functionalized imidazolium cation, can capture 0.5 mol of CO_2 per mol of the IL under ambient pressure (Bates et al., 2002). In addition to that IL provided a low-energy pathway for CO_2 reduction to CO by the formation of an adsorbed complex with CO_2, e.g., CO_2-$[C_2MIm]^+$(C_2MIm=1-ethyl-3-methylimidazolium). It has been suggested that $[C_2MIm]^+$ can stabilize $CO_2^{-\bullet}$ radical anion, therefore, inhibits their dimerization to form oxalate, instead allows the formation of CO. One of the advantages of this method is that a high concentration (0.211 $molL^{-1}$ at 25°C) of CO_2 reactants can be attained using functionalized ILs as absorbent. This helped to overcome the difficulty associated with the low solubility of CO_2 in water (0.033 $molL^{-1}$ at 25°C under 1 atm). IL used in this PEC reduction strategy plays a key role in promoting the conversion of CO_2 to FA while suppressing the reduction of H_2O to H_2. The Faradaic efficiency for FA production was 94.1% and the electro-to-chemical efficiency was 86.2% at an applied voltage of 1.7 V.

Zeng et al. has reported the photocatalytic reduction of CO_2 to methanol on TiO_2-passivated GaP (gallium phosphide) (2014). The TiO_2 passivated layer improved the stability of GaP surface in solution by preventing it from photocorrosion. In addition, the TiO_2 passivation layer also enhanced the photoconversion efficiency through the passivation of surface states and the formation of a charge separating p-n region. Owing to these two aforementioned reasons, the carrier recombination is reduced which in turn lowers the over-potential required to initiate this reaction by approximately 0.5 V. It was observed that the 5 nm thick TiO_2 sample produced 4.9 (±0.02) µmol ofCH₃OH during an 8 h reaction consuming 5.2 Coulombs of charge corresponds to a Faradaic efficiency of 55%. This general approach of passivating narrower bandgap semiconductors with TiO_2 will enable more efficient photocatalysts enabling efficient use of the solar spectrum.

Already we have mentioned that one-electron reduction of CO_2 to $CO_2^{-\bullet}$ is extremely difficult since it happens at a very high reduction potential of−1.9 V vs. SHE. This value is above the CB edge of almost all the semiconductors, therefore, restricting to initiate the one-electron reduction process. In a new approach, Zhang et al. have performed the one-electron reduction of CO_2 to $CO_2^{-\bullet}$ in aqueous medium using solvated electrons (2014). The final reduced product was CO with greater than 90% selectivity with minimal formation of H_2. Diamond substrates and aqueous iodide ion (I^-) were the sources of the solvated electrons. UV-photon illumination of the diamond substrate can directly eject electron into the water followed by the formation of solvated electron which then reduce CO_2 to $CO_2^{-\bullet}$. This approach is able to initiate reduction by emitting electrons directly into solution without adsorption of CO_2 (reactants) on surface which is not accessible in conventional electrochemical or photochemical processes.

Figure 2.8 is displaying the energy level diagram for diamond and some of the relevant redox reactions of importance. The CB of the diamond (terminated with H-atom) lies at −5.2 V versus SHE. Therefore, it obviously lies above the one-electron reduction potential of CO_2, as well as at a higher energy than that of a free electron in space (the vacuum level). The typical bandgap of H-terminated diamond is −5.5 eV, thus it requires a UV photon of $\lambda < 225$ nm in order to excite the electron in CB. These excited electrons are directly emitted into the adjacent aqueous phase followed by a fast relaxation to form solvated electrons and other high-energy species.

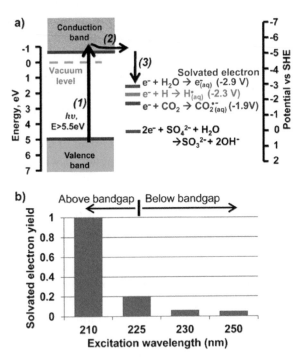

FIGURE 2.8   a) Energy diagram of H-terminated diamond compared with relevant redox potentials on absolute energy scale (left) and electrochemical energy scale (right). E^0 values are shown for the production of the solvated electron ($e^-_{(aq)}$), a hydrogen atom (H'), and CO_2 radical anion ($CO_2^{\cdot-}$). $E(SO_4^{2-}/SO_3^{2-})$ for counter-electrode is shown for pH 3.2. b) Solvated-electron yield versus wavelength from transient absorption measurements, normalized to yield at 210 nm.

Source: Reprinted with permission from Zhang, Zhu, Nathanson, and Hamers (2014). © John Wiley and Sons.

In a quartz electrochemical cell, inexpensive "electrochemical grade" boron-doped diamond acted as an electron emitter, the WE, while platinum was used as a CE. These electrodes were immersed in a Na$_2$SO$_4$/Na$_2$SO$_3$ solution. Sulfite acted as a sacrificial hole scavenger by the oxidation half-reaction SO$_3^{2-}$ + 2OH$^-$ → SO$_4^{2-}$ + H$_2$O + 2e$^-$, (E=−0.27 V under reaction conditions). CO$_2$ pressure in the quartz cell was 2.5 MPa and the diamond sample was irradiated with broadband light from an HgXe lamp. The absolute yield of CO was determined to be 300 mg per 1 cm^2 of the diamond sample which corresponds to a gas phase concentration of CO is about 100 parts per million (ppm). H$_2$ product lies below 5 ppm, i.e., below the detection limit. At this point, it is important to mention that the mechanistic pathways for one electron photo-chemical CO$_2$ reduction to remain only partially understood. Hori et al. hypothesized that CO$_2^-$ first get protonated and could form adsorbed HOCO which then reduced to OH + CO (1994). Costentin et al. (2013) proposed that two CO$_2^-$$_{(ads)}$ anions made a transient intermediate followed by 2CO$_2^-$ → CO + CO$_3^{2-}$ disproportionation in a solution with low proton concentrations (Amatore et al., 1981). Another possibility mentioned was that, CO$_2^-$ in large water clusters can photodissociate into CO and O$^-$ in presence of light at λ = 266 nm or 355 nm (Habteyes et al., 2007; Velarde et al., 2006). Jang et al. fabricated ZnO/ZnTe heterostructures by a dissolution-recrystallization mechanism in an aqueous solution (2014). The ZnO nanowires (NWs) on a zinc metal substrate were transformed into ZnTe after reacting with tellurite and hydrazine in the aqueous solution. This is the simplest way to synthesize ZnTe in an aqueous solution without using any surfactant. Since ZnTe is directly formed on the Zn/ZnO electrode (Zn/ZnO/ZnTe), it was used as a photocathode for CO$_2$ reduction. This p-type semiconductor has a direct bandgap of 2.26 eV with the most negative CB edge at −1.63 eV. The Zn/ZnO/ZnTe photocathode was used for the photo-electrochemical reduction of CO$_2$. The formation of CO and hydrogen was observed at −0.7 to 0.2 V vs. RHE in aqueousKHCO$_3$. The incident-photon-to-current-conversion efficiency value was~60% at −0.7V vs. RHE without any sacrificial agent.

Another hybrid system containing Cu(I)/Cu have also gained wide interests in the electrochemical reduction of CO$_2$ to hydrocarbons, specifically to ethylene (Ogura et al., 2003–2005). It was purported that Cu(I) species anchored on the surface of Cu electrode can act as the adsorption sites for the intermediates involved in CO$_2$ reduction and subsequently improve the overall efficiency along with the product selectivity as compared to that of the bare Cu electrodes (Ba et al., 2014). Ba et al. recently designed a similar hybrid system combining p-type Cu_2O with Cu electrode and used it in the photoelectrochemical reduction of CO_2 (2014). When the bare Cu electrode was used in the electrochemical reduction of CO_2, the formation of methane, as well as ethylene, was found. But, the hybrid system shows much higher selectivity and efficiency in the formation of ethylene over the formation of methane. It is widely known that the CO is the predominant intermediate formed during the course of CO_2 reduction to hydrocarbons. $CO^{·-}$ anion radical is formed initially on the surface of Cu through the transfer of an electron from Cu electrode to the surface adsorbed CO, which is followed by the formation of C-H bond through the further protonation of the $CO^{·-}$. However, there might be chance for the association of two adjacent $CO^{·-}$ anion radical prior to the protonation step. Formation of methane is expected if the mechanism follows the first path, whereas ethylene is expected in the latter case (Hori et al., 1989; Kim et al., 1988; Gattrell et al., 2006; Schouten et al., 2011, 2012). Ba et al. argued that the electrochemical reduction of CO_2 on bare Cu electrode might possibly follow both of the available routes and consequently shows the equal formations of methane and ethylene (Ba et al., 2014). But the presence of Cu_2O possibly suppresses the first route and predominantly forms ethylene. Light illumination (i.e., photoelectrochemical) on this hybrid system further improves the efficiency in the formation of ethylene. In addition, they found that the morphology of Cu_2O plays a vital role in the photoreduction of CO_2. Cu_2O having stone-like morphology was more active as compared to the activity of the same with nanobelt arrays. Moreover, the stability of the Cu/Cu_2O hybrid electrode was found to be appreciably higher as compared to the bare Cu electrode over the course of the photoelectrochemical reduction of CO_2 (Ba et al., 2014).

2.3   ELECTROCHEMICAL OXIDATION OF CO

2.3.1   BACKGROUND

The study of CO electrooxidation (COE) is also a very crucial in the field of energy conversion from the alternative fuels using fuel cells. The fuels are used in fuel cells mostly are hydrogen and methanol, which are oxidized at the anode of the fuel cells (Grgur et al., 2001). Due to the associated technical difficulties for hydrogen storage and transportation, hydrogen is readily produced through the reformation of hydrocarbons or alcohol fuels like methane or methanol (Eqn. 2.11), prior to feed it into the fuel cells. As a result, H_2 is largely contaminated with CO. In the case of methanol as fuel, CO is the major intermediate formed in fuel cells. Platinum with high surface area (platinized platinum) is conventionally used as the catalyst for this anodic oxidation.

$$CH_4 + H_2O \rightarrow CO + H_2 \qquad (2.11)$$

The strong chemistry between CO and metals is well known. The ability of CO to strongly bind metals with low or zero oxidation states can easily be understood from its molecular orbital diagram. It can simultaneously donate its electron pair from its 5σ bonding orbital to metal and accept electron pair in its 2π antibonding orbital from the d orbital of metal. This type of back donation leads to the strong adsorption of CO (CO_{ad}) on Pt and subsequent deactivation of Pt towards its catalytic activity, which is commonly known as CO *poisoning* (Figure 2.9) (Baschuk and Li, 2001). Figure 2.10 shows the voltage-current dependencies of a hydrogen fuel cell (Grgur et al., 2001). Eqs. (2.12) and (2.13) show the oxidation of H_2 and reduction of O_2, which happen separately in anode and cathode of the fuel cell, respectively.

At anode:

$$H_2 = 2H^+ + 2e^- \qquad (2.12)$$

At cathode:

$$O_2 + 4H^+ + 4e^- = 2H_2O \qquad (2.13)$$

Standard potentials for both the reactions are known to be 0 and 1.23 V *vs.* NHE, respectively. The theoretical value of the output cell potential (U_T) is known to be the difference between the standard potentials for the H_2 oxidation and O_2 reduction, which is 1.23 V *vs.* NHE. However, the experimental values lie within 0.8 to 1.1 V due to the slow nature of the O_2 reduction kinetics (Grgur et al., 2001). Preferential adsorption of CO over H_2 on Pt is so high that the presence of CO in the H_2 stream, even in a trace amount (5–10 ppm), makes the complete coverage of CO on the Pt surface, leaving behind no or a little number of free catalytic site(s) for the H_2 to be oxidized. As a result, the output cell potential further reduces to the range of 0.4 to 0.5 V. Therefore, the study of COE has been essential for finding the appropriate catalysts or the ways to mitigate this CO *poisoning* during the fuel cell operations (Grgur et al., 2001).

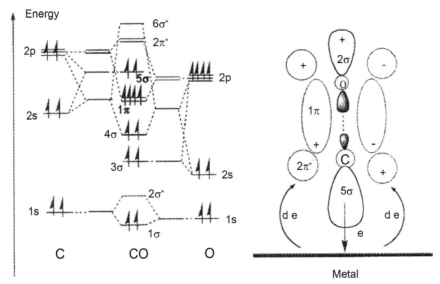

FIGURE 2.9 (a) Molecular orbital energy level diagram of carbon monoxide and (b) the formation of metal-carbon monoxide bonding.
Source: Reprinted with permission from Grgur et al. (2001).

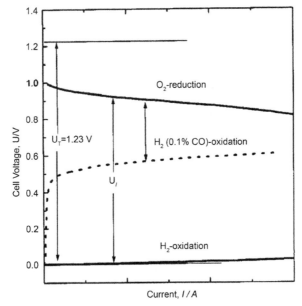

FIGURE 2.10 Plot of voltage *vs.* current for a hydrogen fuel cell.
Source: Reprinted with permission from Grgur et al. (2001).

2.3.2 WORKING PRINCIPLE

CO electrochemical oxidations are typically studied in a three-electrode electrochemical cell (Liu et al., 2016). The WE is basically the anode where the CO oxidation takes place. A glassy carbon electrode may be loaded with the desired electrocatalyst and may serve as the WE. Pt wire is used as the CE where O_2 is reduced. A standard REs is used during the CO electrochemical oxidations to measure the cell output voltage accurately (Liu et al., 2016). Anodic chamber and cathodic chamber are often separated from each other by using a proton exchange membrane (PEM), where the streams of CO and O_2 are separately fed through a connected gas pipelines, respectively. A solution containing NaOH, $HClO_4$, or $NaClO_4$ is generally used as the electrolyte. The pH of the electrolytic solution plays a very crucial role in controlling the electrochemical behavior (Kita and Shimazu, 1988). The rate of CO oxidation is enhanced even up to one to two orders of magnitude just by changing the solution pH from acidic to basic. Before understanding the reason behind the enhanced rate of CO oxidation in alkaline medium properly, certain terms and features associated with a typical cyclic voltammogram (CV) using a polycrystalline Pt electrode in the acidic medium are necessary to be introduced here. A typical CV, Figure 2.11, consists of three distinct regions: *hydrogen region* and *oxygen region,* which are at the two extremes of the CV, are separated by the *double-layer region* (Łukaszewski et al., 2016).

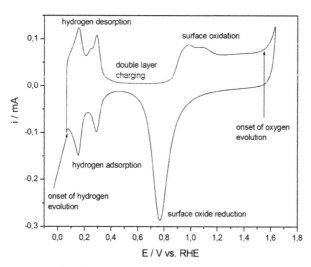

FIGURE 2.11   A typical cyclic voltammogram recorded at a scan rate of 0.1 V s^{-1} using a polycrystalline Pt as the working electrode in 0.5 M H_2SO_4.
Source: Reprinted with permission from Łukaszewski, Soszko, and Czerwiński (2016).

Reversible hydrogen adsorption and desorption happen in the *hydrogen region* during the cathodic and anodic scans, respectively, whereas the formation or reduction of surface oxides occurs in the *oxygen region*. At the *double layer region,* no faradaic process takes place (Łukaszewski et al., 2016). During the oxidation of CO, the required potential for the same grossly depends on which region the adsorptions of CO take place and how strongly it bonded. If the adsorption of CO happens in the hydrogen region, both weakly and strongly bonded CO can be observed, which can be oxidized above 0.4 V and 0.6 V, respectively (Grgur et al., 2001). In the case of the CO adsorbed in the double-layer region, only the strongly bonded CO can be observed which requires even higher potential to be oxidized (Łukaszewski et al., 2016). Now if we look at the Figure 2.12(a) and (b) of the typical CVs of the CO oxidation in 0.1 M NaOH and 0.1 MHClO$_4$, respectively, it is quite obvious that the onset potential for the CO oxidation is ca. 0.5 V more negative in 0.1 M NaOH than that in 0.1 MHClO$_4$. This suggests that the CO binds weakly and hence, it is oxidized easily in alkaline solution. In acidic solution, CO binds strongly and subsequently, oxidation starts at a higher potential (Kita and Shimazu, 1988).

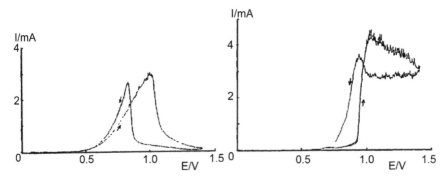

FIGURE 2.12 Typical cyclic voltammograms of CO oxidation using Pt as a working electrode in (a) 0.1 M NaOH and (b) 0.1 M HClO$_4$.
Source: Reprinted with permission from Kita and Shimazu (1988). © Elsevier.

2.3.3 MECHANISM

The CO oxidation on noble metals has extensively been studied since the pioneering report of Langmuir in 1922 (Bourane and Bianchi, 2001; Wojciechowski and Aspey, 2000; Venderbosch et al., 1998; Ertl, 1993).

Conventionally it happens at the gas-solid interface and follows simple overall stoichiometry as stated in the Eqn. (2.14).

$$2CO + O_2 \rightarrow 2CO_2 \qquad\qquad (2.14)$$

However, there have been different contradictory opinions about the detailed mechanism. Though, in most of the cases, the overall rate of the reaction is directly and inversely proportional to the pressures of the oxygen and carbon monoxide (CO), respectively (Allian et al., 2011). This can be understood as follows. The surface of the metalcatalyst first needs to be near completely covered by CO with few vacant sites adjacent to the adsorbed CO, where the oxygen molecules can bind to the surface and react with the adjacent adsorbed CO. In case of sufficiently large pressure of the CO, the surface of the metalcatalyst would be completely covered by the CO, leading to the unavailability of the free catalyst surface for the O_2 to be adsorbed. It is highly unlikely that the free O_2 would react with the CO to form CO_2. Therefore, adsorption of O_2 on the surface of the metalcatalyst is also mandatory, and the chances of filling the free catalyst surface adjacent to the adsorbed CO becomes high with the increasing O_2 pressure. Hence, the overall rate follows the rate equation, $r = k[O_2][CO]^{-1}$ (Gilman, 1964; Allian et al., 2011). The next important elementary step is the dissociation of the adsorbed O_2 assisted by the adjacent adsorbed CO. The dissociation of the adsorbed O_2 happens through the concerted nucleophilic addition to the vicinal adsorbed CO to form $CO_2(g)$. Obviously, there exists another possibility of dissociation of adsorbed O_2 to adsorbed O, followed by the reaction of adsorbed O with adsorbed CO. Here the activation barrier in the former case is much smaller than the latter one (Allian et al., 2011).

In case of CO oxidation electrochemically, which usually starts at a voltage range from 0.6 to 0.9 V (Kita and Shimazu, 1988), mechanism differs slightly from that of gas-phase catalytic oxidation, mentioned above, as the oxidation of CO and reduction of O_2 occur separately at anode and cathode, Eqs. (2.15) and (2.16), respectively in the case of COE (Ammal and Heyden, 2015).

At anode:

$$CO + H_2O = CO_2 + 2H^+ + 2e^- \qquad\qquad (2.15)$$

At cathode:

$$O_2 + 4H^+ + 4e^- = 2H_2O \qquad\qquad (2.16)$$

However, the rate of CO oxidation is still inversely proportional to the CO concentration, i.e., the higher oxidation current is obtained at a low coverage of CO. Reactions occurring at anode broadly consist of three elementary steps, which is known as *reactant pair* mechanism, postulated by Gilman in 1964. Unless there is any presence of O_2 in the CO stream, adsorbed H_2O molecules take part in the *reactant pair* instead of O_2 mentioned earlier and the overall mechanism is all about the dissociation of the adsorbed H_2O molecules, assisted by the neighboring adsorbed CO molecules. Rapid adsorption of CO replacing most of the already adsorbed H_2O molecules on the metal surface is the first among the three elementary steps. This step is important in maintaining the appropriate concentrations of the *reactant pair*. In the second step, ET and deprotonation happen to generate adsorbed OH. This step is the slowest among the three and hence, rate-determining. In the final step, CO_2 is formed through the fast oxidation of adsorbed CO by the neighboring adsorbed OH (Gilman, 1964). The idea of the formation of adsorbed OH can be supported by the experimental facts that the onset potential of H_2O oxidation matches with that of CO oxidation, i.e., 0.6 to 0.9 V. The point is to mention here that the H_2O oxidation also involves the formation of OH adsorbed on the metal catalyst surface (Grgur et al., 2001; Andersonz and Neshev, 2002). Now it is clear that COE to happen more efficiently at a faster rate, the onset potential for the same needs to be shifted towards more negative potentials. This can happen only if the formation of the adsorbed OH radical intermediate from the adsorbed H_2O can happen at a more negative potential. The remaining of this chapter will focus on the discussion about the materials developed till now to lower the potential for the formation of adsorbed OH radical intermediate and subsequently the overall rate of COE.

2.3.4   STATE OF THE ART ELECTROCATALYSTS

2.3.4.1   METALS AND THEIR VARIOUS ALLOYS

Noble metals, specifically Pt have extensively been used as the electrode materials in fuel cells for the oxidations of hydrogen or the small-molecule alcohols including methanol and ethanol (Langmuir, 1922; Bourane and Bianchi, 2001; Wojciechowski and Aspey, 2000; Venderbosch et al., 1998; Ertl, 1993). On the other hand, strong chemistry between Pt and CO hinders the electrocatalytic efficiency of Pt in the oxidations of hydrogen or the small-molecule (Grgur et al., 2001). Therefore, discussion on the Pt and its various alloys as the electrocatalyst for the CO oxidation may be a good starting point. Then focus will be shifted towards the electrocatalysts containing Pt and Au on various types of supports.

2.3.4.1.1 *Metals*

As shown in Figure 2.10, CO tolerance of Pt electrode in a hydrogen fuel cell is very low. Even the presence of CO as low as 10–25 ppm reduces the output cell potential abruptly. However, it was understood from several reports over the last two decades that the surface structure of the Pt controls the oxidation behavior of CO significantly (Chang and Weaver, 1992; Chang et al., 1990; Akemann et al., 1998; Pozniak et al., 2002). Akemann et al. reported that CO oxidation starts at a potential of *ca.* 0.88 V on a Pt (111) surface in a CO saturated acidic medium, whereas the same happens at a 100 mV more negative potential on a Pt surface with step edges under an identical condition. The presence of step edges helps the formation of the adsorbed OH radical intermediate from the adsorbed H_2O at a more negative potential, which is very crucial for COE to happen more efficiently at a faster rate (Akemann et al., 1998). Akemann et al. confirmed the formation of adsorbed OH radical intermediate at a more negative potential through the presence of a vibrational band in optical second harmonic generation spectroscopy which is associated with the adsorbed OH radical (Akemann et al., 1998).

Parallel to the extensive researches on the COE using Pt as the electrocatalysts, copious reports on the COE using Au as the electrocatalysts are also available (Haruta et al., 1987; Edens et al., 1996; Haruta, 1997). The mechanism behind COE on Au differs from the *reactant pair* mechanism for the same on Pt. Interaction of CO with Au is relatively much weaker as compared to that with Pt (Edens et al., 1996). Hence, CO molecules are adsorbed weakly and reversibly on Au. No direct evidence for the formation of adsorbed OH on gold was found. Rather Weaver proposed that weakly adsorbed CO on Au reacts directly with the bulk H_2O molecules in the rate-determining step during the course of its oxidation to CO_2 (Edens et al., 1996). Like in the case of CO oxidation on Pt, surface crystallinity of Au also plays a very important role in controlling the rate of CO oxidation. Chang et al. evaluated the COE rates on Au surfaces with six different crystalline orientations: (111), (533), (100), (221), (210), and (110) (1991).

Enhancement in the oxidation rate was observed much higher, up to 100-fold, by just changing the surface crystallinity of Pt from (111) to (110). What was more astonishing in their interesting findings is that the extent of CO coverage doesn't follow the same trend as in the case of the overall COE. Despite having the moderate CO coverage on the surface, Au (110) was the most reactive. Chang et al. explained that these two apparently contrasting observations were found due to the presence of a higher number of step sites or rows of low-coordination metal atoms in Au (110). H_2O molecules are adsorbed preferentially at those step sites which are the prerequisite to form the adsorbed OH intermediates to further react with adjacent adsorbed CO (Chang et al., 1991). Edens et al. studied the kinetics involved in COE in aqueous solution over a wide pH range (0–13.5) (Edens et al., 1996). Figure 2.13(a) shows the plot of the logarithm of current density vs. solution pH.

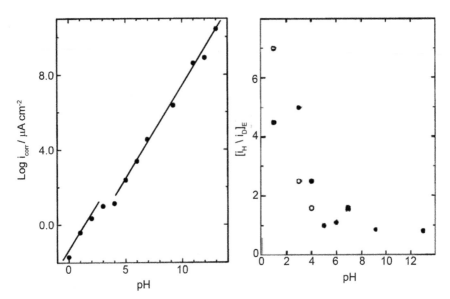

FIGURE 2.13   (a) plot of the logarithm of current density vs. solution pH and (b) the ratio of the current densities obtained in CO electrooxidation in aqueous to deuterated solvent with respect to the solution pH.
Source: Reprinted with permission from Edens, Hamelin, and Weaver (1996). © American Chemical Society

An increase in current density is regarded as the increase in the rate of reaction, whereas the increase of solution pH indicates the increasing concentration of [OH⁻] in the solution. Therefore, the linear dependency of log (rate) over log[OH⁻] clearly suggests that the COE follows the first-order kinetics with respect to [OH$^-$] at the ranges pH ≤ 2 and pH ≥ 4. The rate is independent of [OH$^-$] at the range $2 \leq$ pH ≥ 4. Based on these evidences, Edens et al. proposed two-step mechanisms: the formation of an adsorbed hydroxycarbonyl Au(CO$_2$H) intermediate, (Eqn. (2.17a) or (2.17b)) and subsequently its decomposition to CO$_2$ (Eqn. (2.18)). At the range pH ≥ 4, the formation of an adsorbed hydroxycarbonyl Au(CO$_2$H) intermediate involves the discharge of hydroxide onto adsorbed CO (Au-CO) which is eventually the rate-determining step (Eqn. (2.17a)). At the range $2 \leq$ pH ≥ 4, pH-independent reaction rate indicates that the rate-determining step involves the reaction between adsorbed CO (Au-CO) with water rather than OH$^-$ during the formation of the adsorbed Au(CO$_2$H) intermediate (Eqn. 2.17b). However, it is apparently surprising that the rate law is still first order with respect to [OH$^-$] even at pH as low as below 2. This behavior at low pH was due to the fact that the decomposition, rather than the formation, of adsorbed Au(CO$_2$H) intermediate is the rate-determining step as the reaction associated with the Eqn. (2.17b) will predominantly be shifted towards the reverse direction and hence, the reaction associated with Eqn. (2.17b) becomes much faster in the reverse direction as compared to the decomposition of the adsorbed Au(CO$_2$H) intermediate to CO$_2$ (Eqn. (2.18)). This makes the concentration of the adsorbed Au(CO$_2$H) intermediate proportional to [OH$^-$] and the overall reaction rate follows the first-order kinetics with respect to [OH$^-$] (Edens et al., 1996).

$$Au-CO + OH^- \Leftrightarrow Au(CO_2H) + e^- \tag{2.17a}$$

$$Au-CO + H_2O \Leftrightarrow Au(CO_2H) + e^- + H^+ \tag{2.17b}$$

$$Au(CO_2H) \rightarrow CO_2 + H^+ + e^- \tag{2.18}$$

Figure 2.13b shows the changes in the ratio of the current densities obtained in COE in aqueous to deuterated solvent with respect to the solution pH. The ratio of the current densities remains constant at unity up to the pH≥ 4 which suggests that no deprotonation/protonation of water is involved in the rate-determining step at this pH range. Rather, there is the direct involvement of the OH$^-$ (Eqn. (2.17a)). The ratio of the current densities starts deviating from unity when the solution pH reaches below 4. This clearly indicates that deprotonation/protonation of water is involved during the course of CO oxidation to CO$_2$ at the range pH ≥ 4 (Eqs. (2.17b) and (2.18)).

2.3.4.1.2   Metal Alloys

As mentioned in the previous sections that the formation of the adsorbed OH radical intermediate from the adsorbed H_2O at a more negative potential can enhance the overall rate of the CO oxidation and subsequently reduce CO poisoning, the idea of providing H_2O a different adsorption site where CO is not adsorbed preferably, have been floating for a long time (Baschuk and Li, 2001). Watanabe and Motoo proposed the bifunctional theory of activity enhancement in 1975 by alloying Pt with such a metal (M), which doesn't have an affinity to bind CO as much as Pt does but has lower oxidation potential as compared to that of Pt (Watanabe and Motoo, 1975). As a result, H_2O would preferentially be adsorbed on M and the formation of adsorbed OH can be possible at a less positive potential due to the lower oxidation potential of M. Use of Pt alloy containing Ru as the anode in the hydro-carbon fuel cells to lower the overpotential required for the anodic oxidation by reducing the CO poisoning has been started way back in 1965 (Petry et al., 1965). Schmidt et al. have shown in their report in 1995 that alloying Pt with Ru can lower the potential required for CO oxidation on pure Pt surface and also increase the CO tolerance level of Pt electrocatalyst used in a hydrogen fuel cell using H_2/CO as the fuel, Figure 2.14 (Schmidt et al., 1995). $Pt_{0.5}Ru_{0.5}$ composition showed the best result in shifting the onset of

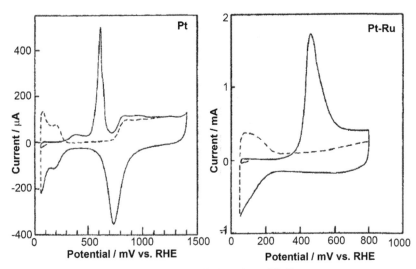

FIGURE 2.14   Electrochemical oxidation of CO on Pt and Pt-Ru.
Source: Reprinted with permission from Ianniello et al. (1994). © Elsevier.

CO oxidation towards more negative potential and in terms of CO tolerance even up to 250 ppm (Schmidt et al., 1995). In addition to the enhancements in the activity due to alloying of Pt with Ru, excellent long-term stability was also found, even over a time period of 3000 h (Ralph et al., 1997). Due to lower oxidation potential of Ru as compared to that of Pt, the preferential formation of the adsorbed OH radical intermediate from the adsorbed H_2O on Ru atoms at a more negative potential, whereas CO is preferentially adsorbed on the Pt atoms due to its stronger affinity towards Pt over Ru and subsequently reacts with the adsorbed OH on the adjacent Ru through the *reactant pair* mechanism to form CO_2 (Ianniello et al., 1994). However, Pt-Ru alloy doesn't follow the bi-functional theory beyond 250 ppm CO anymore as the Ru sites are also covered with CO beyond 250 ppm CO.

A similar enhancement in CO tolerance was found when Pt-Mo alloy was used as the anode in hydrogen fuel cells. Figure 2.15 shows that the best result was obtained with the composition of $Pt_{0.77}Mo_{0.23}$ (Grgur et al., 2001).

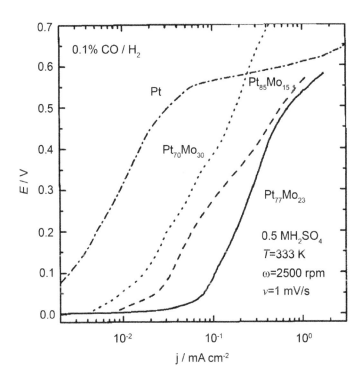

FIGURE 2.15   Electrochemical oxidation of CO on Pt and Pt-Mo alloys.
Source: Reprinted with permission from Grgur et al. (2001).

Another important parameter, which should be kept in mind while selecting appropriate M for alloying with Pt is the stability of the metal under working pH. The oxidation potentials of the transition metals (Fe, Co, Ni, Mn) along with Zn, Cd, Ga, Bi, In, Sb, Ge, etc., are also known to be much lower as compared to that of Pt. However, their poor thermal stability and high corrosion rate make them poor candidates for alloying them with platinum. In this regard, PtSn alloy could easily be the best among the known real catalyst as shown in Figure 2.16. But, the dissolution of the surface Sn under the working condition hinders its applications in reality (Grgur et al., 2001).

Chang et al. reported if bismuth is deposited (predosing) on Pt(100), enhancement in the electro-oxidation of FA to CO_2 was found to be up to 30 to 40 folds (Chang et al., 1992). Predosing of Pt(100) with bismuth reduces the extent of CO poisoning of the catalyst. It was found that the intermediate CO, which is formed during the oxidation of FA, completely covers pure Pt at a particular pressure of CO at which predosing Pt(100) with bismuth with a surface coverage above ca. 0.2 is capable to essentially eliminate the CO coverage (Chang et al., 1992).

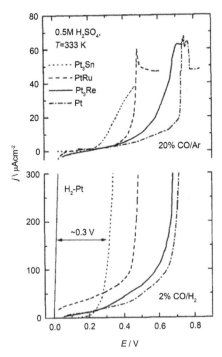

FIGURE 2.16 Electrochemical oxidation of CO on Pt and different Pt alloys.
Source: Reprinted with permission from Grgur et al. (2001).

2.3.4.2   METALS ON VARIOUS SUPPORTS

Pure noble metals, especially gold, are known to be the least reactive. In addition, CO poisoning makes those pure metals less attractive as the catalysts for CO oxidation. However, making alloys of metals shows great improvements in reducing the CO poisoning, the processes involved in the alloying include sintering above 700°C to make the alloy homogeneous. This high-temperature sintering makes this process economically dull. Haruta et al. showed the excellent catalytic activity of gold nanoparticle deposited on selected metal oxides in CO oxidation even at a low temperature far below 0°C (1987). Recently in 2016, Liu et al. devised a novel and durable catalyst system containing nickel-aluminum double layered hydroxide (NiAl-LDH) nanoplates on mildly oxidized carbon nanotubes (CNTs) (Liu et al., 2016). In addition, NiAl-LDH, gold nanoparticle was also deposited on the surface of CNTs. This makes the complex system having extended dual-metal active sites: Au/NiO. Further attachment of bis(trifluoromethylsulfonyl) imide (NTf_2) anion of IL electrolyte was to facilitate the CO/O_2 adsorption as well as the conversion of $Ni(OH)_2$ to NiOOH by making the catalytic interface more electrophilic. Activity of this novel catalytic system surpasses the activity and stability of the commercial and other reported catalysts containing precious metals. Not only the turnover frequency (TOF) of the LDH-Au/CNTs COE catalyst was excellent but its long-term stability even after 1000 cycles makes it very attractive for the practical application in the field of COE (Liu et al., 2016).

2.4   CONCLUSION

Apparently, it may appear puzzling that the reduction of CO_2 to CO/hydrocarbons and oxidation of CO to CO_2 have been discussed in a single chapter. However, it became clear now that these two processes are quite independent of each other for their applications in the field of alternative energy. Both of them are complementary to each other. Without one, the very purpose of the other would not be fulfilled. Photochemical/photoelectrochemical reduction of CO_2 bears immense potential in the field of solar fuel generation by garnering hugely abundant sunlight. In addition, it can put caps on the anthropogenic CO_2 level in Earth's atmosphere. Now, this step would be highly beneficial only if we can extract energy from these solar fuels through their combustion in fuel cells as much efficient as possible.

During the combustions, CO is found to be the most predominant product which basically hinders the efficient combustion of fuels through the CO poisoning of the electrocatalysts used in the fuel cells. Therefore, the cycle would not be complete without paying enough attention for the process of electrochemical oxidation of CO. Photoelectrochemical reduction of CO_2 is extremely challenging as well as potentially rewarding. While one intends to design an efficient catalyst for the photoelectrochemical reduction of CO_2, it should be kept in mind that the catalyst must be a hybrid system containing an appropriate light absorber and an efficient electrocatalyst for reducing the activation barrier involved in the process of CO_2 reduction. Also, the surface of the catalyst must provide enough adsorption sites for CO_2 molecules. On the contrary, ideal catalysts for fuel cells should be less adsorbent of CO. So that CO poisoning can be avoided. This goal can be achieved through alloying conventionally used electrocatalyst Pt with some other appropriate metals which are necessarily less adsorbent of CO as compared to that of Pt. Therefore, this chapter can provide a brief idea of designing electrocatalysts in the field of carbon-neutral energy.

KEYWORDS

- **catalyst poison**
- **electrocatalysts**
- **hydrogen evolution reaction**
- **photoelectrocatalysts**
- **renewable energy**
- **semiconductor nanostructures cocatalyst**

REFERENCES

Akemann, W., Friedrich, K. A., Linke, U., & Stimming, U., (1998). The catalytic oxidation of carbon monoxide at the platinum/electrolyte interface investigated by optical second harmonic generation (SHG): Comparison of Pt(111) and Pt(997) electrode surfaces. *Surf. Sci., 402*, 571.

Alenezi, K., Ibrahim, S. K., Li, P. Y., & Pickett, C. J., (2013). Solar fuels: Photoelectrosynthesis of CO from CO_2 at p-type Si using Fe porphyrin electrocatalysts, *Chem-Eur. J., 19*, 13522–13527.

Allian, D. A., et al., (2011). Chemisorption of CO and mechanism of CO oxidation on supported platinum nanoclusters. *J. Am. Chem. Soc., 133*, 4498–4517.

Amatore, C., & Saveant, J. M., (1981). Mechanism and kinetic characteristics of the electrochemical reduction of carbon-dioxide in media of low proton availability, *J. Am. Chem. Soc., 103*, 5021–5023.

Ammal, S. C., & Heyden, A., (2015). Reaction kinetics of the electrochemical oxidation of CO and syngas fuels on a Sr$_2$Fe$_{1.5}$Mo$_{0.5}$O$_{6-\delta}$ perovskite anode. *J. Mater. Chem. A, 3*, 21618–21629.

Andersonz, A. B., & Neshev, N. M., (2002). Mechanism for the electro-oxidation of carbon monoxide on platinum, including electrode potential dependence theoretical determination. *Journal of the Electrochemical Society, 149*, E383–E388.

Arai, T., Sato, S., Uemura, K., Morikawa, T., Kajino, T., & Motohiro, T., (2010). Photo electrochemical reduction of CO$_2$ in water under visible-light irradiation by a p-type InP photocathode modified with an electro polymerized ruthenium complex. *Chem. Commun., 46*, 6944–6946.

Arai, T., Tajima, S., Sato, S., Uemura, K., Morikawa, T., & Kajino, T., (2011). Selective CO$_2$ conversion to format in water using a CZTS photocathode modified with a ruthenium complex polymer. *Chem. Commun., 47*, 12664–12666.

Ba, X., Yan, L. L., Huang, S., Yu, J., Xia, X. J., & Yu, Y., (2014). New way for CO$_2$ reduction under visible light by a combination of a Cu electrode and semiconductor thin film: Cu$_2$O conduction type and morphology effect. *J. Phys. Chem. C, 118*, 24467.

Barton, E. E., Rampulla, D. M., & Bocarsly, A. B., (2008). Selective solar-driven reduction of CO$_2$ to methanol using a catalyzed p-GaP based photoelectrochemical cell. *J. Am. Chem. Soc., 130*, 6342–6344.

Baschuk, J. J., & Li, X., (2001). Carbon monoxide poisoning of proton exchange membrane fuel cells, *Int. J. Energy Res., 25*, 695.

Bates, E. D., Mayton, R. D., Ntai, I., & Davis, J. H., (2002). CO$_2$ capture by a task-specific ionic liquid. *J. Am. Chem. Soc., 124*, 926–927.

Bott, A. W., (1998). Electrochemistry of semiconductors. *Current Separations, 17*, 87–91.

Bourane, A., & Bianchi, D., (2001). Oxidation of CO on a Pt/Al$_2$O$_3$ catalyst: From the surface elementary steps to light-off tests: I. kinetic study of the oxidation of the linear CO species. *J. Catal., 202*, 34.

Chang, S. C., Hamelin, A., & Weaver, M. J., (1991). Dependence of the electro oxidation rates of carbon monoxide at gold on the surface crystallographic orientation: A combined kinetic-surface infrared spectroscopic study. *The Journal of Physical Chemistry, 95*, 5560.

Chang, S. C., Ho, Y., & Weaver, M. J., (1992). Applications of real-time infrared spectroscopy to electro catalysis at bimetallic surfaces I. Electro oxidation of formic acid and methanol on bismuth-modified Pt(lll) and Pt(100). *Surface Science, 265*, 81–94.

Chang, S. C., Leung, L. W. H., & Weave, M. J., (1990). Metal crystallinity effects in electro catalysis as probed by real-time FTIR spectroscopy: Electro oxidation of formic acid, methanol, and ethanol on ordered low-index platinum surfaces. *J. Phys. Chem., 94*, 6013–6021.

Chang, X. X., Wang, T., & Gong, J. L., (2016). CO$_2$ photo-reduction: Insights into CO$_2$ activation and reaction on surfaces of photo catalysts. *Energ. Environ. Sci., 9*, 2177–2196.

Costentin, C., Robert, M., & Saveant, J. M., (2013). Catalysis of the electrochemical reduction of carbon dioxide, *Chem. Soc. Rev., 42*, 2423–2436.

Edens, G. J., Hamelin, A., & Weaver, M. J., (1996). Mechanism of carbon monoxide electro oxidation on mono crystalline gold surfaces: Identification of a hydroxyl carbonyl intermediate. *J. Phys. Chem., 100*, 2322–2329.

Ertl, G., (1993). Self-organization in reactions at surfaces, *Surf. Sci., 287*, 1.

Gattrell, M., Gupta, N., & Co, A., (2006). A review of the aqueous electrochemical reduction of CO_2 to hydrocarbons at copper. *J. Electroanal. Chem., 694*, 1–19.

Gilman, S., (1964). The mechanism of electrochemical oxidation of carbon monoxide and methanol on platinum: II. The "reactant-pair" mechanism for electrochemical oxidation of carbon monoxide and methanol. *J. Phys. Chem., 68*(1), 70–80.

Grgur, B. N., Markovi, N. M., Lucas, C. A., & Ross, Jr. P. N., (2001). Electrochemical oxidation of carbon monoxide: From platinum single crystals to low temperature, fuel cells catalysts: Part I: Carbon monoxide oxidation onto low index platinum single crystals. *J. Serb. Chem. Soc., 66*(11–12), 785–797.

Habisreutinger, S. N., Schmidt-Mende, L., & Stolarczyk, J. K., (2013). photo catalytic reduction of CO_2 on TiO_2 and other semiconductors. *Angew. Chem. Int. Edit., 52*, 7372–7408.

Habteyes, T., Velarde, L., & Sanov, A., (2007). Photo dissociation of CO_2^- in water clusters via Renner-teller and conical interactions. *J. Chem. Phys., 126*, 154301.

Haruta, M., (1997). Size- and support-dependency in the catalysis of gold. *Catalysis Today, 36*, 153.

Haruta, M., Kobayashi, T., Sano, H., & Yamada, N., (1987). Novel gold catalysts for the oxidation of carbon monoxide at a temperature far below 0°C. *Chemistry Letters*, pp. 405–408.

Hori, Y., Murata, A., & Takahashi, R., (1989). Formation of hydrocarbons in the electrochemical reduction of carbon dioxide at a copper electrode in aqueous solution. *J. Chem. Soc., Faraday Trans., 185*, 2309–2326.

Hori, Y., Wakebe, H., Tsukamoto, T., & Koga, O., (1994). Electro-catalytic process of CO selectivity in electrochemical reduction of CO_2 at metal-electrodes in aqueous-media. *Electrochim. Acta., 39*, 1833–1839.

Ianniello, R., Schmidt, V. M., Stimming, U., Stumper, J., & Wallau, A., (1994). CO adsorption and oxidation on Pt and Pt-Ru alloys: Dependence on substrate composition. *Electrochimica Acta, 39*, 1863.

Ida, S., Yamada, K., Matsunaga, T., Hagiwara, H., Matsumoto, Y., & Ishihara, T., (2010). Preparation of p-type $CaFe_2O_4$ photocathode's for producing hydrogen from water. *J. Am. Chem. Soc., 132*, 17343–17345.

Inoue, T., Fujishima, A., Konishi, S., & Honda, K., (1979). Photo-electro catalytic reduction of carbon-dioxide in aqueous suspensions of semiconductor sowders. *Nature, 277*, 637–638.

Jang, J. W., Cho, S., Magesh, G., Jang, Y. J., Kim, J. Y., Kim, W. Y., Seo, J. K., et al., (2014). Aqueous-solution route to zinc telluride films for application to CO_2 reduction. *Angew. Chem. Int. Edit., 53*, 5852–5857.

Jiang, C. R., Moniz, S. J. A., Wang, A. Q., Zhang, T., & Tang, J. W., (2017). Photoelectrochemical devices for solar water splitting-materials and challenges, *Chem. Soc. Rev., 46*, 4645–4660.

Khodakov, A. Y., Chu, W., & Fongarland, P., (2007). Advances in the development of novel cobalt Fischer-Tropsch catalysts for synthesis of long-chain hydrocarbons and clean fuels. *Chem. Rev., 107*, 1692–1744.

Kim, J. J., Summers, D. P., & Frese, K. W. Jr., (1988). Reduction of CO_2 and CO to methane on Cu foil electrodes. *J. Electroanal. Chem., 245*, 223–230.

Kita, H., & Shimazu, K., (1988). Electrochemical oxidation of CO on Pt in acidic and alkaline solutions: Part I. voltammetric study on the adsorbed species and effects of aging and Sn(IV) pretreatment. *J. Eleciroanal. Chem., 241*, 163–179.

Kumar, B., Smieja, J. M., & Kubiak, C. P., (2010). Photo reduction of CO_2 on p-type silicon using Re(bipy-Bu^i)(CO)$_3$Cl: Photo voltages exceeding 600 mV for the selective reduction of CO_2 to CO, *J. Phys. Chem. C, 114*, 14220–14223.

Kumar, B., Smieja, J. M., Sasayama, A. F., & Kubiak, C. P., (2012). Tunable, light-assisted co-generation of CO and H_2 from CO_2 and H_2O by Re(bipy-tbu)(CO)$_3$Cl and p-Si in non-aqueous medium. *Chem. Commun., 48*, 272–274.

Langmuir, I., (1922). The mechanism of the catalytic action of platinum in the reactions 2CO + O_2 = 2CO$_2$, and 2H$_2$ + O$_2$ = 2H$_2$O. *Trans. Faraday Soc., 17*, 621.

Liu, Z., Huang, F. C., Zhanhu, G., Guangdi, W., Xu, C., & Zhe, W., (2016). Efficient dual-site carbon monoxide electro-catalysts via interfacial nano-engineering. *Sci. Rep., 6*, 33127.

Lu, W., Jia, B., Cui, B., Zhang, Y., Yao, K., Zhao, Y., & Wang, J., (2017). Efficient photoelectrochemical reduction of carbon dioxide to formic acid: A functionalized ionic liquid as an absorbent and electrolyte. *Angew. Chem. Int. Ed., 56*, 11851–11854.

Łukaszewski, M., Soszko, M., & Czerwiński, A., (2016). Electrochemical methods of real surface area determination of noble metal electrodes: An overview. *Int. J. Electrochem. Sci., 11*, 4442–4469.

Maeda, K., Sekizawa, K., & Ishitani, O., (2013). A polymeric-semiconductor-metal-complex hybrid photocatalyst for visible-light CO_2 reduction. *Chem. Commun., 49*, 10127–10129.

Mishra, B., & Chaudhary, Y. S., (2017). In: Yatendra, S. C., (ed.), *Photocatalytic CO_2 Reduction to Fuel in Solar Fuel Generation*. CRC Press, ISBN: 9781498725514.

Ogura, K., Ohara, R., & Kudo, Y., (2005). Reduction of CO_2 to ethylene at three-phase interface effects of electrode substrate and catalytic coating. *J. Electrochem. Soc., 152*, D213−D219.

Ogura, K., Yano, H., & Shirai, F., (2003). Catalytic reduction of CO_2 to ethylene by electrolysis at a three-phase interface. *J. Electrochem. Soc., 150*, D163−D168.

Ogura, K., Yano, H., & Tanaka, T., (2004). Selective formation of ethylene from CO_2 by catalytic electrolysis at a three-phase interface. *Catal. Today, 98*, 515−521.

Petry, O. A., Podlovchenko, B. I., Frumkin, A. N., & Lal, H., (1965). The behavior of platinized-platinum and platinum-ruthenium electrodes in methanol solutions. *J. Electroanal. Chem., 10*, 253.

Pozniak, B., Mo, Y., & Scherson, D. A., (2002). The electrochemical oxidation of carbon monoxide adsorbed on Pt(111) in aqueous electrolytes as monitored by in situ potential step-second harmonic generation. *Faraday Discuss, 121*, 313–322.

Ralph, T. R., Hards, G. A., Keating, J. E., Campbell, S. A., Wilkinson, D. P., Davis, M., St-Pierre, J., & Johnson, M. C., (1997). Low cost electrodes for proton exchange membrane fuel cells: Performance in single cells and Ballard stacks. *Journal of the Electrochemical Society, 144*, 3845–3857.

Sato, S., Arai, T., Morikawa, T., Uemura, K., Suzuki, T. M., Tanaka, H., & Kajino, T., (2011). Selective CO_2 conversion to formate conjugated with H_2O oxidation utilizing semiconductor/complex hybrid photo catalysts. *J. Am. Chem. Soc., 133*, 15240–15243.

Sato, S., Morikawa, T., Saeki, S., Kajino, T., & Motohiro, T., (2010). Visible-light-induced selective CO_2 reduction utilizing a ruthenium complex electro catalyst linked to a p-type nitrogen-doped Ta$_2$O$_5$ semiconductor. *Angew. Chem. Int. Edit., 49*, 5101–5105.

Schmidt, V. M., Ianneillo, R., Oetjen, H. F., Reger, H., Stimming, U., & Trila, F., (1995). Oxidation of H_2/CO in a proton exchange membrane fuel cell. In: Gottesfeld, S., Halpert, G., & Landgrebe, A., (eds.), *Proton Conducting Membrane Fuel Cells I: Electrochemical Society Proceedings* (Vol. 95, pp. 1–11, 23). The Electrochemical Society, Inc.: Pennington.

Schouten, K. J. P., Kwon, Y., Van, D. H. C. J. M., Qin, Z., & Koper, M. T. M., (2011). A new mechanism for the selectivity to C1 and C2 species in the electrochemical reduction of carbon dioxide on copper electrodes. *Chem. Sci., 2*, 1902–1909.

Schouten, K. J. P., Qin, Z., Gallent, E. P., & Koper, M. T. M., (2012). Two pathways for the formation of ethylene in CO reduction on single crystal copper electrodes. *J. Am. Chem. Soc., 134*, 9864–9867.

Takeda, H., Koike, K., Inoue, H., & Ishitani, O., (2008). Development of an efficient photo catalytic system for CO_2 reduction using rhenium(I) complexes based on mechanistic studies. *J. Am. Chem. Soc., 130*, 2023–2031.

Van, D. K. R., & Gratzel, M., (2012). Photoelectrochemical hydrogen production. *Electronic Materials: Science and Technology, 102*, 48. doi: 10.1007/978-1-4614-1380-6_1, Springer Science Business Media, LLC.

Velarde, L., Habteyes, T., & Sanov, A., (2006). Photo detachment and photo fragmentation pathways in the $[(CO_2)_2(H_2O)_m]^-$ cluster anions. *J. Chem. Phys., 125*, 114303.

Venderbosch, R. H., Prins, W., & Van, S. W. P. M., (1998). Platinum catalyzed oxidation of carbon monoxide as a model reaction in mass transfer measurements. *Chem. Eng. Sci., 53*, 3355.

Watanabe, M., & Motoo, S., (1975). Electro catalysis by ad-atoms Part II. Enhancement of the oxidation of methanol on platinum by ruthenium ad-atoms. *Electro-Analytical Chemistry and Interracial Electrochemistry, 60*, 267–273.

Wojciechowski, B. W., & Aspey, S. P., (2000). Kinetic studies using temperature-scanning: The oxidation of carbon monoxide. *Appl. Catal. A: Gen., 190*, 1.

Yamanaka, K., Sato, S., Iwaki, M., Kajino, T., & Morikawa, T., (2011). Photoinduced electron transfer from nitrogen-doped tantalum oxide to adsorbed ruthenium complex. *J. Phys. Chem. C, 115*, 18348–18353.

Yang, Y., Niu, S. W., Han, D. D., Liu, T. Y., Wang, G. M., & Li, Y., (2017). Progress in developing metal oxide nano materials for photo electro chemical water splitting. *Adv. Energy Mater, 7*, 1700555.

Yuan, J. L., & Hao, C. J., (2013). Solar-driven photo electro chemical reduction of carbon dioxide to methanol at $CuInS_2$ thin film photocathode, *Sol. Energ. Mat. Sol. C, 108*, 170–174.

Zeng, G. T., Qiu, J., Li, Z., Pavaskar, P., & Cronin, S. B., (2014). CO_2 reduction to methanol on TiO_2-passivated GaP photocatalysts. *Acs. Catal., 4*, 3512–3516.

Zhang, L. H., Zhu, D., Nathanson, G. M., & Hamers, R. J., (2014). Selective photoelectrochemical reduction of aqueous CO_2 to CO by solvated electron. *Angew. Chem. Int. Edit., 53*, 9746–9750.

Zhao, J., Wang, X., Xu, Z. C., & Loo, J. S. C., (2014). Hybrid catalysts for photo electro chemical reduction of carbon dioxide: A prospective review on semiconductor/metal complex co-catalyst systems. *J. Mater. Chem. A, 2*, 15228–15233.

CHAPTER 3

Hybrid Polymer Nanocomposites for Energy Storage/Conversion Devices: From Synthesis to Applications

ANIL ARYA and A. L. SHARMA

Department of Physical Sciences, Central University of Punjab, Bathinda – 151001, Punjab, India.
E-mail: alsharma@cup.edu.in (A. L. Sharma)

ABSTRACT

Due to the increasing demand for the energy at the global level, the development of an advanced, clean, and sustainable energy source has attracted worldwide attention. Also, depletion of the traditional non-renewable energy sources such as fuel, coal, etc., require the urgent implementation of renewable energy sources that can help, sustain the traditional sources, and fulfill the environmental norms. However, to achieve a continuous supply of energy from these sources possess stiff challenges unless until a robust and high-performance storage solution is developed. In this context, the secondary/rechargeable battery has been one of the most promising candidates for energy storage applications. It comprises four components, two electrodes (cathode and anode), a separator (to provide electrical insulation, while facilitating ion mobility), and an electrolyte. Out of the mentioned components, the electrolyte is seeking more attention, as it plays a significant role in battery performance. The conventional/traditional secondary batteries, the electrolyte is usually poured in the organic/inorganic separator that possesses several drawbacks. In order to eliminate the drawback associated with it, hybrid polymer electrolyte (HPE) seems to be more appropriate, which may be at par to full fill the entire requirement in liquid electrolyte based devices. HPE possesses improved structural, electrochemical, and transport properties. These are advanced polymer electrolytes (PEs) as they combine the properties of both the organic and inorganic components. This chapter starts with a brief introduction to PEs, and classification of HPEs. A special section has been introduced to explore the challenges in suppressing the dendrite growth formation, which has been one of the most detrimental factors in such devices. Then, several important preparation techniques and advanced characterization techniques have been summarized in brief. Finally, we highlight some key updates in the field of HPEs and provide an emphasis on their application as practical devices.

3.1   INTRODUCTION

Globally, there is a rapidly growing demand for a sustainable and safe renewable source of energy due to the depletion of the traditional fuels and related consequences such as global warming. Both raise serious concerns that have forced the researchers to develop new energy resources due to increased dependency on electronic gadgets. Alternatives that seem to be the most feasible are renewable energy sources such as wind energy, hydrothermal energy, and solar energy. Although the above energy sector is promising, the issue of storage of energy from these sources still remains unresolved. The only available energy storage option with efficient performance is the battery system, e.g., lead–acid, nickel-cadmium (Ni-Cd), nickel-metal hydride (Ni-MH), and lithium-ion batteries (LIBs). However, the first two have been used since decades in storage systems and vehicles. But, due to limited capacity and bulky size, their use in the daily electronic gadgets is limited to a great extent. Other important issues are cost-effectiveness, low energy density, poor cyclic stability, and environmental toxicity. So, the only alternative is LIB that has the potential to replace the traditional battery systems in all the electronic gadgets, electric vehicles, and the household sector (Manthiram et al., 2014; Xia et al., 2018; Pritam et al., 2019).

3.1.1   THE BASIC PRINCIPLE OF SECONDARY BATTERIES

Li-ion batteries are the most popular energy storage/conversion devices due to their widespread applications from portable electronics, electric vehicles, to grid storage. The general principle that drives a battery is the conversion of chemical energy stored inside the cell into electrical energy via the redox process (oxidation/reduction). The battery assembly comprises three crucial components, (i) cathode: positive electrode usually consists of layered oxide material, (ii) anode: negative electrode, usually consists of graphite, and (iii) electrolyte cum separator: placed between electrodes to separate electrodes and acts as a medium for ion migration (Figure 3.1) (Muench et al., 2016).

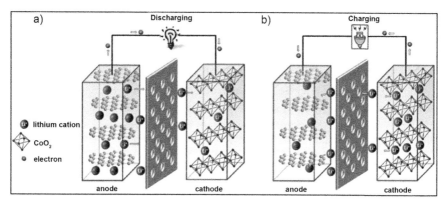

FIGURE 3.1   Schematic representation of the processes during (a) discharging and (b) charging of a Li-ion battery.
Source: With permission from Muench et al. (2016). © American Chemical Society.

3.1.2   HYBRID POLYMER ELECTROLYTES (HPE) AND CLASSIFICATION

Most of the existing batteries have a liquid-based electrolyte, and though. They perform efficiently, testability, and safety can only be achieved within a narrow temperature range. Other limitations are bulky size, safety issues, and flammable nature of the electrolyte. So, polymer electrolyte (PE) has been considered as potential replacement electrolyte since the first report by Fenton in 1973. They reported an ionic conductivity in a polymer matrix comprising of polymer and alkali metal salt. This was further improved and in 1978, Armand reported their practical implementation in the battery system. So, this escalated the research and development of polymer electrolytes having enhanced thermal and electrochemical properties (Arya and Sharma, 2017).

A typical polymer electrolyte comprises of a polymer dissolved in a solvent with an alkali metal salt of low lattice energy. As polymers are mostly insulators, the salt gets dissociated and provides ions in the matrix. The polymer chain provides the cation coordinating sites for the cation migration and cation migrates via the hopping mechanisms. The ionic conduction in polymer electrolytes is linked with relation, $\sigma = \sum n_i z_i \mu_i$ where, n_i, z_i, and μ_i area number of charge carriers, ion charge, and ion mobility respectively. So, the approach is to dissociate the salt effectively for the release of as many free ions possible (Arya and Sharma, 2017). Three important factors that boost the ion migration are (i) number of charge carriers, (ii) polymer flexibility, and (iii) number of percolation pathways. To achieve these parameters, guest additives are added to the polymer salt matrix. The polymer electrolytes are classified into three types depending on the nature of guest species, (i) ionic liquid (IL)-based PE, (ii) gel polymer electrolytes, and (iii) solid polymer electrolytes (SPE) (Croce et al., 1998; Arya and Sharma, 2018).

The IL-based polymer electrolyte system comprises of an IL and a polymer salt matrix (acts as separator). The IL having a bulky and asymmetrical cation is chosen, e.g., imidazolium, pyrrolidinium, pyridinium, ammonium, and phosphonium (Macfarlane et al., 2016). The Gel polymer electrolyte matrix comprises of a polymer host, salt, and a low molecular weight plasticizer (ethylene carbonate (EC), propylene carbonate (PC), diethyl carbonate (DEC), and dimethyl carbonate (DMC)). One advantage with GPE is that it possesses properties of both solid and liquid. But, a problem associated with both types are, (i) interfacial issues with electrodes, (ii) narrow temperature range of operation, (iii) packaging issues leading to poor mechanical properties, and (iv) cyclic stability. So, the above issues were resolved by replacing the liquid/gel electrolyte with a completely free-standing SPE. A SPE matrix comprises of a polymer matrix dissolved with some alkali salt. The advantages associated with the SPE are; desirable flexibility, lightweight, varied shape geometry, and low reactivity towards the electrodes. In conclusion, it eliminates the issues of electrolyte leakage, bulky size and also prevents additional chemical reactions which are undesired during the cell operation. One unique feature is that it plays a dual role in the cell; it acts both as an electrolyte as well as a separator. Still, the issue of low ionic conductivity needs to be resolved completely to fulfill the concept of a solid-state battery that can be used for commercial devices. So, a new branch of SPE with the addition of nanoparticles (NPs) of different shapes (spherical, rod, belt, wire) is developed, which is known as composite polymer electrolyte (CPE). The crucial roles of NPs are, (i) increase salt dissociation rate, (ii) disrupt the polymer recrystallization tendency, (iii) provide additional conducting pathways, and (iv) long continuous conduction path (Meyer, 1998; Sandi et al., 2003).

Nowadays, research is focused on the preparation of the hybrid polymer electrolytes (HPE). In general, a HPE system combines the advantage of each constituent (polymer, nanofiller, and plasticizer), i.e., chemical, thermal, and electrical or ion transport, and avoids the disadvantages faced by a liquid electrolyte. HPE system consists of an inorganic and organic material (Koksbang et al., 1994). These make them alternative for future energy storage/conversion devices. Figure 3.2 shows the classification of different polymer electrolytes (PE) broadly on the basis of source and composition. Further, on the addition of different guest species composition, PE is classified as solid, gel, and HPEs. To enhance the electrical properties of the HPE, one of the most effective approaches is the addition of nanofillers. Nanofillers are classified into two types, i.e., active, and passive, depending on their contribution to the overall conductivity. Some important advantages and disadvantages of both types are summarized in Table 3.1 (Chua et al., 2018).

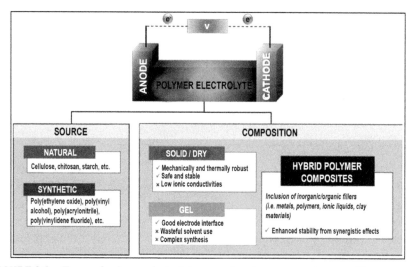

FIGURE 3.2   Types of polymer electrolytes according to source and composition.
Source: Reprinted with permission from Chua et al. (2018). © John Wiley and Sons.

TABLE 3.1   Advantages and Disadvantages of Active and Passive Nanofillers in SPEs

Active Nanofillers	Passive Nanofillers*
Advantages	
Exceptional ionic conductivities (> 10^{-3} S cm^{-1})	Ease of preparation and inexpensive
High lithium transference numbers (> 0.5)	Tuneable characteristics
Disadvantages	
Complex conductivities and synthesis	Low ionic conductivities (<10^{-4} S cm^{-1})
Interfacial incompatibility with lithium metal	Poor interfacial contact with electrodes
*Two-dimensional nanofillers (i.e., LDHs, natural clay, GO)	

Source: Reprinted with permission from Chua et al. (2018). © John Wiley and Sons.

Here, in the case of HPE, the salt content seems to be the most critical constituent that decides the overall ion dynamics. So, the salt must have some characteristics that need to be taken care of during selection. Some of the important characteristics are, (i) fast dissociation in the polymer matrix, (ii) inertness toward cell components, (iii) low lattice energy and bulky anion size, (iv) non-toxic and water-stable, and (v) easy availability (Marcinek et al., 2015). For better comparison and selection of salt, Table 3.2 may be referred. It summarizes the crucial properties and provides a detailed comparison of various salts (Tasaki et al., 2003; Mauger et al., 2018).

TABLE 3.2 Comparison between Salts for Lithium Batteries

Properties	From Best to Worst					
Ion mobility	$LiBF_4$	$LiClO_4$	$LiPF_6$	$LiAsF_6$	LiTf	LiTFSI
Ion pair dissociation	LiTFSI	$LiAsF_6$	$LiPF_6$	$LiClO_4$	$LiBF_4$	LiTf
Solubility	LiTFSI	$LiPF_6$	$LiAsF_6$	$LiBF_4$	LiTf	
Thermal stability	LiTFSI	LiTf	$LiAsF_6$	$LiBF_4$	$LiPF_6$	
Chem. inertness	LiTf	LiTFSI	$LiAsF_6$	$LiBF_4$	$LiPF_6$	
SEI formation	$LiPF_6$	$LiAsF_6$	LiTFSI	$LiBF_4$		
Al corrosion	$LiAsF_6$	$LiPF_6$	$LiBF_4$	$LiClO_4$	LiTf	LiTFSI

3.1.3 ESSENTIAL PROPERTIES OF HYBRID POLYMER ELECTROLYTE (HPE)

The development of an advanced HPE polymeric system, for the application of the battery unit, some properties need to be monitored. Enhancement of these properties is the key goal of the HPE developing community. As faster ion dynamics are the characteristics of an electrolyte, therefore, it is further linked with the structure, morphology, and stability of HPE. The balanced enhancement in the properties is beneficial for the successful development of HPE matrix. A brief overview of the key properties given in the forthcoming section. By optimizing the above-said properties, an electrolyte may yield better performance. The complete details of the above said properties are discussed as follows;

3.1.3.1 STRUCTURAL AND MORPHOLOGICAL PROPERTIES

As it is well known that the ion dynamics in the case of polymer electrolytes occur via the amorphous phase, therefore, the morphology and the structure investigation need to be explored in detail. The crystallite size, interplaner-spacing, inter-chain separation, and crystallinity are some of the key properties. The higher the crystallite size more will be the available free volume, for the ion migration and favors faster ion transport. The high value of crystallite size indicates the low degree of crystallinity (Stoeva et al., 2003). Another important parameter is the value of crystallinity (X_C), as lowering of crystallinity promotes faster ion migration owing to the increased free volume.

3.1.3.2 IONIC RADII OF CATION

In polymer electrolytes, the cation is the only dominating migrating species. So, investigating the ionic radii of the cation is very important prior to the selection of any salt. Salt dissociation is also linked with the ionic radii of ions, as large anion size makes it easier for the salt to dissociate into cations/anions. Another important factor is that the cation with small size migrates at a faster rate in the polymer matrix via the coordinating sites of the polymer chain. It affects the polymer chain flexibility, ionic conductivity, and the performance of the whole device. Table 3.3 summarizes the comparison of various salts (Bandara et al., 2016).

TABLE 3.3 Some of the parameters that influence the ionic conductivity in gel/solid polymer electrolyte containing alkaline metal diodes. Reprinted with permission from Bandara, T. M. W. J.; Fernando, H. D. N. S.; Furlani, M.; Albinsson, I.; Dissanayake, M. A. K. L.; Ratnasekera, J. L.; & Mellander, B. E. Effect of the alkaline cation size on the conductivity in gel polymer electrolytes and their influence on photoelectrochemical solar cells. *Physical Chemistry Chemical Physics*, 2016, 18, 10873–10881.

Alkaline Ion	Li^+	Na^+	K^+	Rb^+	Cs^+	Expected Influences from Left to Right
Radius/pm (six coordinations)	76	102	138	152	167	Increasing cation size increases ionic dissociation, increases the separation of the polymer chains, and decreases the viscosity.
	Increases from Li^+ to Cs^+					
Charge density	Decreases from Li^+ to Cs^+					Weakening coordination bond strength, activation energy decreases. Salt dissociation decreases (but can be compensated by an increasing coordination number, see point 4).

TABLE 3.3 *(Continued)*

Alkaline Ion	Li⁺ Na⁺ K⁺ Rb⁺ Cs⁺	Expected Influences from Left to Right
Acidity (as a Lewis acid)	Decreases from Li^+ to Cs^+	Weaker interactions with electronegative/donor sites in the electrolyte
Coordination number	Increases more or less from Li^+ to Cs^+	Salt solvation improves
Polymer flexibility	Increases from Li^+ to Cs^+	Improves the ion conduction
Local viscous force (of cation moving in the media)	Increases from Li^+ to Cs^+	Reduces the cation conduction
Viscosity change in the electrolyte	Decreases from Li^+ to Cs^+	Anion conductivity improves
Ionic dissociation of salts	Increases from Li^+ to Cs^+	Enhance the carrier density and conductivity
Free volume and voids in the electrolyte	Increases from Li^+ to Cs^+	Improves the ion conduction

Source: Reprinted with permission from Bandara et al. (2016). © Royal Society of Chemistry.

3.1.3.3   FREE IONS AND ION PAIR'S CONTRIBUTION

In HPEs, the number of ions contributing to the conduction is an important parameter that needs to be investigated before its use in any application. The fraction of free ions (FFI) and ion pairs (FIP) is obtained by the deconvolution of the anion mode of salt (obtained from FTIR data) using the peak fitting software, (Peak fit v4). The highest FFI indicates an enhanced ionic conductivity and ion mobility (Edman, 2000; Bar et al., 2017). It is obtained using the following equation:

$$
\begin{cases}
FFI(\%) = \dfrac{A_{free}}{A_{free} + A_{pair}} \\[2mm]
\quad and \\[2mm]
FIP(\%) = \dfrac{A_{pair}}{A_{free} + A_{pair}}
\end{cases}
$$

Here, A_{free} is the area representing free ion peak and A_{pair} area of the peak representing ion pair peak respectively. Figure 3.3a shows the chemical structure of two anions, (i) $\{M[N(SO_2CF_3)_2]$; M=Li, Na, K, TFSI⁻; (trifluoromethanesulfonyl)imide anion, and (ii) $\{[N(SO_2CF_2H)(SO_2CF_3)]^-$,

DFTFSI⁻[(difluoromethanesulfonyl) (trifluoromethanesulfonyl)imide]. Figure 3.3b shows the decrease in the dissociation energy (ΔE_d) with the increasing cation size (Oteo et al., 2019).

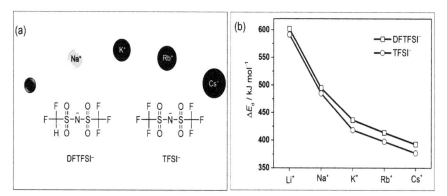

FIGURE 3.3 (a) Chemical structure and (b) computed dissociation energy of the DFTFSI⁻ and TFSI⁻ based alkali metal salts.
Source: Reprinted with permission from Oteo et al. (2019). © John Wiley & Sons.

3.1.3.4 ION DYNAMICS PROPERTIES

The ion dynamics in the polymer electrolytes is influenced by three parameters, ionic conductivity (σ), cation/ion transference number (t_{cation}/t_{ion}), activation energy (E_a), glass transition temperature (T_g) and ion transport parameters (number of charge carriers; n, mobility; μ, diffusion coefficient; D). So, to obtain an electrolyte with balanced properties for stable device performance, above said properties need to be explored.

3.1.3.4.1 Ionic Conductivity

The ionic conductivity of an electrolyte is a key parameter that can provide information about the rate of ion migration in the HPE matrix. For an optimum device performance, a higher value of ionic conductivity is desirable. The ions generated from the salt dissociation migrates via the coordinating sites (i.e., electron-rich group) provided by the polymer chains. The cation migration is also influenced by the amorphous content, ionic radii of cation/anion, and thickness of the polymeric film. In general, the ionic conductivity is linked with the number of free charge carriers, and ion mobility ($\sigma = \sum n_i z_i \mu_i$) where, n_i, z_i, and μ_i are the number of charge carriers, ion charge, and ion mobility respectively (Bandara et al., 2011). The ionic conductivity of HPE is obtained by using the expression; $\sigma = d / (R_b . S)$ where, 'd' is the thickness of the polymer electrolyte films, R_b is the bulk resistance, and S is the area of the stainless-steel (SS) electrodes. The ionic conductivity is effectively enhanced by the temperature, and is attributed to the thermal activation of charge carriers, and enhanced polymer flexibility owing to the disorder.

3.1.3.4.2   Activation Energy

The activation energy parameter is crucial as it enables us to understand the effect of temperature on the hybrid polymer matrix. The increase of temperature disrupts the polymer chain arrangement (i.e., enhanced polymer flexibility), influences the salt dissociation process (i.e., increased number of free ions), and thermal activation of charge carriers occurs which contribute to the overall enhancement of conductivity. Two important mechanisms are observed with temperature variation:

1. **Arrhenius Behavior:** Two key changes in the polymer matrix with an increase in the temperature are, (i) enhancement of amorphous content owing to the disruption of the crystallinity, and (ii) enhancement in the segmental motion of polymer chain wing to improve polymer flexibility. The simultaneous presence of these effects results in the enhancement of ionic conductivity. The temperature-dependent conductivity plot (log σ vs. $1/T$ plot) aggress well with the Arrhenius equation expressed as; $\sigma = \sigma_o \exp(-E_a / k_B T)$, where, σ_o is the pre-exponential, k_B is the Boltzmann constant, and E_a is the activation energy. The value of E_a is obtained by fitting the plot with the Arrhenius equation and extracting the slope. The lowering of the activation energy suggests the reduction of barrier height that ion needs to cross for a successful migration. It also suggests the enhancement of ionic conductivity (Agrawal and Pandey, 2008). The dominating contribution to ion migration is from the amorphous content.

2. **Vogel-Tamman-Fulcher (VTF) Behavior:** VTF is based on the concept that the temperature effectively enhances the segmental motion of the polymer chains and thermal activation of charge

carriers enables us to extract the value of activation energy by fitting the temperature-dependent conductivity plot by VTF equation expressed as; $\sigma = AT^{-1/2}\exp(-B/T-T_o)$; were, σ is the ionic conductivity, A is the pre-exponential factor, B is the pseudo-activation energy for the conductivity, and T_0 is the temperature, which is close to the T_g of the polymer matrix. The non-linear nature of the plot indicates the possibility of long-range ion migration. It also suggests that the increase of temperature enhances the free volume for the ion migration (Agrawal and Pandey, 2008; Ratner et al., 2000).

3.1.3.4.3   Glass Transition Temperature

One of the most important points is the flexibility of the hybrid polymer matrix and is examined in terms of glass transition temperature (T_g). The low value of T_g represents the enhancement in the polymer flexibility, which facilitates cation migration via the segmental motion of polymer chains. At this temperature, the polymer matrix is switched from the rigid phase to the viscous phase. It is influenced by the polymer molecular weight, arrangement of chains, crystallinity, and the thermal behavior of the material (Zhang et al., 2017). The lowering of T_g value is beneficial for ion migration and can be achieved by various approaches, e.g., polymer blending, cross-linking, and the addition of NPs.

3.1.3.4.4   Cation/Ion Transference Number

Three migrating species in the HPE are cations, anions, and electrons. For the ideal electrolyte, the cation transport number must be unity $(t_{Li}^+=1)$. As anions lead to the build-up of polarization, they need to be chemically bound with the backbone of the polymer chain or some anion-receptors (Maranas et al., 2012). However, electron migration is avoided, since electrons flow out of the circuit and their contribution needs to be minimized (Abreha et al., 2016; Sharma and Thakur, 2010). The separation of ions and electron is done by the dc Wagner's polarization technique by sandwiching the HPE film in-between two SS electrodes (SS‖HPE‖SS). The ion transference number (t_{ion}) is obtained from the polarization current vs. time plot using equation, $t_{ion} = [(I_t - I_e)/I_t] \times 10$, where I_t and I_e are the total current and the residual current, respectively. The cation transport number (t_{Li}^{+}) is obtained by combining the ac impedance and dc polarization technique, first time reported by Vincent and co-workers (Evans et al., 1987). The cation transport number is obtained by the equation; $\left[I_s (V - I_i R_i) / I_i (V - I_s R_s) \right]$, where V is the applied voltage to the cell configuration (SS||HPE||SS), I_i and I_s are the initial and steady-state currents, respectively. R_i and R_s are the interfacial resistance before and after polarization, respectively.

3.1.3.5   ION TRANSPORT PARAMETERS

Ion transport parameters in the case of HPE, need to be examined for getting insights on the cation migration and three important parameters, diffusion constant (D), mobility (μ) and charge carrier concentration (n) that influence directly the ionic conductivity. The number density (n), mobility (μ) and diffusion coefficient (D) of the mobile ions were calculated using three approaches (Table 3.4), Bandara and Mellander (B-M) approach, Impedance Spectroscopy Approach, and FTIR method (Bandara et al., 2011; Arof et al., 2014; Petrowsky et al., 2008; Ericson et al., 2000). All three approaches are independent of any conductivity-temperature relationship and the electrical double layer (EDL) capacitance.

TABLE 3.4   Approaches to Obtaining the Number Density (n), Mobility (μ), and Diffusion Coefficient (D).

Units	Method		
	Bandara and Mellander (B-M) Approach	Impedance Spectroscopy Approach	FTIR Method
D (cm²s⁻¹)	$D = \dfrac{d^2}{\tau_2 \delta^2}$	$D = \dfrac{(k_2 \varepsilon_r \varepsilon_o A d)^2}{\tau_2}$	$D = \dfrac{\mu k_B T}{e}$
μ (cm²V⁻¹s⁻¹)	$\mu = \dfrac{ed^2}{kT\tau_2 \delta^2}$	$\mu = \dfrac{e(k_2 \varepsilon_r \varepsilon_o A d)^2}{k_B T \tau_2}$	$\mu = \dfrac{\sigma}{ne}$
N (cm⁻³)	$n = \dfrac{\sigma kT \tau_2 \delta^2}{e^2 d^2}$	$n = \dfrac{\sigma k_B T \tau_2}{(ek_2 \varepsilon_r \varepsilon_o A d)^2}$	$n = \dfrac{M \times N_A}{V_{Total}} \times$ free ion area (%)

TABLE 3.4 *(Continued)*

Units	Method		
	Bandara and Mellander (B-M) Approach	**Impedance Spectroscopy Approach**	**FTIR Method**
Parameters	τ_2 is a time constant corresponding to the maximum dissipative loss curve, $\delta = d/\lambda$, where λ is the thickness of the electrical double layer, d is half-thickness of the polymer electrolyte,	k_2 and k_1 are obtained from the trial and error on the Nyquist plot, The value of τ_2 was taken at the frequency corresponding to a minimum in the imaginary parts of the impedance, Z_i, i.e., at $Z_i \rightarrow 0, k_B$ is the Boltzmann constant (1.38 $\times$ 10^{-23} J K^{-1}) and T is the absolute temperature	M is the number of moles of salt used in each electrolyte, N_A is Avogadro's number (6.02 $\times$ 10^{23} mol^{-1}), where V_{Total} is the total volume of the solid polymer electrolyte, and σ is dc conductivity. e is the electric charge (1.602 $\times$ 10^{-19} C), k_B is the Boltzmann constant (1.38 $\times$ 10^{-23} J K^{-1}). T is the absolute temperature

3.1.3.6 STABILITY PROPERTIES

3.1.3.6.1 Thermal Stability

When any battery device is in operation, then there may be some heat generation internally owing to the combustible organic gases, overcharging, and material decomposition that may result in a short-circuit. This internally generated heat may decompose the cathode and anode as well as electrolyte that may exhaust the battery. So, to prevent the battery from the explosion and/or short-circuit, the thermal stability of the electrolyte must be higher than the recommended safety limit. The thermal stability depends on the interactions in the polymer matrix and also the degradation temperature (Arya and Sharma, 2018; Wu et al., 2006). The thermal stability of an electrolyte is investigated from the thermogravimetric analysis (TGA) and differential scanning calorimetry (DSC). TGA enables us to know about the degradation temperature, weight loss, and the decomposition range while DSC allows us to obtain the thermal transition such as glass transition temperature, crystallization, and melting temperature. The thermal stability of the polymer electrolyte is enhanced by the addition of NPs such as nanofiller, nanoclay, and nanorod.

3.1.3.6.2   Electrochemical Stability

The electrochemical stability window (ESW) is a fundamental parameter that determines the durability and energy output of a lithium cell. The ESW of an electrolyte needs to be determined prior to its use in the battery. ESW is the difference between the oxidation and reduction potential and decides the potential window that needs to be maintained during the charge/discharge activity. The ESW may be determined in terms of molecular properties using the quantum chemical characterization of its frontier energy levels [highest occupied molecular orbital (HOMO), and lowest unoccupied molecular orbital (LUMO)]. The difference between the HOMO and LUMO of a liquid electrolyte or the bottom of the conduction band (CB) and top of the VB of a solid electrolyte is the ESW. It must be greater than that of the open circuit energy ($V_{oc}=(\mu_A-\mu_C)/e$) (difference in Li chemical potential in each of the electrodes) for good stability (Zheng et al., 2017; Etacheri et al., 2011; Goodenough et al., 2013; Goodenough and Kim, 2011).

For better performance of the device, the oxidation potential must be higher than that of the cathode, while the reduction potential must be lower than that of lithium metal in the anode (Long et al., 2016). The linear sweep voltammetry (LSV) technique is used to determine the voltage stability of an electrolyte. In LSV, a voltage signal is applied across the two electrode assembly with a sandwiched electrolyte. Change in the current with respect to the applied potential sweep rate is observed and at a specific potential value, a rapid increase in the current is observed. The voltage at which such an instantaneous increase in current occurs is the decomposition voltage or operational voltage of the electrolyte. The desirable value is ~4–5 V. The electrolyte may be used within this limit safely for the fabrication of battery.

3.1.3.6.3   Mechanical Stability

Polymer electrolyte plays a dual role in the battery systems, i.e., it provides a medium for ion migration and physically separates the electrodes. During the battery operation, the electrolyte may be damaged due to heating or any other undesired cause. The mechanical properties need to be considered with the utmost care during the large-scale production of the energy storage devices. The electrolyte must possess a high mechanical strength so that it may operate safely in adverse conditions and can sustain the pressure or stress during cell assembly, and cell operation. The mechanical properties of the electrolyte are investigated by the stress-strain test (Zhou et al., 2018; Mural et al., 2014). The elongation at break and tensile strength of the samples is determined by using the tensile strength measuring machine. The most suitable approaches to enhance the mechanical properties are the addition of nanofiller, polymer blending, and the addition of nanoclay.

3.1.4 CHALLENGES TO SUPPRESS DENDRITE GROWTH

In the LIB, one of the challenging issues that effectively impacts the performance of the cell and makes cells dead after few cycles are dendrite growth formation. Dendrites are thread-like structures generated on the anode (generally Li) that get penetrated through the electrolytes and damages the battery by short-circuiting it. It also threatens the safe operation of the battery. Some serious hurdles that are resulted from the growth of dendrites are, (i) short-circuiting of the cell due to penetration from anode to the cathode via the electrolyte, (ii) increased surface area due to dendrites may result in undesirable reactions, (iii) reaction between the electrolytes and newly generated dendrites with SEI may result in dead Li that reduces Columbic efficiency, (iv) development of large polarization occurs due to uneven structure of dendrite that hinders cation migration, and (v) large volume change during cell operation at electrodes may occur (Figure 3.4i) (Cheng et al., 2017; Li et al., 2015). So, this issue needs to be taken seriously and should be eliminated to develop a cell with long term stability and better performance.

Zhao prepared an anion-immobilized solid-state composite electrolyte to suppress the dendrite growth (2017). Here, the Garnet-type Al-doped $Li_{6.75}La_3Zr_{1.75}Ta_{0.25}O_{12}$ (LLZTO) ceramic particles were dispersed in the polymer matrix (PEO-LiTFSI). It shows improved stability of the anode as illustrated in Figure 3.4 ii. Figure 3.4 ii shows the comparison of dendrite growth suppression in the case of CPE in comparison to the liquid electrolyte. Here, the polymer-salt part compensates the volume changes with temperature, while the ceramic part suppresses the dendrite growth. Another report by Yang states that, they have tried to suppress the dendrite growth by preparing a polymer/ceramic hybrid membrane based on PEO as polymer and a garnet-type solid-state conductor $Li_7La_3Zr_{1.75}Nb_{0.25}O_{12}$ (LLZNO) (2017).

The Li-plating behavior is examined and shown in Figure 3.4 ii (a-b). Here, both the bare Li and Li-dendrite growth visible with Li deposition (Figure 3.4 ii b-d). When Li metal is coated with a hybrid film (PEO-LLZTO), then the Li deposition is stable and there is no dendrite growth as evidenced by the *in-situ* microscopic images (Figure 3.4 iii e-h).

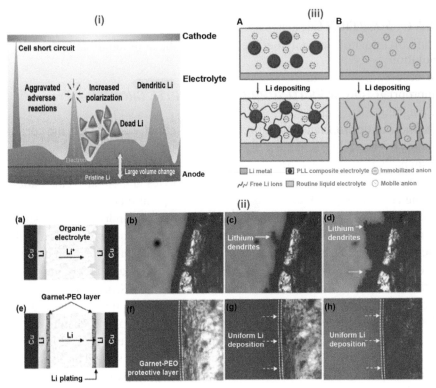

FIGURE 3.4 (i) Scheme of a dilemma for Li metal anode in rechargeable batteries (*Source:* Reprinted with permission from Cheng et al. (2017) © American Chemical Society.) (ii) Schematic of the electrochemical deposition behavior of the Li metal anode with (*A*) the PLL solid electrolyte with immobilized anions and (*B*) the routine liquid electrolyte with mobile anions (*Source:* Reprinted with permission from Yang et al. (2017) © Springer Nature.), (iii) In situ optical microscope observation of Li plating onto Li metal with or without a protective layer in planar symmetric cells. (a) Schematic of Li plating on bare Li metal in a planar battery. Optical microscope images of a bare Li metal anode (b) before Li deposition and after plating with (c) 0.5 mA·h·cm⁻² and (d) 1 mA·h·cm⁻² of Li metal. (e) Schematic of Li plating on Li metal coated by a PEO-garnet hybrid layer in a planar battery. Optical microscope images of a PEO-garnet coated Li metal anode (f) before Li deposition and after plating with (g) 0.5 and (h) 1 mA·h·cm⁻² of Li metal (*Source:* Reprinted with permission from Zhao et al. (2017)).

3.2 PREPARATION TECHNIQUES

For the preparation of the polymer electrolytes, various methods are adopted on the basis of the constituents. The applicability of each method is dependent on the type of polymer electrolyte we want to prepare.

3.2.1 SOLUTION CAST TECHNIQUE

Solution cast technique is the most adaptable technique and is chosen due to ease of preparation. Here, firstly the polymer host is dissolved in the appropriate solvent (low boiling point and high dielectric constant) using a magnetic stirrer until a homogenous solution is obtained followed by the addition of appropriate salt (generally, smaller cation salt with low lattice energy). After that, the addition of varied concentration of guest additive is dispersed in the polymer salt solution. The guest species may be dispersed in a separate beaker and then added to the solution of the polymer salt matrix. In the case of nanoparticle addition, sonication may be done for the uniform dispersion of the NPs depends on the nanoparticle content. Finally, a homogeneous polymer matrix is obtained, and the solution is cast in the well-cleaned Petri dishes (Glass/Polypropylene/Teflon) and kept in the vacuum oven for evaporation of the solvent (below-melting temperature of polymer). After that, the film is carefully peeled off from the Petri-dish and kept in a desiccator with silica gel for performing characterizations (Pritam et al., 2019; Arya and Sharma, 2017).

3.2.2 HOT-PRESS TECHNIQUE

The hot-press technique is a simple and versatile technique as no solvent is used in this. First of all, the constituents (polymer/salt/nanofiller) of polymer electrolytes are mixed together. After that, the powdered mixture is sandwiched between two sheets of material (Mylar) and kept in the heating chamber. Then the electrolyte is pressed with a pressure controller and film is obtained after cooling down the assembly. The prepared film is kept in the vacuum desiccator to avoid contamination with the external environment (Arya and Sharma, 2017; Agrawal and Pandey, 2008).

3.3   CHARACTERIZATION TECHNIQUES

Some characterization techniques need to be performed to check the suitability of prepared polymer electrolytes for application in the battery. The structural, microstructural, morphological, electrochemical, thermal, transport, and mechanical properties are examined using the various characterization techniques (Zhang et al., 2017; Sharma and Thakur, 2010; Arya and Sharma, 2018). Some important techniques are summarized as follows;

3.3.1   STRUCTURAL, MORPHOLOGICAL, AND MICROSTRUCTURAL CHARACTERIZATION

1. **Field Emission Scanning Electron Microscopy (FESEM):** It is an important technique to analyze the surface morphology and particle distribution in the polymer matrix. It also allows us to obtain the particle size, porosity, etc.

2. **X-Ray Diffraction (XRD):** The structural investigations of the prepared are obtained by the x-ray diffraction (XRD) technique. It confirms the complex formation by identifying the characteristics peaks of the constituents of the polymer matrix. If any corresponding peak of the salt is absent in the diffraction spectra, it indicates the complete dissociation of the salt. The d-spacing (d), the crystallite size of nanofiller, interchain separation (R) in nanoclay is obtained to investigate the effect of the guest species on the polymer host.

3. **Fourier Transform Infrared (FTIR) Spectroscopy:** It is an important technique to study the presence of various interactions between the constituents of polymer electrolytes, such as polymer-ion, ion-ion, polymer-nanofiller, ion-nanofiller, polymer-ion-nanofiller, etc. It provides critical information about the complex formation and examined by identifying the shift in the peak position of the characteristics functional group of the host matrix, i.e., polymer. It provides a quantitative analysis of the free ions and is correlated with the ionic conductivity. Besides this, the FFI is used to evaluate the transport parameters (n, μ, D) as discussed in the earlier section.

3.3.2 ELECTRICAL AND ION DYNAMICS CHARACTERIZATION TECHNIQUE

1. **Complex Impedance Spectroscopy (CIS):** The ionic conductivity is obtained by performing the Impedance study and Nyquist plot (plot of the imaginary part of impedance against the real part of impedance) enables us to obtain the bulk resistance. For this, a film of the polymer electrolyte is sandwiched between the SS blocking electrodes and an AC signal (10–50 mV) is applied across the cell configuration. A Nyquist plot comprises of a semicircle at high frequency followed by a spike in the low-frequency side. The diameter of the semi-circle is the bulk resistance and in the absence of the semi-circle, an intercept is drawn from the spike by extrapolating it and the value of the bulk resistance is calculated from the position where it cuts the abscissa.

2. **i-t Characteristics (Transference Number):** In the case of HPEs, ionic, and electronic contributions need to be separated to check the dominant charge carrier contributing to transport. The i-t characteristic technique is performed to obtain the ionic (t_{ion}) and electronic ($t_{electronic}$) transference number. Another parameter is the cation transference number (t_{cation}) and for an ideal conductor need to be close to unity.

3. **Differential Scanning Calorimetry (DSC):** This technique allows us to obtain various transitions in the polymer electrolyte with temperature, and the difference in heat flow between the sample and reference is observed with temperature. The ionic conductivity of the polymer electrolyte is linked with the glass transition temperature (T_g) of the polymer matrix and the low value of T_g suggests the increased segmental motion of the polymer chain, hence ionic conductivity. The DSC provides many important parameters; glass transition temperature, melting temperature, crystallinity.

4. **Nuclear Magnetic Resonance (NMR):** It is an important technique to analyze the ion dynamics in polymer electrolytes. The line-width of the NMR spectra is linked with the ion migration in the polymer matrix and is correlated with the ionic conductivity.

3.3.3 STABILITY CHARACTERIZATION TECHNIQUES

1. **Linear Sweep Voltammetry (LSV):** This technique is performed to check the decomposition voltage of the HPE. A HPE membrane is sandwiched between the two SS electrodes and its i-V characteristics has been performed within a specified potential range, and the corresponding change in the output current value was recorded. At a particular voltage, the rapid increase in the current is observed, which can be associated with the breakdown of the electrolyte. This voltage is known as ESW and is used during the charging/discharging measurement of the fabricated cell.

2. **Thermogravimetric Analysis (TGA):** The thermal stability of electrolyte needs to be examined before its use in a battery. TGA is performed to check the behavior of polymer electrolyte with temperature variation and weight loss within abroad temperature range. It provides sufficient information regarding the safe limit of the electrolyte for application in the battery.

3. **Mechanical Properties:** These are examined by performing the stress-strain analysis, and dynamical mechanical analysis. It provides information about the mechanical stability of the electrolyte that allows ease of fabrication in different geometries such as cylindrical and pouch cell.

3.4 STATE-OF-THE-ART STATUS

This section of the chapter discusses some advanced HPEs prepared recently. A detailed explanation has been given to understand the effectiveness of HPE as compared to other systems. The most crucial part is that all the HPEs that have been used in the solid-state battery systems and the corresponding device performance have been discussed in detail.

The addition of nanofiller is an effective approach to improve the electrical properties of the polymer electrolytes, but the possibility of agglomeration can have a negative impact on the ion dynamics. To tackle this issue, Chet reported the preparation of HPE ((PEO)/Li_3PS_4) by *in-situ* synthesis (ISS), and the results were compared with HPE prepared by mechanical-mixing method (MMM) (2018). The XRD analysis indicated an enhancement in the amorphous content. The Li_3PS_4 were found to be uniformly distributed with a diameter of about 400–700 nm and ionic conductivity was of the order of 4.03×10^{-4} S cm^{-1} (at room temperature). The obtained HPE was semi-transparent and flexible with a thickness in the range of about 100–150 μm (Figure 3.5). The ionic conductivity for 2% vol Li$_3$PS$_4$ is about 6.98×10^{-4} S cm^{-1} (at 60°C) for HPE prepared by MMM while HPE prepared by ISS depicts higher conductivity, i.e., 8.01×10^{-4} S cm^{-1} (at 60°C). The enhancement is attributed to the ion-conducting paths provided by Li$_3$PS$_4$ (acts as active nanofiller), and uniform distribution which results in the enhancement of amorphous content. The thermal stability of the prepared HPE was close to 373°C. HPE prepared by ISS shows the lowest value of crystallinity and DSC results suggest that Li$_3$PS$_4$ acts as a plasticizer and plays two key roles, i.e., (i) disruption of polymer re-crystallization tendency, and (ii) enhancement of

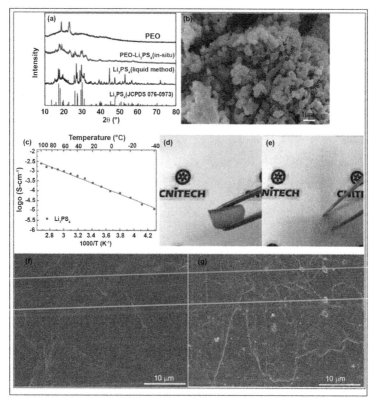

FIGURE 3.5   (a) XRD of PEO, Li$_3$PS$_4$ and in-situ prepared PEO/Li$_3$PS$_4$ hybrid electrolyte; (b) FESEM micrograph of the Li$_3$PS$_4$ sample prepared in liquid method; (c) Ionic conductivities of Li$_3$PS$_4$ glass-ceramic; photos of (d) in-situ prepared hybrid electrolyte and (e) mechanical-mixing prepared hybrid electrolyte.
Source: Reprinted with permission from Chen et al. (2018).

the free volume besides improving the ion mobility. Also, the ISS prepared HPE shows improved mechanical properties than that obtained through the MMM technique. The cation transport number is 0.28 and 0.33 for the HPE (PEO-2% vol Li_3PS_4) prepared by MMM and ISS, respectively. The voltage stability window was found to be about 5.1 V for ISS electrolyte, which was higher than MMM (4.9 V) and pure PEO (4.6 V). A solid-state Li-cell was fabricated using Li metal as anode, LFP as a cathode, and in-situ prepared PEO-2% vol Li_3PS_4 as the electrolyte. The specific capacity obtained from the configuration is 125 mAh g^{-1} after 100 cycles with capacity retention of about 86.1% at 0.2 C for ISS prepared HPE, while at 0.5 C, the discharge capacity at the 325th cycle is 116 mAh g^{-1} with capacity retention of about 80.9%.

Recently, Cho reported the preparation of hybrid solid electrolyte (HSE) using PEO mixed with LiTFSI ([EO]:[Li] = 8:1) and LATP ($Li_{1.3}Al_{0.3}Ti_{0.7}(PO_4)_3$) (2018). The morphological analysis shows the reticulated PEO connected with LATP particles which creates ion-conducting pathways. It decreases the boundary resistance in the solid electrolyte. The ionic conductivity of the prepared electrolyte with SN addition is about 2.0×10^{-4} S cm^{-1} (at 23°C), and 1.6×10^{-3} S cm^{-1} (at 55°C). The voltage stability window of the prepared electrolyte was close to 6 V. A HSC/HSE/Li metal cell (HSC stands for the hybrid solid cathode) was fabricated which shows a charge capacity of 82/62 mAh g^{-1} (at 23°C) and 123.4/102.7 mAh g^{-1} (at 55°C). One unique advantage with this electrolyte is that even in the absence of a protection layer, there was no Ti reduction (Ti^{4+} to Ti^{3+}).

Another feasible approach to develop the HSE is the integration of ceramic Li^+ ion-conducting electrolyte in the polymer matrix. Keller reported the preparation of the HSE comprising of PEO-LiTFSI+ LLZO ($Li_7La_3Zr_2O_{12}$). DSC analysis evidenced the decrease in the melting temperature with addition of salt, while, there was no effect of LLZO addition on the same (2017). The thermal stability was close to 300°C. The temperature-dependent ionic conductivity for SPE and HSE (containing pristine and annealed LLZO) is shown in Figure 3.6a. The conductivity of the PE and HSE is lower than the bulk LLZO (10^{-4} S cm^{-1}). Two conductivity trends are observed for two separate temperature ranges, i.e., 20°C–40°C, and 60°C–80°C. At high temperature, the lowest value of activation energy is attributed to the enhanced polymer flexibility and faster ion mobility. It was concluded that the LLZO is completely inert. Figure 3.6b shows the ion conduction model in HSE. The decrease of conductivity with the addition of LLZO in PS system reduces cross-sectional area and high tortuosity leads to longer pathways for cation migration. The grain boundary resistance between polymer and nanoparticle hinders the cation migration via the ceramic-polymer and ceramic-ceramic interfaces. Another key reason is that cation may get trapped somewhere between the interface and/or within the LLZO surface layer (composed of LiOH and Li_2CO_3), resulting in a decreased conductivity.

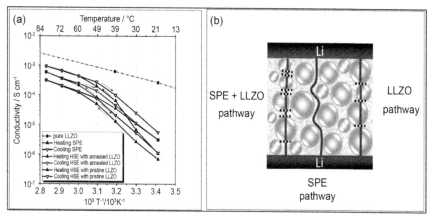

FIGURE 3.6 (a) Arrhenius conductivity vs. temperature plot of the SPE and HSE (containing pristine and annealed LLZO) samples, and (b) schematic model of possible lithium-conducting pathways through the HSE samples.
Source: Reprinted with permission from Keller et al. (2017). © Elsevier.

Zheng investigated the LLZO–PEO (LiTFSI) composite electrolytes by a solution casting method (2018). The average thickness of the film is 30 to 50 μm. The ^{6}Li NMR spectra show a broad shoulder peak located at 1.8 ppm associated with LLZO–PEO interface. While the peak located at 1.3 ppm is associated with the LLZO decomposition, the peak height increases with the addition of TEGDME that indicates the increase in the possibility of LLZO breakdown. The addition of LLZO broadened the peak associated with LiTFSI (Figure 3.7).

The Li-ion mobility is examined by the $_7$Li T_1 relaxation time. With the addition of LLZO, relaxation time increases, which can be attributed to the rigid nature of LLZO (Table 3.5). Figure 3.7 shows the schematic of Li-ion pathways in PEO-LiTFSi + LLZO composite electrolytes. At low LLZO content, the present system behaves like normal polymer electrolyte, while an increase in the LLZO content creates an inter-connecting path, which acts as a ceramic electrolyte. At critical LLZO content, the percolation LLZO network formation occurs which results in faster ion dynamics. Further addition of TEGDME results in Li-ion migration via the TEGDME-modified polymer phase.

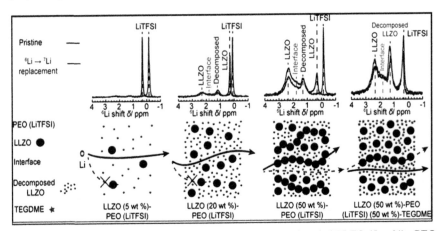

FIGURE 3.7   At the top: ⁶Li NMR comparison of pristine and cycled LLZO (5 wt%) –PEO (LiTFSI), LLZO (20 wt%) –PEO (LiTFSI), LLZO (50 wt%) –PEO (LiTFSI), and LLZO (50 wt%) –PEO (LiTFSI) (50 wt%) –TEGDME composite electrolytes. At the bottom: schematic of Li-ion pathways with these composites.
Source: Reprinted with permission from Zheng and Hu (2018). © American Chemical Society.

TABLE 3.5   ₇Li T₁ Results of PEO (LiTFSI), LLZO (5 wt%)–PEO (LiTFSI), LLZO (20 wt%)–PEO (LiTFSI), LLZO (50 wt%)–PEO (LiTFSI), and LLZO (50 wt%)–PEO (LiTFSI) (50 wt%)–TEGDME. Reprinted with permission from Zheng, J.; Hu, Y. Y. New Insights into the Compositional Dependence of Li-Ion Transport in Polymer-Ceramic Composite Electrolytes. *ACS applied materials & interfaces*, 2018, 10, 4113–4120.

T_1/s	PEO (LiTFSI)	LLZO (5 wt%)–PEO (LiTFSI)	LLZO (20 wt%)–PEO (LiTFSI)	LLZO (50 wt%)–PEO (LiTFSI)	LLZO (50 wt%)–PEO (LiTFSI) (50 wt%)–TEGDME
LiTFSI	0.28	0.35	0.52	0.73	0.69
LLZO and interface			0.75	1.33	1.18
Decomposed LLZO			11.36	19.28	11.16

Source: Reprinted with permission from Zheng et al. (2018). © American Chemical Society.

Another important issue in the case of an all-solid-state battery is the reaction with Li-metal and dendrite growth. To resolve this issue, Zhou reported the preparation of a sandwiched electrolyte in configuration polymer/ceramic membrane/polymer sandwich electrolyte (PCPSE) (2016). The host polymer is cross-linked poly(ethylene glycol) methyl ether acrylate (CPMEA), the electrochemical performance of which is examined by fabricating a coin cell of Li/CPMEA/SS (Fe). The impedance plot of the Li/CPMEA/Fe depicts improved stability for 300 h with a continuous dc bias voltage of 4.5 V as compared to PEO. The ionic conductivity increases with temperature (1×10^{-4} S cm^{-1} (at 65°C) $\rightarrow$ 2×10^{-4} S cm^{-1} (at 100°C) owing primarily to the softening of polymer chains. The ESW of the CPMEA−LATP electrolyte was close to 5 V. Figure 3.8 shows the cell performance of the two solid-state cells LiFePO$_4$/CPMEA/Li, and LiFePO$_4$/CPMEA−LATP/Li. The charge-discharge profile is shown in Figure 3.8 a, and 3.8 b corresponding to 0.2 C, and 0.5 C, respectively. The discharge capacity is 130 mAh g^{-1} (at 0.2 C) and 120 mAh g^{-1} (at 0.5 C). The flat region is dominant for the PCPSE electrolyte than CPMEA. Figure 3.8 c shows the cycling performance for 650 cycles. For 100 cycles, both the systems show similar performances, while after 200 cycles, capacity fading is more in the case of CPMEA electrolyte (70 mAh g^{-1} after 325th cycle). It is important to note that even after 640th cycle, the capacity retention for PCPSE electrolyte is 102 mAh g^{-1} (at 0.6 C). The Coulombic efficiency was also stable for the PCPSE electrolyte, while there were large fluctuations in the case of CPMEA electrolyte. The long cycle stability suggests the suppression of the dendrite growth with PCPE electrolyte.

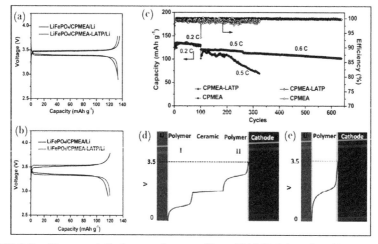

FIGURE 3.8 Charge and discharge voltage profiles of Li/LiFePO$_4$ cells with CPMEA and CPMEA−LATP-based PCPSE at 0.2C (a) and 0.5C (b). (c) Cycling and C rate performance of the Li/LiFePO$_4$ cells with CPMEA and CPMEA−LATP-based PCPSE. Illustration of the electric potential profile across the sandwich electrolyte (d) and individual polymer electrolyte (e) in the charge process of a Li/LiFePO$_4$ cell.

Source: Reprinted with permission from Zhou et al. (2016). © American Chemical Society.

Figures 3.8d and 3.8e show the electric potential profile across phase boundaries. In Figure 3.8d, between two conducting phases, redistribution of charge carrier creates an electric double layer (i.e., membrane with a potential difference). The ceramic electrolyte blocks anion movement, while cation migration is enhanced owing to the reduction in the trapped positive charge. In brief, it results in a stabilized system with a low electric field across the interface. Figure 3.8 e shows the case when anion is not blocked, making the electric field across the interface is stronger. Here, high electric field lowers the LUMO energy of interface polymer relative to anode Fermi energy which indicates an electron transfer (ET) for an interfacial chemical reaction (Gireaud et al., 2006).

Another interesting approach is the reinforcement of the inorganic nano-filler in the polymer matrix. It combines the properties of both nanofiller as well as polymer. The concept is that the LATP acts as a reinforcing phase which improves both the thermal and electrochemical properties.

As flexibility is a crucial requirement for the faster ion conduction. So, the flexibility of the polymer chain makes it easier to push the cation via coordinating sites of polymer chains. So, keeping this in mind, Ban reported the preparation technique of all-solid-state battery based on (PEO)-LiClO$_4$–Li$_{1.3}$ Al$_{0.3}$ Ti$_{1.7}$ (PO$_4$)$_3$ (LATP) matrix (2018). XRD analysis indicates the suppression of crystallinity and good stability of both polymer matrix and LATP (Figure 3.9a). The prepared film was flexible and LATP is uniformly dispersed in the polymer matrix (Figure 3.9b-d). The ionic conductivity increases with the addition of LATP and maximum conductivity was found to be 1.6×10^{-3} S cm^{-1} for the 25 wt% LATP (at 80°C). The voltage stability window is close to 4.6 V (with t$_+$=0.216) which is high as compared to PEO (3.8 V; with t$_+$=0.163). The thermal stability of the prepared PE is about 300°C and tensile strength decreases with the addition of LATP (Figure 3.9d-e). The charging-discharging performance has been examined for the cell, LiFePO$_4$|PEO-50 wt% LATP CSE|Li. The cell performance observed at 80°C shows a discharge capacity of about 109.3 mAh g^{-1} (at 1 C) after 500 cycles with capacity retention of 76.3%. Figure 3.9g-h shows the plot of discharge capacity (at different bending states of the pouch cell) and the demonstration of a LED powered by the cell. It was concluded that the prepared system has the potential to make flexible batteries with high energy and power density.

A garnet based HSE is advantageous as compared to the nanofiller based PE due to factors such as (i) high ionic conductivity, (ii) broad electrochemically stable potential window, (iii) low solid/solid interface resistance, and

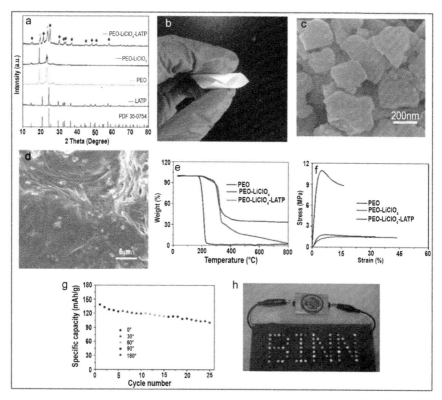

FIGURE 3.9 (a) XRD patterns of the composite solid electrolyte (pink), PEO-LiClO$_4$ electrolyte (blue), pure PEO electrolyte (green), LATP powers (red). (b) Optical photo of the composite solid electrolyte membrane, showing the flexibility. (c) SEM image of the LATP powders. (d) SEM images of the composite solid electrolyte membrane. (e) TGA results of the PEO-LiClO$_4$-LATP and PEO-LiClO$_4$ composite electrolyte membranes. (f) The stress–strain curves of PEO, PEO-LiClO$_4$, and PEO-LiClO$_4$-LATP composite electrolyte membranes. (g) The discharge capacity of the LFP|PEO-50 wt% LATP|Li pouch battery bending with different angles at 80°C and 0.2C rate, (h) Demonstration of a LED screen powered by this solid-state LFP| PEO-50 wt% LATP|Li pouch cell.
Source: Reprinted with permission from Ban et al. (2018). © American Chemical Society.

(iv) high thermal stability. Based on this approach, a polymer-ceramic composite electrolyte is prepared via a tape casting method (Zhang et al., 2018). Here, the HPE matrix comprises of the PVDF-HFP as a polymer matrix and LLZO particles as an additive. XRD analysis confirmed that the LLZO has no effect on the crystallinity after mixing with PVDF-HFP. Figure 3.10 shows the SEM micrograph of the hybrid system and depicts the uniform dispersion of LLZO and film is of a flexible nature. Figure 3.10 b shows the ion conduction mechanism and increase in cation transport number, which is attributed to the dominance of the hopping process between ceramic particles. Also, the flexibility of the polymer matrix supports the cation migration. The mechanical stability of the prepared system is sufficient for solid-state battery applications. The ionic conductivity of the HSE infiltrated with 20 μL liquid electrolyte is about 1.1×10^{-4} S cm^{-1} (at 25°C), and 7.63×10^{-4} S cm^{-1} (at 100°C). The voltage stability window for the HSE membrane is 5.3 V (vs. Li$^+$/Li) and is much higher as compared to PVDF-HFP which shows a potential difference of 4.5 V. The cation transport number is 0.61 and is within the desirable limit. This increase is attributed to the relaxation of polymer chain motion and enhanced segmental motion which facilitates cation migration with the addition of LLZO. Authors conclude that the high value of cation transport number for HSE as compared to liquid electrolytes is associated with the contribution from LLZO and anion trapping. The cell (Li|HSE|LiFePO$_4$) performance is examined and the initial discharge capacity is found to be 140 mAh g^{-1} (at 0.1 C). The Coulombic efficiency of the cell is very high (100%) after 180 cycles with a capacity retention of 92.5%. Triboelectric nanogenerators (TENGs) are used for harvesting mechanical energy from the environment and here, the combination of TENG, power management circuit, and a solid-state lithium battery is demonstrated to examine the effect of different charging frequencies of the pulsed output of the TENG on the performance of the batteries (Figure 3.10 c). The output current is shown in Figure 3.10 d. It is observed from the charge-discharge profile that recharging current increases with an increase in current of TENG as well as the rotational speed. Figure 3.10 f shows the impedance plot, and both the R$_f$ and R$_{ct}$ of the Li|HSE|LFP cell are found to increase during cell operation. It was concluded that the above-demonstrated cell can store the pulsed energy harvested by TENG.

3.5   PATENTS IN THE AREA

From the detailed discussion in the present chapter, it can be concluded that the electrolyte plays a very important role as it participates during charging as well as discharging also. The key role of the separator is to physically separate both electrodes (to prevent short-circuit), and (ii) provides a path for ion migration. To provide a glimpse of the existing companies working in the development of the electrolytes, some important inventions are summarized in Table 3.6.

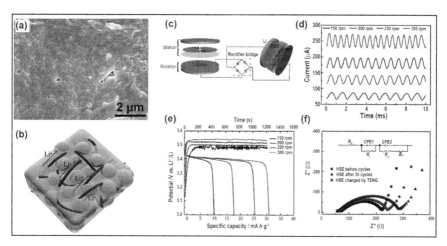

FIGURE 3.10 (a) SEM images of the hybrid solid electrolyte membrane. (b) Schematic image illustrating the Li-ion conducting paths in the HSE membrane. (c) Schematic image of storing pulse energy generated by the TENG in an all-solid-state lithium-ion battery. (d) Output currents of the TENG at different rotation rates. (e) The charge and discharge profiles of the Li|HSE|LFP cell charged by a TENG at different rotation rates with a fixed time of 20 min and discharged at a current density of 40 μA. (f) Electrochemical impedance spectra of the Li|HSE|LFP cell before cycles and after 30 cycles at different rates and after charged by TENG.

Source: Reprinted with permission from Zhang et al. (2018). © Elsevier.

TABLE 3.6   Some Available Patents on Electrolytes

Inventor	Polymer Used	Patent No.	Year	Conductivity
Bauer et al.	LiClO$_4$ in a 400 MW PEG	4,654,279	1987	4×10^{-4} S/cm at 25°C
Kuzhikalail et al.	PAN, EC, PC, and LiPF$_6$/LiAsF$_6$	5510209	1995	OCV: 2.85 V
Nitash Pervez Balsara et al.	Block copolymer	US 8,889,301 B2	2014	1×10^{-4} S cm at 25°C
Wunder et al.	PEO-POSS-phenyl 7(BF$_3$Li)$_3$	9680182 B2	2017	1×10^{-4} S cm at 25°C. (for O/Li=14)
Michael A. Zimmerman	PPS, PPO, PEEK, PPA	2017/0018781 A1	2017	1×10^{-5} S/cm (At RT)
Russell Clayton Pratt et al.	perfluoropolyether electrolytes terminated with urethane	9923245	2018	3.6×10^{-5} (@40°C) 1.1×10^{-4} (@80°C)

TABLE 3.6   *(Continued)*

Inventor	Polymer Used	Patent No.	Year	Conductivity
Mohit Singh et al.	Ceramic electrolyte	20110281173	2018	Stability up to 500 cycles
Nanotek Instruments, Inc.	Hybrid solid-state electrolyte	20180166759	2018	>10^{-3} S/cm, dendrite penetration resistant
Michael A. Zimmerman, Randy Leising	Lithium Metal Battery With Solid Polymer Electrolyte	20180151914	2018	1×10^{-3} S/cm at 80°C/1×10^{-5} S/cm at −40°C.
Michael A. Zimmerman, Randy Leising	Solid-State Bipolar Battery	20180151910	2018	1×10^{-4} S/cm at RT/1×10^{-3} S/cm at 80°C/cycling efficiency 99%

3.6   CONCLUSIONS

Energy storage devices, specifically batteries, are the most promising candidate all over the world. The polymer electrolyte based batteries are the best alternative to the existing batteries consisting of liquid-based electrolytes. The properties associated with the polymer make it easier to develop flexible and lightweight batteries. The IL and gel polymer electrolytes have desirable properties, but poor mechanical properties prevent their use in commercial applications. Therefore, hybrid electrolytes are the most promising candidates due to improved electrical and mechanical properties. HPEs also suppress the issue of dendrite growth and have unique physicochemical properties that are crucial for developing a safe and high-performance battery. In brief, the HPE based solid-state batteries have great potential and may be adopted as the future electrolytes for energy storage/conversion devices.

ACKNOWLEDGMENTS

The authors thank the Central University of Punjab for providing fellowship.

KEYWORDS

- **hybrid polymer electrolyte**
- **ionic conductivity**
- **polymer electrolytes**
- **secondary batteries**
- **solid polymer electrolyte**
- **voltage window**

REFERENCES

Abreha, M., Subrahmanyam, A. R., & Kumar, J. S., (2016). Ionic conductivity and transport properties of poly (vinylidene fluoride-co-hexafluoropropylene)-based solid polymer electrolytes. *Chemical Physics Letters, 658*, 240–247.

Agrawal, R. C., & Pandey, G. P., (2008). Solid polymer electrolytes: Materials designing and all-solid-state battery applications: An overview. *Journal of Physics D: Applied Physics, 41*, 223001.

Arof, A. K., Amirudin, S., Yusof, S. Z., & Noor, I. M., (2014). A method based on impedance spectroscopy to determine transport properties of polymer electrolytes. *Physical Chemistry Chemical Physics, 16*(5), 1856–1867.

Arya, A., & Sharma, A. L., (2017). Insights into the use of polyethylene oxide in energy storage/conversion devices: A critical review. *Journal of Physics D: Applied Physics, 50*, 443002.

Arya, A., & Sharma, A. L., (2017). Polymer electrolytes for lithium-ion batteries: A critical study. *Ionics, 23*, 497–540.

Arya, A., & Sharma, A. L., (2018). Effect of salt concentration on dielectric properties of Li-ion conducting blend polymer electrolytes. *Journal of Materials Science: Materials in Electronics, 29*, 17903–17920.

Arya, A., & Sharma, A. L., (2018). Structural, microstructural and electrochemical properties of dispersed-type polymer nano composite films. *Journal of Physics D: Applied Physics, 51*, 045504.

Ban, X., Zhang, W., Chen, N., & Sun, C., (2018). A high-performance and durable poly(ethylene oxide)-based composite solid electrolyte for all solid-state lithium battery. *The Journal of Physical Chemistry C, 122*, 9852–9858.

Bandara, T. M. W. J., Dissanayake, M. A. K. L., Albinsson, I., & Mellander, B. E., (2011). Mobile charge carrier concentration and mobility of a polymer electrolyte containing PEO and $Pr_4N^+ I^-$ using electrical and dielectric measurements. *Solid State Ionics, 189*, 63–68.

Bandara, T. M. W. J., Fernando, H. D. N. S., Furlani, M., Albinsson, I., Dissanayake, M. A. K. L., Ratnasekera, J. L., & Mellander, B. E., (2016). Effect of the alkaline cation size on the conductivity in gel polymer electrolytes and their influence on photo electrochemical solar cells. *Physical Chemistry Chemical Physics, 18*, 10873–10881.

Bar, N., Basak, P., & Tsur, Y., (2017). Vibrational and impedance spectroscopic analyses of semi-interpenetrating polymer networks as solid polymer electrolytes. *Physical Chemistry Chemical Physics, 9,* 14615–14624.

Chen, S., Wang, J., Zhang, Z., Wu, L., Yao, L., Wei, Z., Deng, Y., et al., (2018). In-situ preparation of poly (ethylene oxide)/Li_3PS_4 hybrid polymer electrolyte with good nano filler distribution for rechargeable solid-state lithium batteries. *Journal of Power Sources, 387,* 72–80.

Cheng, X. B., Zhang, R., Zhao, C. Z., & Zhang, Q., (2017). Toward safe lithium metal anode in rechargeable batteries: A review. *Chemical Reviews, 117,* 10403–10473.

Cho, S., Kim, S., Kim, W., Kim, S., & Ahn, S., (2018). All-Solid-state lithium battery working without an additional separator in a polymeric electrolyte. *Polymers, 10,* 1364.

Chua, S., Fang, R., Sun, Z., Wu, M., Gu, Z., Wang, Y., Hart, J. N., et al., (2018). Hybrid solid polymer electrolytes with two-dimensional inorganic nanofillers. *Chemistry: A European Journal, 24,* 18180–18203.

Croce, F., Appetecchi, G. B., Persi, L., & Scrosati, B., (1998). Nano composite polymer electrolytes for lithium batteries. *Nature, 394,* 456.

Edman, L., (2000). Ion association and ion solvation effects at the crystalline-amorphous phase transition in PEO-LiTFSI. *The Journal of Physical Chemistry B, 104,* 7254–7258.,

Ericson, H., Svanberg, C., Brodin, A., Grillone, A. M., Panero, S., Scrosati, B., & Jacobsson, P., (2000). Poly(methyl methacrylate)-based protonic gel electrolytes: A spectroscopic study. *Electrochimica Acta, 45*(8–9), 1409–1414.

Etacheri, V., Marom, R., Elazari, R., Salitra, G., & Aurbach, D., (2011). Challenges in the development of advanced Li-ion batteries: A review. *Energy and Environmental Science, 4*(9), 3243–3262.

Evans, J., Vincent, C. A., & Bruce, P. G., (1987). Electrochemical measurement of transference numbers in polymer electrolytes. *Polymer, 28*(13), 2324–2328.

Gireaud, L., Grugeon, S., Laruelle, S., Yrieix, B., & Tarascon, J. M., (2006). Lithium metal stripping/plating mechanisms studies: A metallurgical approach. *Electrochemistry Communications, 8,* 1639–1649.

Goodenough, J. B., & Kim, Y., (2011). Challenges for rechargeable batteries. *J. Power Sources, 196,* 6688–6694

Goodenough, J. B., & Park, K. S., (2013). The Li-ion rechargeable battery: A perspective. *Journal of the American Chemical Society, 135,* 1167–1176.

Keller, M., Appetecchi, G. B., Kim, G. T., Sharova, V., Schneider, M., Schuhmacher, J., Roters, A., & Passerini, S., (2017). Electrochemical performance of a solvent-free hybrid ceramic-polymer electrolyte based on $Li_7La_3Zr_2O_{12}$ in P $(EO)_{15}$ LiTFSI. *Journal of Power Sources, 353,* 287–297.

Koksbang, R., Olsen, I. I., & Shackle, D., (1994). Review of hybrid polymer electrolytes and rechargeable lithium batteries *Solid State Ionics, 69,* 320–335

Li, W., Yao, H., Yan, K., Zheng, G., Liang, Z., Chiang, Y. M., & Cui, Y., (2015). The synergetic effect of lithium polysulfide and lithium nitrate to prevent lithium dendrite growth. *Nature Communications, 6,* 7436.

Long, L., Wang, S., Xiao, M., & Meng, Y., (2016). Polymer electrolytes for lithium polymer batteries. *Journal of Materials Chemistry A, 4,* 10038–10069.

MacFarlane, D. R., Forsyth, M., Howlett, P. C., Kar, M., Passerini, S., Pringle, J. M., & Zhang, S., (2016). Ionic liquids and their solid-state analogs as materials for energy generation and storage. *Nature Reviews Materials, 1,* 15005.

Manthiram, A., Fu, Y., Chung, S. H., Zu, C., & Su, Y. S., (2014). Rechargeable lithium-sulfur batteries. *Chemical Reviews, 114*, 11751–11787.

Maranas, J. K., (2012). In polymers for energy storage and delivery: Poly electrolytes for batteries and fuel cells. *ACS Symposium Series, 1096*, 1–17.

Marcinek, M., Syzdek, J., Marczewski, M., Piszcz, M., Niedzicki, L., Kalita, M., & Kasprzyk, M., (2015). Electrolytes for Li-ion transport-review. *Solid State Ionics, 276*, 107–126

Mauger, A., Julien, C. M., Paolella, A., Armand, M., & Zaghib, K., (2018). A comprehensive review of lithium salts and beyond for rechargeable batteries: Progress and perspectives. *Materials Science and Engineering: R: Reports, 134*, 1–21.

Meyer, W. H., (1998). Polymer electrolytes for lithium-ion batteries. *Advanced Materials, 10*, 439–448.

Muench, S., Wild, A., Friebe, C., Häupler, B., Janoschka, T., & Schubert, U. S., (2016). Polymer-based organic batteries. *Chemical Reviews, 116*, 9438–9484.

Mural, P. K. S., Rana, M. S., Madras, G., & Bose, S., (2014). PE/PEO blends compatibilized by PE brush immobilized on MWNTs: Improved interfacial and structural properties. *RSC Advances, 4*, 16250–16259.

Oteo, U., Martinez-Ibañez, M., Aldalur, I., Sanchez-Diez, E., Carrasco, J., Armand, M., & Zhang, H., (2019). Improvement of the cationic transport in polymer electrolytes with (Difluoromethanesulfonyl) (trifluoromethanesulfonyl) imide salts. *Chem. Electro. Chem., 6*, 1019–1022.

Petrowsky, M., & Frech, R., (2008). Concentration dependence of ionic transport in dilute organic electrolyte solutions. *The Journal of Physical Chemistry B, 112*(28), 8285–8290.

Pritam, Arya, A., & Sharma, A. L., (2019). Dielectric relaxations and transport properties parameter analysis of novel blended solid polymer electrolyte for sodium-ion rechargeable batteries. *Journal of Materials Science, 54*, 7131–7155.

Ratner, M. A., Johansson, P., & Shriver, D. F., (2000). Polymer electrolytes: Ionic transport mechanisms and relaxation coupling. *Mrs. Bulletin, 25*, 31–37.

Sandí, G., Carrado, K. A., Joachin, H., Lu, W., & Prakash, J., (2003). Polymer nanocomposites for lithium battery applications. *Journal of Power Sources, 119*, 492–496.

Sharma, A. L., & Thakur, A. K., (2010). Improvement in voltage, thermal, mechanical stability, and ion transport properties in polymer-clay nano composites. *Journal of Applied Polymer Science, 118*(5), 2743–2753.

Stoeva, Z., Martin-Litas, I., Staunton, E., Andreev, Y. G., & Bruce, P. G., (2003). Ionic conductivity in the crystalline polymer electrolytes PEO_6: $LiXF_6$, X= P, As, Sb. *Journal of the American Chemical Society, 125*, 4619–4626.

Tasaki, K., Kanda, K., Nakamura, S., & Ue, M., (2003). Decomposition of $LiPF_6$ and stability of PF_5 in Li-ion battery electrolytes density functional theory and molecular dynamics studies. *Journal of the Electrochemical Society, 150*, A1628–A1636.

Wu, G. M., Lin, S. J., & Yang, C. C., (2006). Preparation and characterization of PVA/PAA membranes for solid polymer electrolytes. *Journal of Membrane Science, 275*(1–2), 127–133.

Xia, S., Wu, X., Zhang, Z., Cui, Y., & Liu, W., (2018). Practical challenges and future perspectives of all-solid-state lithium-metal batteries. *Chem.* doi: 10.1016/j.chempr.2018.11.013.

Yang, C., Liu, B., Jiang, F., Zhang, Y., Xie, H., Hitz, E., & Hu, L., (2017). Garnet/polymer hybrid ion-conducting protective layer for stable lithium metal anode. *Nano Research, 10*, 4256–4265.

Zhang, Q., Liu, K., Ding, F., & Liu, X., (2017). Recent advances in solid polymer electrolytes for lithium batteries. *Nano Research, 10,* 4139–4174.

Zhang, Q., Liu, K., Ding, F., & Liu, X., (2017). Recent advances in solid polymer electrolytes for lithium batteries. *Nano Research, 10*(12), 4139–4174.

Zhang, W., Nie, J., Li, F., Wang, Z. L., & Sun, C., (2018). A durable and safe solid-state lithium battery with a hybrid electrolyte membrane. *Nano Energy, 45,* 413–419.

Zhao, C. Z., Zhang, X. Q., Cheng, X. B., Zhang, R., Xu, R., Chen, P. Y., Peng, H. J., Huang, J. Q., & Zhang, Q., (2017). An anion-immobilized composite electrolyte for dendrite-free lithium metal anodes. *Proceedings of the National Academy of Sciences, 114,* 11069–11074.

Zheng, J., & Hu, Y. Y., (2018). New insights into the compositional dependence of Li-ion transport in polymer-ceramic composite electrolytes. *ACS Applied Materials and Interfaces, 10,* 4113–4120.

Zhou, B., He, D., Hu, J., Ye, Y., Peng, H., Zhou, X., & Xue, Z., (2018). A flexible, self-healing, and highly stretchable polymer electrolyte via quadruple hydrogen bonding for lithium-ion batteries. *Journal of Materials Chemistry A, 6,* 11725–11733.

Zhou, W., Wang, S., Li, Y., Xin, S., Manthiram, A., & Goodenough, J. B., (2016). Plating a dendrite-free lithium anode with a polymer/ceramic/polymer sandwich electrolyte. *Journal of the American Chemical Society, 138,* 9385–9388.

CHAPTER 4

Process, Design, and Technological Integration of Flexible Microsupercapacitors

SYED MUKULIKA DINARA

National Post-Doctoral Fellow, Indian Institute of Technology, School of Basic Sciences, Bhubaneswar, Odisha, E-mail: dr.smdinara@gmail.com

ABSTRACT

Nowadays, flexible electronic components are gaining significant attention in the modern energy storage system. The flexibility of electronics has triggered a tremendous advantage to miniature a micro-sized energy storage cell with very high energy consumption. Flexible electronics leads a sparked revolution in electronics such as smartphone, laptop, electronic skin, flexible display, light-emitting diode (LED), implantable medical devices, etc. Micro-sized energy storage are being paved a great attention due to their various unique properties like ultra-high power density, long cycling lifetime, excellent flexibility, smaller size, high electrochemical performance, and interdigited architecture. In present technological demand micro-sized energy storage devices, i.e., microsupercapacitors (MSCs) have a pronounced potential demand rather than conventional batteries or electrolytic capacitors for various types of energy and power sources. The rapid development on micro-supercapacitor technology have been researched successively through various mode like material evolution, morphology evolution, structural evolution, patterning evolution, growth technique evolution, etc. The materials used for the micro-supercapacitor study are thinned to its micrometer or nanometer scale to exhibit various novel and exceptional physical and electronic properties. Basically, the potentiality of MSCs depends upon the physical, mechanical, and electrochemical properties of the electrode material. Hence, to find out the appropriate electrode materials is a great challenge for the fabrication of flexible and stable MSCs.

4.1   INTRODUCTION

Nowadays flexible electronic components are gaining significant attention in the energy storage system (Liu et al., 2017; Qi et al., 2017; Tyagi et al., 2015). Also, it leads to a sparked revolution in electronics such as smartphones, laptops, electronic skin, flexible display, light-emitting diode (LED), implantable medical devices, etc. Micro-sized energy storage systems are being paved a great attention due to their unique properties like ultra-high power density, long cycling life, excellent flexibility, smaller size, high-performance and interdigited architecture for new generation portable electronic devices (Patrice and Yury, 2010; SIMON and GOGOTSI, 2009; Wang and Shi, 2015; Xue et al., 2011; Zhang and Zhao, 2009). In present technological demand micro-sized energy storage devices namely micro-supercapacitors (MSCs) have a pronounced potential impact rather than conventional batteries or electrolytic capacitors for various energy and power sources (SIMON and GOGOTSI, 2009; Zhang and Zhao, 2009). The super-capacitor is a promising energy storage unit with a power density intermediate between those of the conventional capacitor and battery cell. These are classified as an electric double-layer capacitor (EDLC) and pseudo capacitor depending on the charge storage mechanism (Zhang and Zhao, 2009). In EDLC, charge storage is governed by desorption and adsorption of electrolyte ions on the high surface area of the electrode whereas the fast and reversible redox reactions stores the charge in the pseudo capacitor (Augustyn et al., 2014). Among both capacitors, EDLC is more superior due to its high power density and excellent life cycles. However, pseudocapacitors exhibit higher specific capacitance (10–100 times) than EDLCs (Conway, 1991; Dubal et al., 2014; Manaf et al., 2013; Yu et al., 2013; Zhang and Zhao, 2009; Zhang et al., 2009). Although the carbon-based materials are used as, the active electrode for EDLCs, but their limited capacitance value lowers the energy density. In recent years, on-chip planer micro-supercapacitor have taken a significant role to overcome the energy density limitation and low operating life cycles of both capacitors by miniaturizing the electrode material through interdigited finger liked geometry with high surface area (Beidaghi and Gogotsi, 2014; Huang et al., 2016; Liu et al., 2013). The electrode patterning of MSCs are designed like a sandwiched geometry or rolled like geometry (Liu et al., 2017; Qi et al., 2017). The aforementioned geometry of MSCs offers ultra-high responsive behavior and energy/power density.

As a 2D material, layered materials have been extensively studied in literature concerning their physical characteristics for supercapacitor applications (Peng et al., 2014; Wu et al., 2013). Among various 2D material graphene is one of the most promising candidates for MSCs due to its unique physical end electronic properties like excellent electrical conductivity, large theoretical surface area ($\sim$2620 m^2 g^{-1}), high intrinsic double-layer capacitance, high surface to volume ratio, excellent mechanical flexibility/stability and wide potential window (Ivanovskii, 2012; Raccichini et al., 2014). The intensive research on graphene demonstrates that when exfoliated graphene is formed into a monolayer (ML) or few MLs, the properties of the material are extensively varied due to nanoscale confinement effect. Among the carbon family, graphene is considered as an ideal electrode material due to its outstanding and exceptional high specific surface area. These aforementioned exciting properties of graphene enable for high value of charge storage application. The graphene oxide (GO) is a functionalized form of graphene containing oxygen groups that assists for various fascinating properties (Gao et al., 2011). Whenever the sheet resistance of GO is reduced by several order of magnitude then GO is transformed to reduced graphene oxide (RGO) that belongs to the group of semiconductor or semimetal (Beidaghi and Wang, 2012). Also GO has been well reported for conventional energy storage devices due to their large surface area, tunable electrical and optical properties, excellent mechanical stability, high chemically active defect density, excellent scalability and low cost (Ivanovskii, 2012; Raccichini et al., 2014). RGO acts as an effective functionalization material with low cost and high scalability and excellent electrochemical performance for flexible supercapacitor application (Shao et al., 2015). Over the past few years, some researchers have successfully designed RGO and RGO-based composites and explored their energy storage applications. The excellent electrochemical performance of RGO film was reported as a superior electrode with ultra-high energy and power density into the new generation micro-supercapacitor family (Beidaghi and Wang, 2012). Beidaghi et al. stated that the RGO/CNT composite is also a promising electrode material for an on-chip energy storage device with highest energy density ($\sim$ 0.68 mWhcm^{-3}) and highest volumetric power density ($\sim$77 Wcm^{-3}) than compared to bare RGOs (Beidaghi and Wang, 2012). Many other groups have been reported various graphene-based composite materials through various processing routes and technology that would be elaborately discussed in the next section of materials evolution.

Beyond the wide spectrum of layered material, some other materials namely transition-metal dichalcogenides (TMDs) have gained significant attention for MSCs due to their superior electrochemical properties (Cao et al., 2013; Chia et al., 2015). The electrochemical properties of TMDs have a great role to electron transfer (ET) kinetics through the edge site of the material surface towards redox probes (Chia et al., 2015). Typically conductive polymers, transition-metal oxides, transition metal hydroxides, and their composite can be directly deposited and patterned upon the interdigited metal current collector to fabricate on-chip MSC (Du et al., 2013; Hercule et al., 2013; Li et al., 2014; Liu et al., 2014b; Niu et al., 2012a; Yan et al., 2014). A massive research effort has been made to get constantly increasing high energy density of electrochemical capacitor (ECs). These ECs store charges through non-faradaic charge separation and reversible faradaic redox process. To develop effective ECs, two important properties, i.e., high surface area and high electrical conductivity must be satisfied by electrode material. The TMD materials at nanometer scale having active redox centers and show pseudocapacitive behavior (Conway, 2013). Moreover, recently planar microsupercapacitors was fabricated by MoS_2 using a laser patterning method (Cao et al., 2013). The high conductivity is a key parameter of an electrode material for better charge storage performance. In this regards composite materials are developed by combining the high conductive material of higher surface area with the nanostructured TMDs. Therefore, various materials and their composite as electrode materials have been proposed with excellent capability to store charge such as CNT, graphene, transition metal oxides with CNTs, activated carbon (AC), etc. (Hsia et al., 2014; In et al., 2015; Kim et al., 2013, 2014, 2016; Lee et al., 2014, 2015; Sun et al., 2016; Yu et al., 2015).

Recent trends on on-chip micropower energy storage cells with interdigited microelectrode patterned geometry along with high areal and volumetric densities at micro/nanoscale are a great demand in the electronics industry. Various processes have been reported to develop and fabricate electrode materials like electrochemical/electrophoretic deposition, electrodeposition, chemical vapor deposition (CVD), laser reduction, spin coating/transferring/etching, spray deposition, ink-jet printing, etc. (Beidaghi and Wang, 2012; Hsia et al., 2014; Kurra et al., 2015; Lee et al., 2016; Liu et al., 2016; Niu et al., 2013; Song et al., 2015; Wen et al., 2014). However, the material synthesis processes and micro-electrode fabrication techniques have a key role to develop a high-performance MSCs. In literature various research groups have made significant contribution to develop high-performance MSCs through various process technologies like laser patterning, inkjet, and screen printing through lithographic processes, selective area etching process, and photolithography (In et al., 2015; Kim et al., 2013, 2016; Yu et al., 2015). Among various processing technologies, laser writing is a non-contact-mask free direct pattern transfer process to fabricate micropatterned geometry. The different approaches on fabrication technology have been elaborately discussed in microfabrication section.

4.2   GENERAL CHARACTERISTICS AND CONFIGURATION OF FLEXIBLE MSCS

Typically, SCs are consisted of four main components such as electrode materials, electrolyte, current collector, and a separator. Generally, SCs packaging is done into a cell followed by sandwich or spiral-wound configuration (Simon and Gogotsi, 2009). But configuration of conventional MSCs are more simple and compact in contrast with SCs where an interdigited microelectrode pattering promote an ultra-high level energy performance in charge storage applications. The inter-digital patterning in micrometer range effectively offers the device to operate in the millimeter-centimeter scaled in any electronic configuration. Moreover, MSCs show more superior electrode performance due to full access of binder-free active electrode surface area. Importantly microelectrode material should be stable and sustainable in micro-supercapacitor fabrication process. In this regard, a special structure with some unique designs is developed in microelectronic chip to conduct device applications. So material choice is also an important part of micro-super capacitor electrode patterning. Moreover, the thickness of microelectrode material is a driving parameter that impacts significantly on device performance. In this regard, thin-film deposition techniques were used to develop the material in nano-scale and electrode processing in micro-scale. Whereas the thickness of the electrode affects the performance of MSCs in such a way:

1. By reducing the thickness of electrode material, the gravimetric and volumetric capacitances are increased.
2. By reducing the thickness of electrode material, areal capacitance is decreased due to the small quantity of electrode material.

Microelectrode fabrication technique is gaining incredible development in process technology where the size and shape of microelectrode can be tailored differently and the space between two electrodes can be designed into a micro range which exhibit very shorten ion diffusion path and a direct access of high density of ions to the electrode surface. Moreover, the electrochemical activity is significantly monitored by material's internal property, thickness, shape, and size of the electrode as well as the space between consecutive electrodes. So there are many factors involved to influence the microelectrode performances and importantly different architectures can be taken to get better powerful and high-density device structure. Due to smaller size and low structural density, micro-supercapacitor can be integrated easily to any kind of micro-chip to get flexible and lightweight micro-electronic devices. To satisfy the requirements of flexible electronic devices, the flexible MSCs take a great challenge through the choices upon substrate electrode materials and electrolytes. With respect to device stability performance, nano-fiber or nano-sheet like electrodes perform better long life cycle than any nanoparticle-based MSCs. The MSC electrode configuration is compatible and offers a compact electronic configuration packaging system to other electronic devices. To fulfill the flexibility requirement of MSCs the structural configuration has been developed/mounted onto flexible substrates and bounded by gel/solid electrolytes. The gel and solid electrolyte offer more flexibility and high mechanical integrity rather than liquid electrolyte. This type of configuration offers desirable electrochemical properties and can retain large mechanical strain without any deformation of their structural and mechanical properties. The general features of MSCs are:

1. The configuration of MSCs is like a micro-electrode array used to integrate with other devices.
2. The substrates used to develop microelectrodes are glass, plastic film, Si/SiO_2, and other flexible materials like film or paper.
3. Owing to unique features of micro-electrodes, the materials are limited in MSC application like graphene, graphene composite, porous carbon film, CNT, PANI, nano-structured PEDOT, VSe_2, and some transition metal composite.
4. As a current collector Au nano-film are generally used in MSC configuration.
5. Gel and solid electrolyte are used into the MSC to conduct electrochemical performances.

4.3   FABRICATION PROCESSES AND DESIGN OF FLEXIBLE MSCS

Electrode patterning in finger-like orientation can be carried out by two ways, i.e., direct patterning and indirect patterning. The former one involves the direct patterning of electrode material into finger like geometry while the later can be done by using a sacrificial template. In this section, some technology processes are summarized to pattern the microelectrode configuration.

4.3.1 PHOTOLITHOGRAPHY

Photolithography is a widely used method in nano-electronics and electro-mechanical system due to low cost, simple technology, and high resolution. To achieve desired microelectrode geometry template is used to patterning the electroactive material for micro-supercapacitor. A standard photolithography technique is used for micro-patterning of electrode material (Figure 4.1). First, a photoresist film is developed on the surface of desired electrode material by spin coating. Specially, the photoresist film is patterned from the photo-mask by exposing the UV light. After rinsing the photoresist film, a thin layer of Au is developed on the patterned structure by thermal evaporation under optimized growth rate and pressure to acquire a certain thickness. Then photoresist film is lifted off by treating with acetone followed by ultra-sonication method. Finally, the Au microelectrode pattern of the electrode material on-chip is successfully developed after plasma etching process.

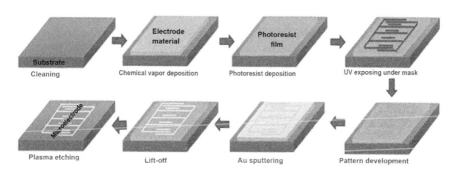

FIGURE 4.1 Schematic illustration of fabricated VSe_2/graphene based interdigital microelectrode pattern on Ni foam.

4.3.2 PLASMA ETCHING PROCESS

Plasma etching is another common technique for preparation of interdigited electrode finger arrays. Initially, the active material is deposited onto a suitable substrate (glass, Si wafer, polyethylene terephthalate (PET), etc.) followed by the formation of current collector arrays through successive procedure of masking, photolithography, and sputtering/evaporation (Chmiola et al., 2010; Liu et al., 2015; Wu et al., 2014a, b). Finally, the plasma-etching process is used to remove the exposed parts of the activated film. For instance, an activated material like graphene or heteroatom doped graphene electrode material is deposited by layer-by-layer or spin coating onto the substrate and the metal contact Au is sputtered upon it through sputtering technique. Afterward, electrode finger arrays are generated by plasma etching of the thin film material which is uncovered by Au collector (Wu et al., 2013, 2014b).

4.3.3 LASER SCRIBING

The photolithography and plasma etching technologies have been extensively explored for microelectrode fabrication process. But somewhere these technologies are limited due to lack of actual sacrificial template during patterning and to yield a proper integrated on a chip device. Hence, as a scalable and cost-effective simple technology, the laser scribing method has been significantly utilized to develop on-chip micro-interdigited patterning without any sacrificial template (Cai and Watanabe, 2016; El-Kady et al., 2012, 2015; El-Kady and Kaner, 2013; Gao et al., 2011; In et al., 2015; Wen et al., 2014). The laser irradiation method can effectively convert the graphite oxide into graphene due to the photothermal effect. So this technique can be directly employed to pattern any graphite oxide into reduced graphene by numerous geometrical patterning micrometer resolutions (Cote et al., 2009; Gilje et al., 2010; Wei et al., 2010). Therefore this technology can buildup both sandwich and in-plane geometrical structure upon graphite oxide film. It is highly desirable that without any external sources like template, electrolyte, and etching solution, the laser irradiation method is directly employed to pattern the microelectrode under different geometry with excellent electrochemical performance. The geometry of microelectrodes has a great influence upon the electrochemical activities due to change of carrier mobility and diffusion length as per the anisotropic nature. As a particular report, direct laser-reduction of GO after fabrication into in-plane MSCs demonstrates fast ion diffusion with specific capacitance (0.51 mF cm^{-2}) was presented (Wei et al., 2010). However, due to relatively low conductivity of electrode, charge/discharge rate, and power density of these MSCs are still limited. So, a facile approach was developed by using a femtolaser in $situ$ reduction of hydrated GO and chloroauric acid (HAuCl$_4$) nanocomposite simultaneously to pattern both rGO electrodes and Au current collectors through a single step. After femtolaser reduction process the fabricated patterned protocol achieved a one hundred-fold increase of electrode conductivities (up to 1.1×106 S/m), which deliver superior capability of charge transfer rate (charging rate increase about 50% from 0.1 V/s to 100 V/s). Moreover the MSC structure yield sufficiently high-frequency responses (362 Hz, 2.76 ms time constant), and high-value specific capacitance of 0.77 mF/cm^2 (Li et al., 2016).

4.3.4 INKJET PRINTING

Inkjet-printing is another fabrication technology of direct writing of electrode finger arrays and is very much useful for the electronic device fabrication process (Dua et al., 2010; Lee et al., 2012; Yan et al., 2009). The inkjet printing can be used precisely and directly onto various substrates by using commercial inkjet printers (i.e., from desktop to roll-to-roll). Moreover, in this typical process a minimum amount of active material is required for the phase transformation from liquid inks to nanoscale solid-state structures. The prime advantage of inkjet printer is its capability to form droplets without any obstruction into the nozzle during patterning the phenomena. Moreover, it greatly depends on some physical properties like particle size, viscosity as well as the surface tension of activated ink material. For instance, the micro supercapacitor patterning was developed by AC based ink material where stable ink material was made by mixing AC with binder poly (tetrafluoroethylene) (PTFE) polymer in ethylene glycol medium (Pech et al., 2010). Also, a graphene/polyaniline (NGP/PANI) ink was reported in inkjet printing technology to produce thin-film electrode with good electrochemical performance (Xu et al., 2014).

Additionally, there are several other methods reported for micro-electrode patterning like directly scraping (onto the thin-film electrode materials), the focused-ion-beam technique, microfluidic etching as well as selective wet etching, etc. In realization of direct integration MSC to any electronic devices, the in-plane MSCs on-chip fabrication technology can be directly patterned into an electrode array without impacting other components of that integrated prototype. Among the currently reported micro-fabrication technologies, the printing method is comparatively favored for MSC fabrication due to ability of direct patterning at room-temperature and not requiring of any sacrificial templates to develop desired micropatterning. Significantly, with microfabrication technology, the electrode active materials also actively influence the electrochemical performances and lead the overall performance of MSCs.

4.4   MATERIAL EVOLUTION FOR MSC APPLICATION

The electrode materials should have some important features to yield a flexible and stable MSC like good mechanical properties, high electrochemical stability, excellent conductivity, and high volumetric area. Therefore, the material choice became an important part to fulfill the requirements of mechanical flexibility and durability to configure the MSC. Different materials thus developed are discussed briefly in the following sections. Because of some special character like unique one-dimensional structure, high carrier mobility (0.79–1.2×10^5 cm$^2 \cdot$V$^{-1} \cdot$s^{-1}) and high flexibility upon large mechanical stress, the CNT acts as a promising material for flexible micro supercapacitor applications (Liu et al., 2016; Niu et al., 2012b). Generally, CNTs are synthesized by hydrothermal route or CVD technique (Hsia et al., 2014). But in large scale, CNTs are synthesized through hydrothermal method as it is inexpensive. So CNT based flexible MSCs are fabricated and designed by combining both hydrothermal and microelectronic processes (Chen et al., 2014; Kim et al., 2013, 2014, 2016; Lee et al., 2014, 2015; Liu et al., 2016; Sun et al., 2016; Yu et al., 2015). Various type of CNT based microelectrodes with their configuration and special characteristics are tabulated from various literatures.

Various technology were developed to make CNT thin film on substrate-like direct dipping, direct printing, photolithography with patterned mask, etc. Microelectrode patterning on different substrate with various geometrical configurations can be achieved through successive microelectrode patterning like lithography, coating, as well as transferring. The spray coating has more accuracy than dipping procedure to control the film thickness precisely. The homogeneity in microelectrode patterning is influenced by the density of CNT materials, substrate surface, and thickness of photosensitive film. CNT thin film is in sticky nature with the substrate surface and maintains a strong adhesion with substrate surface owing to its high surface to volume ratio. So after bending or any mechanical stress the CNT remain in safe from any exfoliation. The wetting-induced micropatterning is another versatile protocol to design microelectrode pattern through selective area solution dispersion method. The CNT films cannot be directly synthesized on desired flexible microelectrode substrate. So in that case after CNT deposition on any arbitrary substrate, the film can be transferred on the flexible substrate for micro supercapacitor application through a laser-assisted dry transfer method. For instance, the CVD synthesized CNT on Si substrate was sputtered by Ni onto the CNT surface and then simultaneously patterned and transferred onto the flexible polycarbonate substrate through laser transfer method as shown by schematic presentation in Figure 4.2 (Hsia et al., 2014). At the last step, an ionogel (ionic liquid (IL)) is employed as a semi-solid matrix that acts as an electrolyte and to develop a complete solid-state MSC.

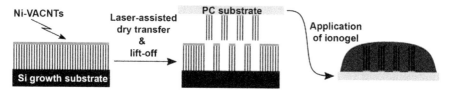

FIGURE 4.2   Schematic illustration of device fabrication (side-view): VACNTs are grown by chemical vapor deposition on a Si substrate and a Ni layer is sputtered on top of the VACNTs. Interdigitated Ni-VACNT electrodes are simultaneously patterned and transferred onto flexible polycarbonate substrate by a laser-assisted dry transfer. In the last step, the ionogel is applied to the patterned electrode structures, completing the micro-supercapacitor fabrication.

Source: Reprinted with permission from Hsia et al. (2014).

Transition metal and transition metal/CNT composites are observed to be an excellent electrode material for flexible micro-supercapacitor application. In this context, various type of CNT composite with transition metal compound has been reported under different array-like configurations. A planar-type MSC array was reported with Au coated functionalized multi-walled carbon nanotube (MWCNT) on MWCNT-COOH/MnOx nanoparticle (NP) composite (Lee et al., 2014). First, as current collectors the patterned Ti/Au electrodes were fabricated through photolithography technique followed by selected area lift-off technique the LbL-MWCNT/Mn_3O_4 film-based electrodes pattern was developed. The functionalized MWCNTs and Mn_3O_4 nanoparticles (NPs) were deposited on NH_2 treated substrate via an LbL assembly method shown in Figure 4.3c. In such configuration layer by layer (LBL) multi-walled CNT thin film assembly with top layer Mn_3O_4 NPs was used as a flexible electrode for MSC using poly(methyl methacrylate)-propylene carbonate (PC)-lithium perchlorate gel electrolyte under a stretchable and patchable array configuration. In this process initially, the patterned MSC arrays were fabricated on PET substrate followed by the same photolithography technique as mentioned earlier (shown in Figure 4.3a). Then by using dry transfer method, the individual MSC are transferred from the PET substrate onto the Ecoflex deformable substrate and the electrical interconnections between each electrode were carried out by Au nanowire solution by dropping method. The top surface of each MSC was drop-casted by PMMA-PC-LiClO$_4$ electrolyte and covered with squarely cutted cured Ecoflex. Also, the four corners of cured Ecoflex were stuck with uncured Ecoflex. The molecular structure of the individual component materials in PMMA-PC-LiClO$_4$ gel electrolyte is shown in Figure 4.3b. Next, the MSC array was encapsulated for protection from external environments and for direct attachment to the human body without contacting the gel-type electrolyte. The top surface of the drop-casted PMMA-PC-LiClO$_4$ electrolyte on each MSC was covered with a square piece of cured Ecoflex, and the four corners of Ecoflex were pasted with uncured Ecoflex (Lee et al., 2015). Then, the entire MSC array was annealed to complete the encapsulation. The cross-sectional view of the encapsulated MSC is shown in Figure 4.3d, with its optical image in inset. Interestingly it was also noticed that metal-free electrode like single-wall carbon nanotube (SWCNT)/carbon/MnO$_2$ was used as hybrid electrodes wherein SWNT/carbon acts as current collectors.

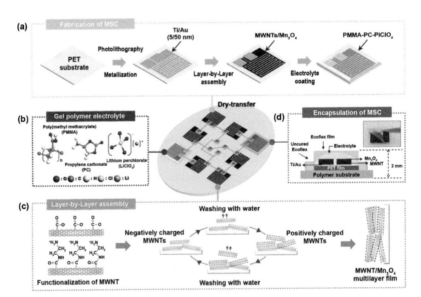

FIGURE 4.3   Schematic illustration of the fabrication process of planar-type LbL-MWCNT/ Mn$_3$O$_4$ MSC array on a stretchable substrate. The central figure shows the dry-transferred MSCs on a deformable Ecoflex substrate with embedded liquid metal interconnections. (a) Fabrication of a MSC on a PET substrate. (b) Molecular structures of the component materials in PMMA-PC-LiClO$_4$ gel electrolyte. (c) Fabrication process of LbL-assembled MWCNT/ Mn$_3$O$_4$ film. (d) Cross-sectional scheme of an encapsulated MSC with Ecoflex film. The inset shows the optical image of the encapsulated MSC.

Source: Reprinted with permission from Lee et al. (2015). © Royal Society of Chemistry.

Graphene is another important 2D material for flexible micro super-capacitor application with thin film and flat morphological structure that efficiently implemented for energy storage system. The in-plane architecture is favorable for graphene as the ionic electrolyte can directly interact with whole surface area of graphene in a short ion diffusion distance. The special characteristics like excellent electron mobility (15000 cm^2 V^{-1} s^{-1}), high theoretical double-layer capacitance ($\approx$550 F g^{-1}) a notable Young's modulus (~1.0 TPa) and high surface area ($\approx$2620 m^2 g^{-1}) favored to implement the graphene as a flexible and suitable material for MSC applications (Liu et al., 2014a; Niu et al., 2014). The in plane ultrathin MSC fabrication was reported on both pristine graphene as well as multilayer reduced graphene oxide with very high value of capacitance compared with conventional supercapacitors. For instance ultrathin patterned rGO microelectrodes laterally designed and fabricated by photolithography and selective electrophoretic build-up process by using gel electrolyte (phosphoric acid/polyvinyl alcohol) (Niu et al., 2013). Most significantly, this typical rGO structure can be easily designed to interconnect the electrode patterning in series or parallel connection on a chip. These advantages help the rGO structure to lead more effective utilization for energy storage application with a low cost and facile fabrication technique.

But pure graphene structure is limited in electrode performance due to reduction of accessible surface area because of aggregation and restacking of the individual layers during the processing (Wen et al., 2014). Therefore, spacers like CNT are employed to restrict the restacking and to achieve better performance in 2D layer structure. This mechanism enlarges the surface area of the electrodes as well as the nanochannels creating pathways for efficient ion transportation parallel with graphene planes. Hence, the graphene/CNT structure significantly enhances the electrochemical performance of planer MSCs. In this system binder, free microelectrode patterned is developed through the combined process of electrostatic spray deposition (ESD) and photolithography lift-off technique. The RGO was directly formed by ESD method without any heat treatment. At first masked Ti/Au, interdigital microelectrodes were patterned by photolithography process. Then a homogeneous solution of rGO/CNT composite was deposited onto the masked Ti/Au interdigital microelectrodes by ESD method. Finally, to remove the mask, etching process was followed to get the desired interdigital rGO-CNT composite electrodes. In this process, a desired amount of COOH-functionalized multiwalled CNT was added into the solution to carry the different rGO: CNT with different weight ratios (Beidaghi and Wang, 2012).

The schematic view of microelectrode fabrication process is shown in Figure 4.4a. Figures 4.4b and 4.4c show the SEM images of the fabricated rGO/CNT composite based micro-supercapacitor with a well-defined and defect-free patterns.

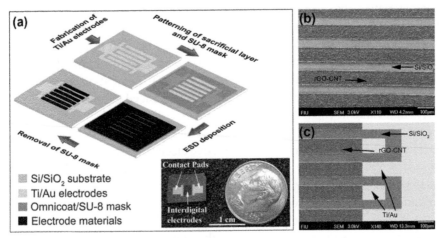

FIGURE 4.4 (a) Schematic drawing of fabrication procedures of microsupercapacitors (inset shows a digital photograph of a fabricated device),(b, c) Top view SEM images of rGO-CNT-based interdigital microelectrode arrays.

Source: Reprinted with permission from Beidaghi and Wang (2012). © Elsevier.

The low conductivity of graphene layer is another limitation for high-performance MSC application. An effective way to increase the conductivity of graphene is the methane (CH_4) plasma treatment. The reported data on CH_4 treated graphene showed more superior performance with a very high value of area and stack capacitance (80.7 µF cm^{-2} and 17.9 F cm^{-3}, respectively) compared with the un-tread graphene (32.6 µF cm^{-2} and 7.3 F cm^{-3}). This phenomenon demonstrates that the electrode materials performance is greatly influenced by material-electronic conductivity. AC is a common electrode material for energy storage application. But due to complexity to fabricate interdigital patterning, the utilization of AC is limited for MSC application. Sometimes pyrolysis was taken to form porous carbon from patterned photoresist followed by a lithography process. The porous carbon-based electrode is rarely used as electrode material for MSC with moderate specific surface area (Hsia et al., 2013; Kim et al., 2014; Wang et al., 2014). But to achieve flexible MSCs, polymer substrates are used with attached by a thin porous carbon layer that can be converted correspondingly into porous carbon-based flexible electrode trough successive pyrolysis and laser irradiation processes (Kim et al., 2014). Some nanostructure-based metal oxides have taken a great importance for fabrication of flexible interdigited patterned substrate for MSC application. For instance, MnO_2 nanoparticle-based interdigital microelectrode finger patterning with 10 nm space prepared by microfluidic etching with H_3PO_4-PVA thin films based solid-state electrolyte (Xue et al., 2011). The entire MnO_2 NPs based MSC shows excellent electrochemical performances and the device durability maintained even under the conditions of the repeated bending process. The mesoporous RuO_x micro-supercapacitor was also reported with good electrochemical performance with specific capacitance of 12.6 mF cm^2 (Makino et al., 2013).

An asymmetric micro supercapacitor reported by designing with different polarity materials wherein MnO_2 acts as a positive electrode material and carbon material acts reversely. The self-supported composite structure contains both MnO_2 and nanoporous AC with a separated interdigital patterning by using micro-electro-mechanical systems (MEMS) fabrication process (Shen et al., 2013). This prototype asymmetric microsupercapacitor shows a well capacitive performance with high value of working voltage in the range of 1 V to 1.5 V in aqueous electrolyte medium. Figure 4.5a shows the 3D interdigital structure where positive and negative electrodes are separated at micrometer scale. In this configuration MnO_2 nanoparticles used for the energy storage unit and porous AC is used to create a fast power transfer process by charge storing activity. The whole protocol is sealed by a PDMS (poly(dimethylsiloxane)) cap after injecting the electrolyte throughout the effective area of interdigited patterning. The schematic cross-sectional view of fabricated MSC is illustrated in Figure 4.5b where interdigital channels are created under optimized distance by selective area etching process. The device is scalable upon the adjustment of the width of each electrode and the gap between the electrodes. Most importantly, the gap between the electrodes has a significant impact upon the electrochemical performance of MSC material. The performance can be improved better by lowering the gap between the electrodes and to create this type of geometrical configuration is limited due to lack of actual precision upon fabrication technology.

Pseudocapacitive polyaniline nanowires (NWs) are also an important electrode material with high flexibility and high electrochemical activities. The PANI electrodes reported for MSC application with high-value specific capacitance of 45.2 mF cm^{-2}/105 F cm^{-3} along with high-value energy and power densities of 7.4 mWh cm^{-3} and 128 W cm^{-3}, respectively (Meng et al., 2014). In this protocol, a thin film of liquid crystal polymer (LCP) is used as a flexible substrate which is mechanically inert and compatible under micromachining treatment to configure the 3D micropatterning. The whole fabrication process is done by three ways (shown in Figure 4.6): (1) top-down micropatterning approach to develop current collector, backside electrical contact pads, and the whole connection between the two contacts (2) "bottom-up" approach to develop nanostructured PANI as supercapacitive electrodes through electrochemical polymerization treatment; and (3) fabrication of whole device using H_2SO_4-PVA polymer electrolyte.

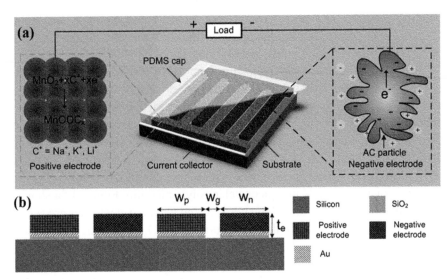

FIGURE 4.5   Schematic of the asymmetric 3D micro supercapacitor using different active materials in interdigital electrodes. The charge storage and delivery mechanism of positive electrode (shown on the left) is based on reversible redox reactions of MnO_2, while that of negative electrode (shown on the right) is based on the EDL effect of highly porous activated carbon, (b) Cross-section view of fabricated the asymmetric micro supercapacitor after selective area etching process and the resulting widths of positive and negative electrode are correspondingly w_p and w_n; the gap between electrodes is w_g, and the thickness of electrodes is t_e.
Source: Reprinted with permission from Shen et al. (2013). © Elsevier.

FIGURE 4.6   Fabrication of integrated 3D microsupercapacitors with through-via bottom electrode contact (cross-sectional view).
Source: Reprinted with permission from Meng et al. (2014). © John Wiley and Sons.

Various types of thin-film and nano-composite under different nano-structural orientations are being also explored for flexible MSC application with high value of electrochemical performances. For instance, the explored materials are onion-like carbon, VS_2, Graphene quantum dots, doped graphene Graphene/MnO_2/Ag NW, etc. A comparison of data on different materials and their composites for MSC application are given in Table 4.1.

4.5 INTEGRATION TECHNOLOGY OF MSCS WITH ELECTRONIC DEVICES

To become flexible and family user all electronic system must be in compact configuration following lightweight, small volume, and flexible nature. So external power sources play an important role to configure the demand in current electronic industry and to fulfill such criteria microsupercapacitor are taking now a great attention. Microsupercapacitor acts as space and energy conservation unit in any electronics devices. Therefore, MSC designs are getting significant attention to establish green energy society by integrating itself to any kinds of electronics system as a self-powered and sustainable energy unit. In this context, some efforts are briefly discussed to design MSC with different kind of electronic systems.

4.5.1 MSC-PHOTO DETECTOR INTEGRATED SYSTEM

Photodetectors are used to convert light into electrical signal that have various kinds of applications in different fields like sensors, actuators, and monitoring systems. Upon integration with photodetector, MSCs can create a self-powered system with exciting characteristics like ultra-high energy density, high-speed operating system, tiny, and light weight configuration as well as very handy microcell. A flexible solid state 3,4-ethylenedioxythio-phene (PEDOT) MSC was fabricated by conventional photolithography and electro deposition method that employed as a power source of light-emitting diode (LED) (Kurra et al., 2015). Glass and plastic polyethylene naphthalate (PEN) were used as substrates to fabricate microsupercapacitor inter-electrode patterning. At first photoresist was spin coated on the substrate and then by UV exposure microelectrode patterning is generated through desired glass mask. Electrode contact on the patterned photoresist was done by Au through sputtering method. After that, the electrode material PEDOT was developed throughout the surface by standard 3-electrode deposition method. Finally,

TABLE 4.1 Various Electrode Materials Along with Their Various Deposition and Fabrication Processes for MSC Application with Exciting Electrochemical Properties

Electrode Materials	Method	Substrate	Electrolyte	Specific Capacitance	Energy Density	Power Density	References
MWCNTs[a]	Dipping	Polyimide films	PVA/H$_2$SO$_4$	—	2.5–10^{-3} Wh·cm^{-3}	5 × 10^{-2}–58 W·cm^{-3}	Yu et al., 2015
	Spray coating	PET film	PEGDA/[EMIM][TFSI]	0.06–0.51 mF·cm^{-2}	3 × 10^{-7}–10^{-8} Wh·cm^{-2}	8 × 10^{-3}–2 mW·cm^{-2}	Kim et al., 2016
	Selective wetting-induced	—	PVA/H$_3$PO$_4$	5.5 F·cm^{-3} at 10 mV·s^{-1}	—	—	Kim et al., 2014
	CVD/transferring	PC film	TMOS/FA/ EMITFSI	430 F·cm^{-2}/86 mF·m^{-3}	0.1–0.5 Wh·cm^{-2}	10 mW·cm^{-2}	Hsia et al., 2014
SWCNTs	Spray coating	PDMS	PMMA/[EMIM][NTf2]	—	20.7 Wh·kg^{-1}	7.4–70.5 kW·kg^{-1}	Kim et al., 2013
MWCNT/ MnO$_x$	LbL/dipping	PET film	PVA/H$_3$PO$_4$	50 F·cm–3 at 10 mV·s^{-1}	4.45 mWh·cm^{-3}	12.3 W·m^{-3}	Lee et al., 2014
PEDOT:PSS/ CNTs	Inkjet printing	Paper	PVA/H$_3$PO$_4$	23.6 F·cm^{-3}	42.1 mWh·cm^{-3}	89.1 mW·cm^{-3}	Liu et al., 2016
SWCNT/ carbon/MnO$_2$	Spin coating/ plasma/ electro-deposition	Plastic film	PVA/H$_3$PO$_4$	20.4 F·cm^{-3} at 20 mV·s^{-1}	—	—	Sun et al., 2016
RGO	Electrophoretic deposition	Plastic film	PVA/H$_3$PO$_4$	359 F·cm^{-3}	31.9 mWh·cm^{-3}	324 W·cm^{-3}	Niu et al., 2013
	Mounting	Paper/ PDMS	PVA/H$_2$SO$_4$	7.6 mF·cm^{-2}	—	—	Weng et al., 2011

TABLE 4.1   (Continued)

Electrode Materials	Method	Substrate	Electrolyte	Specific Capacitance	Energy Density	Power Density	References
	Spin coating/transferring/etching	PET	PVA/H_2SO_4	80.7 mF·cm^{-2} 17.9 F·cm^{-3}	—	—	Wu et al., 2013
	Laser reduction	GO film	H_2O	3.1 F·cm^{-3}	4.3×10–4 Wh·cm^{-3}	1.7 W·cm^{-3}	Gao et al., 2011
Graphene/RGO	Mechanical shaping process	Plastic film	PVA/H_3PO_4	80–394 µF·cm^{-2}	—	—	Yoo et al., 2011
RGO/MWCNTs	Laser reduction	PET film	PVA/H_3PO_4	3.1 F·cm^{-3} at 1 V·s^{-1}	0.84 Wh·cm^{-3}	1 W·cm^{-3}	Wen et al., 2014
RGO/PANI	Spin coating/transferring/etching	Kapton FPC Film	PVA/H_2SO_4	16.55 F·cm^{-3} at 10 mV·s^{-1}	1.51 mWh·cm^{-3}	—	Song et al., 2015
CNTs/RGO	Electrostatic spray deposition	—	Liquid electrolyte	4.4 F·cm^{-3} at 20 mV·s^{-1}	0.68 mWh·cm^{-3}	77 W·cm^{-3}	Beidaghi and Wang, 2012
Photoresist derived carbon	Pyrolysis	Plastic film	0.5 M H_2SO_4	1.7–11 F·cm^{-3}	0.8–1 mWh·cm^{-3}	0.05–56 W·cm^{-3}	Kim et al., 2014
Polyimide derived carbon	Laser direct writing	Plastic film	PVA/H_2SO_4 PVA/H_3PO_4	0.8–19 mF·cm^{-2} at 10 mV·s^{-1}	0.1–1 mWh·cm^{-3}	8–157 mW·cm^{-3}	Cai et al., 2016; In et al., 2015

TABLE 4.1 *(Continued)*

Electrode Materials	Method	Substrate	Electrolyte	Specific Capacitance	Energy Density	Power Density	References
VS$_2$ nanosheets	Mechanical shaping process	Plastic film	PVA/BMIMBF$_4$	394 F·cm^{-3}	—	—	Feng et al., 2011
PEDOT	Electrochemical deposition	Plastic film	PVA/H$_2$SO$_4$	33 F·cm^{-3}	7.7 mWh·cm^{-3}	<80 mW·cm^{-3}	Kurra et al., 2015
PANI	Electrochemical deposition	Plastic film	PVA/H$_2$SO$_4$	67–588 F·cm^{-3} at 0.1 mA·cm^{-2}	5–82 mWh·cm^{-3}	0.45–25 W·cm^{-3}	Hu et al., 2014; Wang et al., 2011

[a] MWCNTs: multi-walled carbon nanotubes.
[b] PEGDA/[EMIM][TFSI]: poly(ethylene glycol) diacrylate/1-ethyl-3-methylimidazolium bis-(trifluoromethylsulfonyl)imide.
[c] TMOS/FA/EMITFSI: Tetramethyl orthosilicate/formic acid/1-ethyl-3-methylimidazolium bis(trifluoromethylsulfonyl)imide.
[d] PMMA/[EMIM][NTf2]: polymethyl methacrylate/1-ethyl-3-methylimidazolium bis(trifluoromethylsulfonyl)imide.
[e] PC: propylene carbonate.

by using lift-off protocol PEDOT/Au interdigited microelectrode finger-like patterning was generated by successive rinsing procedure by acetone and DI water. The schematic view of the successive photolithography technique is shown in Figure 4.7a.

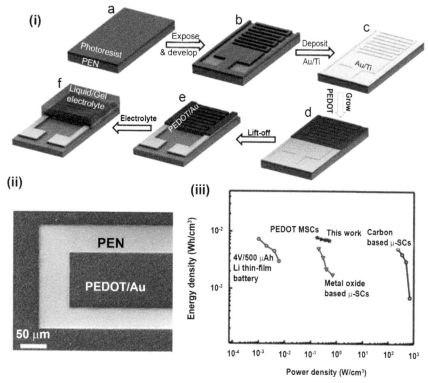

FIGURE 4.7   (i) Schematic of the fabrication of PEDOT-based MSCs by photolithography and electrochemical deposition, (ii) SEM micrographs of PEDOT/Au interdigited finger electrodes, and (iii) Ragoneplot showing the comparison of energy and power density of Li thin-film batteries, carbon and metaloxide based MSCs with respect to PEDOTMSC.
Source: Reprinted with permission from Kurra, Hota, and Alshareef (2015). © Elsevier.

The electrochemical characterizations of PEODT MSCs were done by PVA/H$_2$SO$_4$ gel electrolyte under two electrode configurations. The FESEM image of interdigited PEDOT/Au microelectrode is shown Figure 4.7b. A comparable data on Ragone plot (Figure 4.7c) analysis on PEDOT MSC with respect to Li thin-film batteries and carbon/metal oxide-based MSCs showed that PEDOT MSC exhibits the maximum energy density of 7.7 mW h/cm^3 with a power density of 175 mW/cm^3 rather than other. This energy density value is higher than that of Li thin-film batteries and oxide-based MSCs. Reduced grapheneoxide/Fe_2O_3 hybrid nanostructures was used as an integrated MSC by on-chip fabrication an integration with photo detecting system through a microelectronic photo-lithography process (Gu et al., 2016). The micro-electrode material reduced grapheneoxide/Fe_2O_3 hybrid nanostructures was synthesized by the hydro thermal method under specific growth conditions. The above-mentioned photolithography was used to develop microelectrode on-chip patterning on poly-ethylene terephthalate substrate with Ni current collector that deposited by thermal evaporation technique. The schematic illustration of the successive steps to develop microelectrode integrated patterning is shown in Figure 4.8a. In this integration process, CdS (cadmium sulfides) was used as photo-detector material and nano needle-like CdS structure was synthesized by vapor deposition method. After developing microelectrode patterning CdS NWs was spread at each gap between the successive square electrodes. In that compact on-chip, integrated configuration of CdS nanowire was connected by middle one electrode and on arm of microelectrodes (Figure 4.8d). The 3D schematic diagram of the fabricated on-chip MSCs and the corresponding digital photograph of integrated on-chip MSCs onto the single PET substrate is shown in Figure 4.8b and c respectively. Here the MSC was used as source and drain contact of CdS photodetector, which demonstrated a flexible and self-powered integrated photodetector system in large scale integrated applications.

4.5.2 MSC-SENSOR INTEGRATED SYSTEM

To get an environment-friendly sensor device, MSCs are designed to integrate as a self-powered unit into the same platform with sensor-based material in a large scale application. Now pollution control has taken a great attention to create a pollution-free environment through various kinds of research on gas sensors, UV sensors, mechanically monitoring devices, etc. (Liu et al., 2017). Among of them, the gas sensor was taken considerable attention to control and sense different types of poisonous gasses in the environment. In this context graphene, the based gas sensor was developed to sense the NO_2 gas in rainy environment (Yun et al., 2016). The fabricated graphene gas sensor was powered by MWCNT MSC arrays and the whole integrated system exhibited stable sensing for 50 min with 50% uniaxial stretching.

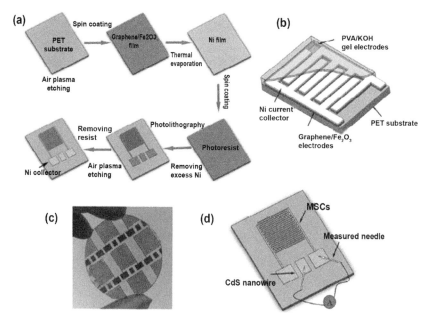

FIGURE 4.8   (a) Schematic illustration of the fabrication process of the flexible on-chip MSCs on a PET substrate, (b) 3D schematic illustration of the fabricated on-chip MSCs, (c) Digital photograph of integrated on-chip MSCs on a single PET substrate and (d) Schematic illustration of electric circuit on integration between MSCs and CdS photodetector.
Source: Reprinted with permission from Gu et al. (2016). © Springer Nature.

To suppress the strain effect of gas sensor a deformable soft Ecoflex substrate was used by a stiff SU-8 platform underneath and electrical inter-connections made by polymer-encapsulated Authin film. The fabricated graphene sensor was then patterned and integrated with MWCNT MSC array on the same substrate (Figure 4.9) through the tape transfer method.

4.5.3   MSC-NANOGENERATOR INTEGRATED SYSTEM

The piezoelectric nanogenerator is an energy harvesting system that converts mechanical energy into electricity. So after integrating the nanogenerator with MSC arrays, the combined self-powered unit readily converts the mechanical energy into electric energy and stored into the system. The fabrication of self-charging micro-supercapacitor power unit (SCMPU) with integrating triboelectric nanogenerator (TENG) and self-powered micro-supercapacitor

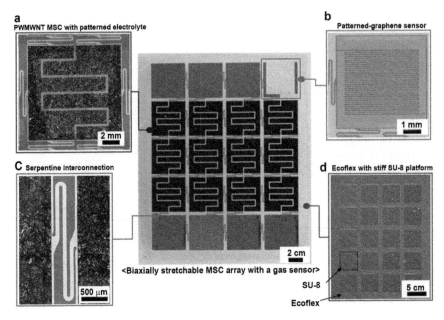

FIGURE 4.9   Design concept of biaxially stretchable patterned-graphene sensor integrated with a PWMWNT MSC array. (Center) Optical image of the fabricated device, (a) Zoomed image of a PWMWNT micro-supercapacitor (MSC) with selectively patterned ion-gel electrolyte, (b) Schematic of the patterned-graphene sensor, (c) Magnified optical image of a serpentine interconnection with polymer-encapsulated Au, and (d) Image of Ecoflex substrate with an embedded stiff SU-8 platform array.
Source: Reprinted with permission from Yun et al. (2016). © Elsevier.

(MSC) arrays was conducted by a simple laser engraving technique (Luo et al., 2015). The whole system is shown in Figure 4.10, where the two sides of the laser-induced graphene (LIG) electrodes are used to design the TENG and MSC array separately. In this system, the MSC unit could be self-powered by mechanical movements and this type of characteristic facilitates the system to couple with other optoelectronic and an electrothermal systems like LED, hygrothermograph, etc. In this system, mechanical energy is converted into electrical energy by applying external stress on piezoelectric material and then the converted electrical energy is stored into the MSC arrays. The PI substrate was taken to fabricate the integrated system into two parts as PI1 and PI2. The top surface of PI1 was used to configure the MSC arrays by four LIG in a series pattern developed by the laser writing method. The bottom surface of PI1 was written by LIG that acts as the top electrode to the integrated unit.

Then poly (tetrafluoroethylene) film was attached on the LIG layer. As the bottom electrode, LIG was written on the single side of PI2 substrate. Finally, both PI1 and PI2 substrate was attached to cover the face to face configuration between PTFE surface and LIG electrode and sealed to form a compact arched structure. The schematic view of the energy converter unit is shown in Figure 4.10. The whole system contains two parts of an MSC array and a patterned TENG unit. These two components were integrated into a single system through the double-faced laser engraving technique. After getting sufficient charging, the SCMPU precedes the stored energy to the energy-supply mode. The circuit diagram of total configuration is shown in Figure 4.10b.

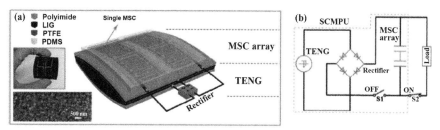

FIGURE 4.10   (a) Schematic depicting the detailed structure of the flexible integrated system of a triboelectric nanogenerator and an MSC array, and (b) Circuit diagram of the energy supply mode.
Source: Reprinted with permission from Luo et al. (2015). © Springer Nature.

4.6   CONCLUSION

The current technological developments on designing of flexible MSCs have been discussed elaborately upon different approaches such as material evolution, process technological evolution, and various integration techniques with other electronics. Graphene, AC, CNTs, conductive polymers, functionalized graphene, and CNT, transition-metal oxides, and some composites are mostly explored as active electrode materials to design flexible MSCs. The electrochemical performance of MSCs depends on intrinsic properties of the active electrode materials, the electrode architecture including the thickness of electrode material, the size of the micro-electrodes and the space between each consecutive electrode into an electrofinger arrays. There are various processing technologies have been developed to fabricate interdigited electrodes through bottom-up and top-down approaches with various structural orientation. The architecture of the interdigited micro-electrodes can be precisely controlled through various well-developed microfabrication techniques. The electrode geometry has a great impact on MSC performances and so forth the smaller size of microelectrodes with micro-scale spacing between the electrodes provides a shorten ion diffusion path length as well as increase the effective surface area of the electrode material. So micro-geometry of electrode material leads to execute better performance than conventional SCs owing to their high value of specific capacitance, columbic efficiency, charge transfer rate, and knee frequency. However, the conductivity of thin film based MSC is limited due to the very small thickness of electrode material in order to nanoscale. In this context gold interdigited patterning as a current collector is essentially required to achieve the desired conductivity of MSCs. Additionally, the lower amount/area of micro-electrodes yield lower areal energy density than conventional SCs.

However, the fabrication processes are getting massive attention in research to design flexible MSCs with excellent electrochemical performance to the requirements of industry electronics. So to feed desired energy and power density to the coupled electronic system, the MSCs must be optimized to match the integrated electronic component. The flexible MSCs also lead power and energy consumption by operating as fully self-powered electric unit upon integration flexible electric devices. However, the reported flexible MSCs other electronic devices are generally operated independently, leading to space and energy consumption from their external connection system. Importantly the miniaturized geometry of MSCs offers space consumption to any large scale integrated system. Yet there is a great challenge in research to fabricate and design flexible MSCs by considering the effect of external forces. However, theoretical simulation is not well established to design flexible MSCs under different bending modes. Still, the in-situ characterization during the electrochemical performance is lacking to understand the device performance during structural deformation. Owing to the lack of both theoretical simulation and in situ characterization during electrochemical operation, to fabricate reliable MSCs integrated system is still under progress. In brief, more attention should be paid to fabricate and design a flexible micro-supercapacitor with the good electrochemical performance during stretching, bending, and release processes, and as a flexible unit, MSCs can be easily integrated to serve as a desired power supply unit under coupling to any stretchable electronic system.

KEYWORDS

- **design and integration**
- **electric double layer capacitor**
- **fabrication**
- **light-emitting diode**
- **micro-supercapacitor**
- **transition-metal dichalcogenides**

REFERENCES

Augustyn, V., Simon, P., & Dunn, B., (2014). Pseudo capacitive oxide materials for high-rate electrochemical energy storage. *Energy Environ. Sci., 7,* 1597–1614.

Beidaghi, M., & Gogotsi, Y., (2014). Capacitive energy storage in micro-scale devices: Recent advances in design and fabrication of microsupercapacitors. *Energy Environ. Sci., 7,* 867–884.

Beidaghi, M., & Wang, C., (2012). Micro-super capacitors based on inter-digital electrodes of reduced graphene oxide and carbon nanotube composites with ultrahigh power handling performance. *Adv. Funct. Mater., 22,* 4501–4510.

Cai, J., & Watanabe, A., (2016). Flexible carbon micro-super capacitors prepared by direct CW-laser writing. In: *Proc. SPIE.*

Cai, J., Lv, C., & Watanabe, A., (2016). Cost-effective fabrication of high-performance flexible all-solid-state carbon microsupercapacitors by blue-violet laser direct writing and further surface treatment. *J. Mater. Chem. A, 4,* 1671–1679.

Cao, L., Yang, S., Gao, W., Liu, Z., Gong, Y., Ma, L., Shi, G., et al., (2013). Direct laser-patterned micro-super capacitors from paintable MoS_2 films. *Small, 9,* 2905–2910.

Chen, J., Jia, C., & Wan, Z., (2014). The preparation and electrochemical properties of MnO_2/poly(3,4-ethylenedioxythiophene)/multiwalled carbon nanotubes hybrid nanocomposite and its application in a novel flexible micro-super capacitor. *Electrochim. Acta, 121,* 49–56.

Chia, X., Eng, A. Y. S., Ambrosi, A., Tan, S. M., & Pumera, M., (2015). Electrochemistry of nanostructured layered transition-metal dichalcogenides. *Chem. Rev., 115,* 11941–11966.

Chmiola, J., Largeot, C., Taberna, P. L., Simon, P., & Gogotsi, Y., (2010). Monolithic carbide-derived carbon films for micro-super capacitors. *Science, 328,* 480–483.

Conway, B. E., (1991). Transition from "supercapacitor" to "battery" behavior in electrochemical energy storage. *J. Electrochem. Soc., 138,* 1539–1548.

Conway, B. E., (2013). *No Ti Electrochemical Super Capacitors: Scientific Fundamentals and Technological Applications.* Springer Science and Business Media.tle.

Cote, L. J., Cruz-Silva, R., & Huang, J., (2009). Flash reduction and patterning of graphite oxide and its polymer composite. *J. Am. Chem. Soc., 131,* 11027–11032.

Du, H., Jiao, L., Wang, Q., Yang, J., Guo, L., Si, Y., Wang, Y., & Yuan, H., (2013). Facile carbonaceous microsphere templated synthesis of Co_3O_4 hollow spheres and their electrochemical performance in supercapacitors. *Nano Res., 6,* 87–98.

Dua, V., Surwade, S. P., Ammu, S., Agnihotra, S. R., Jain, S., Roberts, K. E., Park, S., Ruoff, R. S., & Manohar, S. K., (2010). All-organic vapor sensor using inkjet-printed reduced graphene oxide. *Angew. Chemie Int. Ed., 49*, 2154–2157.

Dubal, D. P., Kim, J. G., Kim, Y., Holze, R., Lokhande, C. D., & Kim, W. B., (2014). Supercapacitors based on flexible substrates: An overview. *Energy Technol., 2*, 325–341.

El-Kady, M. F., & Kaner, R. B., (2013). Scalable fabrication of high-power graphene microsupercapacitors for flexible and on-chip energy storage. *Nat. Commun., 4*, 1475.

El-Kady, M. F., Ihns, M., Li, M., Hwang, J. Y., Mousavi, M. F., Chaney, L., Lech, A. T., & Kaner, R. B., (2015). Engineering three-dimensional hybrid supercapacitors and microsupercapacitors for high-performance integrated energy storage. *Proc. Natl. Acad. Sci., 112*, 4233–4238.

El-Kady, M. F., Strong, V., Dubin, S., & Kaner, R. B., (2012). Laser scribing of high-performance and flexible graphene-based electrochemical capacitors. *Science, 335*, 1326–1330.

Feng, J., Sun, X., Wu, C., Peng, L., Lin, C., Hu, S., Yang, J., & Xie, Y., (2011). Metallic few-layered VS$_2$ ultrathin nanosheets: High two-dimensional conductivity for in-plane supercapacitors. *J. Am. Chem. Soc., 133*, 17832–17838.

Gao, W., Singh, N., Song, L., Liu, Z., Reddy, A. L. M., Ci, L., Vajtai, R., et al., (2011). Direct laser writing of microsupercapacitors on hydrated graphite oxide films. *Nat. Nanotechnol., 6*, 496.

Gilje, S., Dubin, S., Badakhshan, A., Farrar, J., Danczyk, S. A., & Kaner, R. B., (2010). Photothermal deoxygenation of graphene oxide for patterning and distributed ignition applications. *Adv. Mater., 22*, 419–423.

Gu, S., Lou, Z., Li, L., Chen, Z., Ma, X., & Shen, G., (2016). Fabrication of flexible reduced graphene oxide/Fe$_2$O$_3$ hollow nanospheres based on-chip micro-super capacitors for integrated photodetecting applications. *Nano Res., 9*, 424–434.

Hercule, K. M., Wei, Q., Khan, A. M., Zhao, Y., Tian, X., & Mai, L., (2013). Synergistic effect of hierarchical nanostructured MoO$_2$/Co(OH)$_2$ with largely enhanced pseudo capacitor cyclability. *Nano Lett., 13*, 5685–5691.

Hsia, B., Kim, M. S., Vincent, M., Carraro, C., & Maboudian, R., (2013). Photoresist-derived porous carbon for on-chip micro-super capacitors. *Carbon N. Y., 57*, 395–400.

Hsia, B., Marschewski, J., Wang, S., In, J. B., Carraro, C., Poulikakos, D., Grigoropoulos, C. P., & Maboudian, R., (2014). Highly flexible, all solid-state microsupercapacitors from vertically aligned carbon nanotubes. *Nanotechnology, 25*, 55401.

Hu, H., Zhang, K., Li, S., Ji, S., & Ye, C., (2014). Flexible, in-plane, and all-solid-state microsupercapacitors based on printed interdigital Au/polyaniline network hybrid electrodes on a chip. *J. Mater. Chem. A, 2*, 20916–20922.

Huang, P., Lethien, C., Pinaud, S., Brousse, K., Laloo, R., Turq, V., et al., (2016). On-chip and freestanding elastic carbon films for microsupercapacitors. *Science, 351*, 691–695.

In, J. B., Hsia, B., Yoo, J. H., Hyun, S., Carraro, C., Maboudian, R., & Grigoropoulos, C. P., (2015). Facile fabrication of flexible all solid-state micro-supercapacitor by direct laser writing of porous carbon in polyimide. *Carbon N. Y., 83*, 144–151.

Ivanovskii, A. L., (2012). Graphene-based and graphene-like materials. *Russ. Chem. Rev., 81*, 571–605.

Kim, D., Shin, G., Kang, Y. J., Kim, W., & Ha, J. S., (2013). Fabrication of a stretchable solid-state micro-supercapacitor array. *ACS Nano, 7*, 7975–7982.

Kim, H., Yoon, J., Lee, G., Paik, S., Choi, G., Kim, D., Kim, B. M., Zi, G., & Ha, J. S., (2016). Encapsulated, high-performance, stretchable array of stacked planar micro-super capacitors as waterproof wearable energy storage devices. *ACS Appl. Mater. Interfaces, 8,* 16016–16025.

Kim, M. S., Hsia, B., Carraro, C., & Maboudian, R., (2014). Flexible micro-super capacitors with high energy density from simple transfer of photo resist-derived porous carbon electrodes. *Carbon N. Y., 74,* 163–169.

Kim, S. K., Koo, H. J., Lee, A., & Braun, P. V., (2014). Selective wetting-induced micro-electrode patterning for flexible micro-super capacitors. *Adv. Mater., 26,* 5108–5112.

Kurra, N., Hota, M. K., & Alshareef, H. N., (2015). Conducting polymer microsupercapacitors for flexible energy storage and Ac line-filtering. *Nano Energy, 13,* 500–508.

Lee, G., Kim, D., Kim, D., Oh, S., Yun, J., Kim, J., Lee, S. S., & Ha, J. S., (2015). Fabrication of a stretchable and patchable array of high-performance micro-super capacitors using a non-aqueous solvent based gel electrolyte. *Energy Environ. Sci., 8,* 1764–1774.

Lee, G., Kim, D., Yun, J., Ko, Y., Cho, J., & Ha, J. S., (2014). High-performance all-solid-state flexible micro-supercapacitor arrays with layer-by-layer assembled MWNT/MnOx nanocomposite electrodes. *Nanoscale, 6,* 9655–9664.

Lee, H. M., Lee, H. B., Jung, D. S., Yun, J. Y., Ko, S. H., & Park, S., (2012). Bin, solution processed aluminum paper for flexible electronics. *Langmuir, 28,* 13127–13135.

Lee, S. C., Patil, U. M., Kim, S. J., Ahn, S., Kang, S. W., & Jun, S. C., (2016). All-solid-state flexible asymmetric micro super capacitors based on cobalt hydroxide and reduced graphene oxide electrodes. *RSC Adv., 6,* 43844–43854.

Li, P., Shi, E., Yang, Y., Shang, Y., Peng, Q., Wu, S., Wei, J., et al., (2014). Carbon nanotube-polypyrrole core-shell sponge and its application as highly compressible super capacitor electrode. *Nano Res., 7,* 209–218.

Li, R. Z., Peng, R., Kihm, K. D., Bai, S., Bridges, D., Tumuluri, U., Wu, Z., et al., (2016). High-rate in-plane micro-super capacitors scribed onto photo paper using in situ femtolaser-reduced graphene oxide/Au nanoparticle microelectrodes. *Energy Environ. Sci., 9,* 1458–1467.

Liu, L., Niu, Z., & Chen, J., (2016). Unconventional supercapacitors from nanocarbon-based electrode materials to device configurations. *Chem. Soc. Rev., 45,* 4340–4363.

Liu, L., Niu, Z., & Chen, J., (2017). Design and integration of flexible planar microsupercapacitors. *Nano Res., 10,* 1524–1544.

Liu, L., Niu, Z., Zhang, L., & Chen, X., (2014a). Structural diversity of bulky graphene materials. *Small, 10,* 2200–2214.

Liu, L., Niu, Z., Zhang, L., Zhou, W., Chen, X., & Xie, S., (2014b). Nanostructured graphene composite papers for highly flexible and foldable supercapacitors. *Adv. Mater., 26,* 4855–4862.

Liu, W. W., Feng, Y. Q., Yan, X. B., Chen, J. T., & Xue, Q. J., (2013). Superior microsupercapacitors based on graphene quantum dots. *Adv. Funct. Mater., 23,* 4111–4122.

Liu, W., Lu, C., Li, H., Tay, R. Y., Sun, L., Wang, X., Chow, W. L., et al., (2016). Paper-based all-solid-state flexible micro-super capacitors with ultra-high rate and rapid frequency response capabilities. *J. Mater. Chem. A, 4,* 3754–3764.

Liu, W., Lu, C., Wang, X., Tay, R. Y., & Tay, B. K., (2015). High-performance micro super capacitors based on two-dimensional graphene/manganese dioxide/silver nanowire ternary hybrid film. *ACS Nano, 9,* 1528–1542.

Luo, J., Fan, F. R., Jiang, T., Wang, Z., Tang, W., Zhang, C., Liu, M., Cao, G., & Wang, Z. L., (2015). Integration of microsupercapacitors with triboelectric nanogenerators for a flexible self-charging power unit. *Nano Res., 8*, 3934–3943.

Makino, S., Yamauchi, Y., & Sugimoto, W., (2013). Synthesis of electro-deposited ordered mesoporous RuOx using lyotropic liquid crystal and application toward microsupercapacitors. *J. Power Sources, 227*, 153–160.

Manaf, N. S. A., Bistamam, M. S. A., & Azam, M. A., (2013). Development of high-performance electrochemical capacitor: A systematic review of electrode fabrication technique based on different carbon materials. *ECS J. Solid State Sci. Technol., 2*, M3101–M3119.

Meng, C., Maeng, J., John, S. W. M., & Irazoqui, P. P., (2014). Ultra small integrated 3D microsupercapacitors solve energy storage for miniature devices. *Adv. Energy Mater, 4*, 1301269.

Niu, Z., Liu, L., Zhang, L., & Chen, X., (2014). Porous graphene materials for water remediation. *Small, 10*, 3434–3441.

Niu, Z., Luan, P., Shao, Q., Dong, H., Li, J., Chen, J., Zhao, D., et al., (2012a). A "skeleton/skin" strategy for preparing ultrathin free-standing single-walled carbon nanotube/polyaniline films for high-performance super capacitor electrodes. *Energy Environ. Sci., 5*, 8726–8733.

Niu, Z., Ma, W., Li, J., Dong, H., Ren, Y., Zhao, D., Zhou, W., & Xie, S., (2012b). High-strength laminated copper matrix nanocomposites developed from a single-walled carbon nanotube film with continuous reticulate architecture. *Adv. Funct. Mater, 22*, 5209–5215.

Niu, Z., Zhang, L., Liu, L., Zhu, B., Dong, H., & Chen, X., (2013). All-solid-state flexible ultrathin microsupercapacitors based on graphene. *Adv. Mater, 25*, 4035–4042.

Patrice, S., & Yury, G., (2010). Charge storage mechanism in nanoporous carbons and its consequence for electrical double layer capacitors. *Philos. Trans. R. Soc. A Math. Phys. Eng. Sci., 368*, 3457–3467.

Pech, D., Brunet, M., Taberna, P. L., Simon, P., Fabre, N., Mesnilgrente, F., Conédéra, V., & Durou, H., (2010). Elaboration of a microstructured inkjet-printed carbon electrochemical capacitor. *J. Power Sources, 195*, 1266–1269.

Peng, X., Peng, L., Wu, C., & Xie, Y., (2014). Two dimensional nanomaterials for flexible supercapacitors. *Chem. Soc. Rev., 43*, 3303–3323.

Qi, D., Liu, Y., Liu, Z., Zhang, L., & Chen, X., (2017). Design of architectures and materials in in-plane microsupercapacitors: Current status and future challenges. *Adv. Mater., 29*, 1602802.

Raccichini, R., Varzi, A., Passerini, S., & Scrosati, B., (2014). The role of graphene for electrochemical energy storage. *Nat. Mater., 14*, 271.

Shao, Y., El-Kady, M. F., Wang, L. J., Zhang, Q., Li, Y., Wang, H., Mousavi, M. F., & Kaner, R. B., (2015). Graphene-based materials for flexible supercapacitors. *Chem. Soc. Rev., 44*, 3639–3665.

Shen, C., Wang, X., Li, S., Wang, J., Zhang, W., & Kang, F., (2013). A high-energy-density micro supercapacitor of asymmetric MnO$_2$-carbon configuration by using micro-fabrication technologies. *J. Power Sources, 234*, 302–309.

Simon, P., & Gogotsi, Y., (2009). Materials for electrochemical capacitors. In: *Nanoscience and Technology* (pp. 320–329). Co-Published with Macmillan Publishers Ltd, UK.

Song, B., Li, L., Lin, Z., Wu, Z. K., Moon, K., & Wong, C. P., (2015). Water-dispersible graphene/polyaniline composites for flexible microsupercapacitors with high energy densities. *Nano Energy, 16*, 470–478.

Sun, L., Wang, X., Zhang, K., Zou, J., & Zhang, Q., (2016). Metal-free SWNT/carbon/MnO$_2$ hybrid electrode for high-performance coplanar microsupercapacitors. *Nano Energy, 22,* 11–18.

Tyagi, A., Tripathi, K. M., & Gupta, R. K., (2015). Recent progress in micro-scale energy storage devices and future aspects. *J. Mater. Chem. A, 3,* 22507–22541.

Wang, K., Zou, W., Quan, B., Yu, A., Wu, H., Jiang, P., & Wei, Z., (2011). An all-solid-state flexible micro-supercapacitor on a chip. *Adv. Energy Mater, 1,* 1068–1072.

Wang, S., Hsia, B., Carraro, C., & Maboudian, R., (2014). High-performance all solid-state micro-supercapacitor based on patterned photoresist-derived porous carbon electrodes and an ionogel electrolyte. *J. Mater. Chem. A, 2,* 7997–8002.

Wang, X., & Shi, G., (2015). Flexible graphene devices related to energy conversion and storage. *Energy Environ. Sci., 8,* 790–823.

Wei, Z., Wang, D., Kim, S., Kim, S. Y., Hu, Y., Yakes, M. K., Laracuente, A. R., et al., (2010). Nanoscale tunable reduction of graphene oxide for graphene electronics. *Science, 328,* 1373–1376.

Wen, F., Hao, C., Xiang, J., Wang, L., Hou, H., Su, Z., Hu, W., & Liu, Z., (2014). Enhanced laser scribed flexible graphene-based micro-supercapacitor performance with reduction of carbon nanotubes diameter. *Carbon N.Y., 75,* 236–243.

Weng, Z., Su, Y., Wang, D. W., Li, F., Du, J., & Cheng, H. M., (2011). Graphene-cellulose paper flexible supercapacitors. *Adv. Energy Mater, 1,* 917–922.

Wu, Z. S., Parvez, K., Feng, X., & Müllen, K., (2014a.) Photolithographic fabrication of high-performance all-solid-state graphene-based planar microsupercapacitors with different interdigital fingers. *J. Mater. Chem. A, 2,* 8288–8293.

Wu, Z. S., Parvez, K., Winter, A., Vieker, H., Liu, X., Han, S., Turchanin, A., Feng, X., & Müllen, K., (2014b). Layer-by-layer assembled heteroatom-doped graphene films with ultrahigh volumetric capacitance and rate capability for micro-super capacitors. *Adv. Mater., 26,* 4552–4558.

Wu, Z., Parvez, K., Feng, X., & Müllen, K., (2013). Graphene-based in-plane microsupercapacitors with high power and energy densities. *Nat. Commun., 4,* 2487.

Xu, Y., Hennig, I., Freyberg, D., James, S. A., Georg, S. M., Weitz, T., & Chih-Pei, C. K., (2014). Inkjet-printed energy storage device using graphene/polyaniline inks. *J. Power Sources, 248,* 483–488.

Xue, M., Xie, Z., Zhang, L., Ma, X., Wu, X., Guo, Y., Song, W., Li, Z., & Cao, T., (2011). Micro fluidic etching for fabrication of flexible and all-solid-state micro super capacitor based on MnO$_2$ nanoparticles. *Nanoscale, 3,* 2703–2708.

Yan, H., Chen, Z., Zheng, Y., Newman, C., Quinn, J. R., Dötz, F., Kastler, M., & Facchetti, A., (2009). A high-mobility electron-transporting polymer for printed transistors. *Nature, 457,* 679.

Yan, J., Wang, Q., Wei, T., & Fan, Z., (2014). Recent advances in design and fabrication of electrochemical supercapacitors with high energy densities. *Adv. Energy Mater, 4,* 1300816.

Yoo, J. J., Balakrishnan, K., Huang, J., Meunier, V., Sumpter, B. G., Srivastava, A., et al., (2011). Ultrathin planar graphene supercapacitors. *Nano Lett., 11,* 1423–1427.

Yu, G., Xie, X., Pan, L., Bao, Z., & Cui, Y., (2013). Hybrid nanostructured materials for high-performance electrochemical capacitors. *Nano Energy, 2,* 213–234.

Yu, Y., Zhang, J., Wu, X., & Zhu, Z., (2015). Facile ion-exchange synthesis of silver films as flexible current collectors for microsupercapacitors. *J. Mater. Chem. A, 3,* 21009–21015.

Yun, J., Lim, Y., Jang, G. N., Kim, D., Lee, S. J., Park, H., Hong, S. Y., et al., (2016). Stretchable patterned graphene gas sensor driven by integrated micro-supercapacitor array. *Nano Energy, 19*, 401–414.

Zhang, L. L., & Zhao, X. S., (2009). Carbon-based materials as super capacitor electrodes. *Chem. Soc. Rev., 38*, 2520–2531.

Zhang, Y., Feng, H., Wu, X., Wang, L., Zhang, A., Xia, T., Dong, H., Li, X., & Zhang, L., (2009). Progress of electrochemical capacitor electrode materials: A review. *Int. J. Hydrogen Energy, 34*, 4889–4899.

CHAPTER 5

Catalytic Activities of Carbon-Based Nanostructures for Electrochemical CO_2 Reduction: A Density Functional Approach

MIHIR RANJAN SAHOO

School of Basic Sciences, Indian Institute of Technology, Bhubaneswar, Odisha, India, E-mail: mrs10@iitbbs.ac.in

ABSTRACT

The level of carbon dioxide (CO_2), the prime greenhouse gas is rising rapidly in our atmosphere which causes undesirable global warming and climate changes which is very difficult to control due to faster growth of human society. For the sustainable growth, removal of excess CO_2 from the atmosphere in a potential and efficient way is mostly required. The electrochemical CO_2 reduction (ECR) provides a promising solution which can tackle both environmental and energy issue by reducing CO_2 into energy-rich fuels and necessary industrial chemicals. Due to unique electrical conductivity, low-cost, high stability, and plentiful active sites carbon-based nanostructures have gained remarkable attention to be used as catalysts for electrochemical reduction processes. This chapter basically focuses on the study of catalytic activities nano-carbon materials such as graphene and nanotubes through first-principles density functional theory (DFT). Along with the theoretical approach to study the mechanism behind ECR for different catalysts, the reaction pathway towards different end products is clearly discussed. This review will provide a clear insight towards the further improvement of catalysts for better activity.

5.1   INTRODUCTION

No life form can have a sustainable growth without a suitable energy resource. If the sources of energy would get depleted, then not only the economy will heavily disrupt, but also there will be severe undesired consequences on public health, infrastructural growth, and social order. Since renewable energies are the only way to stabilize the energy surge that is inevitable in a few years to come, hence the storage and conversion of energy from various renewable sources have been prioritized the world over. Fossil fuels, such as coal, oil, and natural gas, are known as traditional or non-renewable energy resources, and are available in limited quantities considering their depleting reserves rapidly fueled by the high rate of consumption both in industrial and domestic activities. Combustion of these fuels produces several air pollutants, which play a major role in environmental pollution, global warming, and jeopardizing public health. Not surprisingly, the development of alternative, renewable, abundant, eco-friendly, and efficient energy sources has generated a great research interest within the scientific community recently. Renewable energy sources like sun, wind, and water have enormous potential to be implemented as alternatives, as they don't release hazardous greenhouse gas to the atmosphere. However, in a fast-growing populated world, these renewable sources cannot supply energy at a large scale because of the difficulties and high cost associated with the storage and transportation of the generated power from these intermittent sources. The generated power from these sources is strongly dependent on environmental conditions, which could vary rapidly with time rendering it highly inconsistent with the power demand of the society (Lu and Jiao, 2016). One of the effective ways to tackle this problem of inconsistency between the supply and demand of energy is to store the surplus energy and utilize it when needed. To make practical use of renewable energy resources, electrochemical processes are the most effective methods in which this intermittent electrical energy can be stored in the form of chemical energy in energy-rich materials (Liu, Bai, Zhang, and Peng, 2018). The main purpose this technology is to harvest clean energy which is controlled by a series of electro-catalytic processes such as hydrogen evolution reactions (HER), oxygen evolution reactions (OER), oxygen reduction reactions (ORR), carbon dioxide reduction reactions (CRR), and hydrogen oxidation reactions (HOR).

Most of the primary sources of energy cannot be used directly for various purposes, and need to be converted to secondary sources, which will eventually be stored through storage devices and further exploited through conversion devices (Balogun et al., 2017). Recently developed storage technologies such as electrochemical capacitors (ECs), and metal-ion (e.g., sodium-ion) batteries (Slater, Kim, Lee, and Johnson, 2013; Yabuuchi, Kubota, Dahbi, and Komaba, 2014), including already established lithium-ion technology (Balogun et al., 2016; Etacheri, Marom, Elazari, Salitra, and Aurbach, 2011), are the most prominent energy storage devices that are currently available. These devices possess higher energy and power densities and longer life span in contrast to traditional storage technologies (Balogun et al., 2016; Zhu, Yang, Chao, Mai, and Fan, 2015), and have drawn significant attention due to their effectiveness in portable electronic devices, electrical vehicles and grid-scale energy storage (Liu, Li, Pasta, and Cui, 2014). To develop a next-generation advanced energy conversion device, a solid understanding of underlying theory, and fundamental principles of electro-catalysis is highly essential. Generally, the electrochemical reactions are accelerated on the surfaces of electrodes through heterogeneous electro-catalysis processes (Jin et al., 2018). Using renewable energy as an input, hydrocarbon electro-fuel (Lu and Jiao, 2016) and essential industrial chemicals can be produced by converting carbon dioxide and water (Kuhl et al., 2014). The liquid fuels, produced by the electrochemical process, are called electro-fuel. In an electrochemical reactor, electricity is generated from renewable sources, which then catalyzes the process of production of electro-fuel from CO_2 and H_2O. The overall process is shown as a schematic in Figure 5.1. This is one of the most efficient ways to store energy in the form of organic chemicals. The storage, transportation, and distribution of electro-fuel is quite easy and also economically viable (Lu and Jiao, 2016).

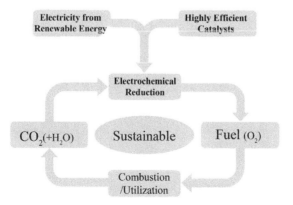

FIGURE 5.1 Schematic showing conversion of CO_2 into electro-fuel.
Source: Reprinted with permission from Gao, Cai, Wang, and Bao (2017). © Elsevier.

Primarily, there are two energy cycles such as the water cycle and the carbon cycle that are observed during the electro-catalysis processes, are required for the conversion and storage of energy. The water cycle consists of a series of electrochemical processes, where hydrogen and oxygen are associated and play a major role in clean energy conversion technologies (Jiao, Zheng, Jaroniec, and Qiao, 2015; Xu, Kraft, and Xu, 2016; Zou and Zhang, 2015). In a hydrogen-oxygen fuel cell, ORR occurs at the cathode, whereas HOR takes place at the anode. But, in an electrolytic cell, HER, and OER occur at cathode and anode respectively (Jiao et al., 2015) as shown in Figure 5.2. As a result, hydrogen, and oxygen molecules are produced in the electrolytic cell at cathode and anode, respectively. So, in the whole water cycle, splitting of water is carried out through HER and OER to produce fuels (hydrogen and oxygen), and then power is generated from the fuel through HOR and ORR in a fuel cell (Zheng, Liu, Liang, Jaroniec, and Qiao, 2012). In the carbon cycle, CO_2 reduction reaction is the central part in which CO_2 can be reduced electrochemically and made available for energy applications. Hence, to establish sustainable and effective clean energy infrastructures, these devices should be the vital components and a subject of both extensive fundamental and application-oriented researches in the field of renewable energy technology (Jiao et al., 2015).

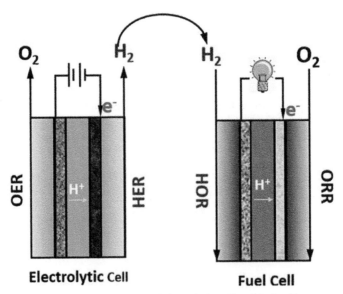

FIGURE 5.2   Schematic representations of electrolytic cell and fuel cell.
Source: Reprinted with permission from Jin et al. (2018). © American Chemical Society.

To enhance the efficiency of the aforementioned energy devices, several active and advanced electrode materials should be developed so that the electrochemical reactions to be catalyzed on the surfaces of those electrodes should have lower overpotentials and higher current densities, for better conversion efficiency (Balogun et al., 2017). There are some highly expensive metals like Pt, Ir, and Ru, which are the state-of-art catalysts for the above-described reactions (Gasteiger, Kocha, Sompalli, and Wagner, 2005; Jin et al., 2018). Due to their high cost and poor stability, applications in large-scale is not possible. Thus the development of high-performance and cost-effective material is necessary to replace these precious metals. Materials with micrometer dimension encounter performance limitations as compared to their bulk counterparts, and can hardly fulfill the ever-growing consumer needs (Leite, 2009). Therefore, new materials with smaller size and better performance are critical for the development of new and advanced technologies (Guo, Hu, and Wan, 2008). Most of the research focus is therefore invested in nanostructures and nanomaterials for their extraordinary physical, electronic, and electrical characteristics, as compared to their bulk counterparts. Reduced dimension, in a specific direction, brings significant change to the properties of these nanostructures, which make them suitable for different technical applications. When, the size of a semiconductor is reduced along a specific dimension, its bandgap increases due to the quantum confinement effect. Change in dimension, therefore leads to the change in the band gap, which can provide control over the optical properties like absorption and emission spectra of a material, for applications in different technological requirements (Alivisatos, 1996; Zhang, Uchaker, Candelaria, and Cao, 2013). Since the number of defects in the nanostructures is very less, their mechanical strength is much higher compared to the bulk structures. Their smaller size offers many advantages in new technologies (Endo, Strano, and Ajayan, 2007).

For electrochemical energy conversions and storage, nanomaterials offer many advantages than other bulk or micro-meter sized materials. The physical and chemical interactions involved in energy storage and conversion process occur primarily at the surface/interfaces of nanostructures. Hence, surface energy and specific surface area play a major role in these processes (Zhang et al., 2013). Two types of size-dependent effects of nanomaterials have been found to be of use in energy devices. First, one is called the trivial size effect in which the surface to volume ratio is increased. The second one is known as the true size effect in which the local material properties change (Guo et al., 2008).

5.2   ELECTROCHEMICAL CO_2 REDUCTION (ECR)

The level of carbon dioxide, a major greenhouse gas, is increasing rapidly in the atmosphere day-by-day through various human activities and contributes significantly towards air pollution and global warming. The rise in the concentration of CO_2 in the atmosphere is difficult to control, as there is a rapid development and growth of human society, which puts both the environment and the health in a compromised state. Therefore, potential, and efficient way to remove the excess CO_2 from the atmosphere is highly necessary for the sustainable growth. There have been several attempts by the researchers to remove the excess amount of CO_2 and utilize it in a proper way for the benefit of our society. Since CO_2 is nontoxic, abundant, and inexpensive, it can readily be used as a building block to produce hydrocarbon fuels and organic chemicals, materials, and carbohydrates (Ma et al., 2009; Wang, Wang, Ma, and Gong, 2011). Though the process of CO_2 reduction is a time-consuming task, it is basically followed by two steps, i.e., capturing CO_2 and transforming or reducing it into desired organic chemicals.

There are primarily three ways to reduce CO_2, which include electrochemical, photochemical, and thermochemical methods. For thermochemical CO_2 reduction process, high reaction temperature and pressure is required along with large amount hydrogen as reducing agent, which would make the process less efficient for large scale energy system. Photochemical CO_2 reduction reaction is economically challenging due to the low rate of production and selectivity (Lu, Rosen, and Jiao, 2015). On the other hand, electrochemical CO_2 reduction (ECR) is most efficient because the reaction can be operated in ambient conditions and controlling the rate of reaction can be easily done by tuning the electrode potential and reaction temperature (Qiao, Liu, Hong, and Zhang, 2014; Wang, Li, and Yamauchi, 2016). In general, ECR gains much more attraction because, (1) it uses atmospheric excess carbon dioxide, (2) the electrolytes which are used in the process can be recycled, leading to much less consumption of chemicals, and is environmentally benign, since water is formed as the by-product, (3) provides a much-needed solution for energy storage with high energy density using renewable sources like solar, wind, hydroelectric, geothermal, tidal, etc., without the regeneration of CO_2, (4) can provide industrial chemicals and organic petrochemical feedstocks, and (5) because the electrochemical cells are transportable, compact, and can be designed to the convenient shape (Qiao et al., 2014; Vasileff, Zheng, and Qiao, 2017). These specific products obtained includes low carbon chemical fuels including carbon monoxide (CO), formic acid (HCOOH), methanol

(CH_3OH), ethylene (C_2H_4), and methane (CH_4) (Centi and Perathoner, 2009; Qiao et al., 2014), which makes CO_2 suitable for potential energy harvesting applications.

An electrochemical reactor or electrolyzer, where CO_2 reduction takes place, consists of two electrodes, i.e., anode, and cathode, separated by a membrane, which allows ions to pass through it. Both the electrodes are placed in individual chambers as shown in Figure 5.3. An external electric potential is provided to the electrochemical cell containing dissolved CO_2 in the aqueous solution, because this neutral or slightly alkaline solution provides a better environment for the formation of formate (HCO_2^-) ion, a reaction intermediate in the hydrogenation of carbon dioxide reaction pathway (Simakov, 2017). At anode, electrolytic dissociation of water occurs in which molecular oxygen is formed that is capable of supplying protons which move towards the cathode through proton exchange membrane (PEM), whereas CO_2 reduction occurs at the cathode to give hydrocarbon liquid fuels. The products obtained at the two electrodes during the electrochemical process can be easily separated at different individual reaction chamber and thus reduces the cost of separation after the reaction (Lu and Jiao, 2016). An external potential is required to drive this thermodynamically uphill reaction. Multi electron reactions associated with protons are energetically more favorable than the single electron associated reduction reaction, because products obtained in the proton-coupled multi electron transfer (ET) processes are thermodynamically more stable. The thermodynamic potential to drive single-electron reduction of CO_2 to CO_2^- is $E_0 = -1.90$ V vs. SHE (standard hydrogen electrode, 7 pH aqueous solution, 25°C and 1 atm pressure), which indicates that the reaction process is energetically unfavorable presenting a large kinetic barrier. This high value of potential is due to the reorganization energy required to convert a linear molecule to a bent radical ion (Benson, Kubiak, Sathrum, and Smieja, 2009). On the other hand, potential in the range of only- 0.2 to 0.6 V vs. SHE is required in multi-electron transfer reduction processes. Eqns. (5.1) to (5.6) summarize the reduction reactions that occur at the cathode along with their potential values (with respect to SHE), whereas Eqn. (5.7) represents a reduction reaction at the anode.

$$CO_2 + 8H^+ + 8e^- \rightarrow CH_4 + 2H_2O, E_0 = -0.24V \qquad (5.1)$$

$$CO_2 + 4H^+ + 4e^- \rightarrow HCHO + H_2O, E_0 = -0.48V \qquad (5.2)$$

$$CO_2 + 2H^+ + 2e^- \rightarrow CO + H_2O, \ E_0 = -0.53V \tag{5.3}$$

$$CO_2 + 2H^+ + 2e^- \rightarrow HCO_2H, E_0 = -0.61V \tag{5.4}$$

$$CO_2 + 6H^+ + 6e^- \rightarrow CH_3OH + H_2O, \ E_0 = -0.38V \tag{5.5}$$

$$CO_2 + e^- \rightarrow CO_2^-, E_0 = -1.90V \tag{5.6}$$

$$2H^+ + 2e^- \rightarrow H_2 \ E_0 = -0.41V \tag{5.7}$$

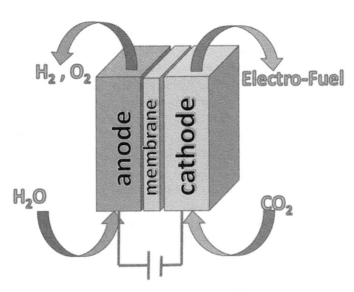

FIGURE 5.3   A schematic diagram of ion-conducting membrane-based CO_2 electrolyzer.

As CO_2 is a very stable molecule, converting CO_2 into organic fuel is quite challenging in which extremely high-quality catalysis is required to get specific target chemicals (Kondratenko, Mul, Baltrusaitis, Larrazábal, and Pérez-Ramírez, 2013). Since the proton-coupled multi-electron pathways lead to a wide range of organic products by converting CO_2, it is quite difficult to produce, selectively, one desired chemical, because all the reaction pathways have almost similar redox potential values (Lu and Jiao, 2016). There is another issue that arises due to the use of an aqueous electrolyte in the electrolyzer, as there is only a slight difference between the reduction potential of CO_2 cathode, and the reduction potential of water at the anode. The reduction of water occurs at a potential of $E_0 = -0.41$ V (Eqn. (5.7)) with the evolution of hydrogen gas. So there is a clear competition in the electrolyzer between the two reduction mechanisms, i.e., H_2 generation through water electrolysis and CO_2 reduction into desired chemicals like methane, methanol, or formic acid (FA) (Simakov, 2017). Also, ECR reaction begins with the adsorption of CO_2 molecule at the electrode-electrolyte interface. Furthermore, the low solubility index of CO_2 in aqueous solutions causes a constraint in diffusion and mass transfer, which slows down the reaction rate (Li and Oloman, 2005; Vasileff et al., 2017). Hence, the electrocatalyst which helps in obtaining the selective desired product, suppressing the production of undesired chemicals, and preventing the side reactions, is highly necessary for designing a successful and efficient electrochemical cell, for the production of electro-fuel through CO_2 reduction.

5.3   CATALYSTS FOR ELECTROCHEMICAL CO_2 REDUCTION (ECR)

As discussed earlier, highly efficient electrocatalyst should be developed, which could enhance the electrochemical reduction of CO_2 to form electro-fuels. At the electrode surface, the ET reaction should take place in the presence of an electrocatalyst, which also accelerates the electrochemical reaction on the electrode surface (Benson et al., 2009; Simakov, 2017). For an efficient catalyst, both the reactions must be accelerated simultaneously. The redox potentials, (E_0), of both the processes, should be matched thermodynamically to get an optimal electrocatalyst. The ideal catalyst should have a few desired characteristics, which could help to reduce CO_2 even under ambient conditions (Vasileff et al., 2017). The catalysts should show low activity towards the competing HER and high activity towards CO_2 reduction reaction. High selectivity is another important characteristic in which desired chemicals should be obtained at the end of the reaction, suppressing the production of other possible chemicals. Besides these, the catalysts must be strong enough to be stable for a longer period and should be of non-precious origin, which would make the cell suitable for practical use and commercialization.

Transition metals and their complexes, in both bulk and nanostructured forms, are the most commonly used catalysts for CO_2 reduction. The active d-orbital electrons in these metals play a critical role in enhancing the binding energy between CO_2 and the metal electrode (Qiao et al., 2014). These metal catalysts are classified into three groups based on the nature of the output products. They are: (1) FA generating metals (Sn, Hg, Pb, In, Bi, Tl, Cd), (2) CO generating metals (Ag, Au, Zn), and (3) Cu which is a special catalyst to produce a wide range of hydrocarbons including CO, ethylene, FA and methanol through ECR (Jones, Prakash, and Olah, 2014; Lu et al., 2015; Vasileff et al., 2017). From the previous experimental investigations by Hori et al. significant production (current efficiency ~ 70%) of hydrocarbons (CH_4, C_2H_4) and alcohol (CH_3OH) can be obtained from the reduction of CO_2 in the presence of Cu based electrodes as compared to other metal electrodes described above (Hori et al., 2005; Yoshio Hori, Koga, Yamazaki, and Matsuo, 1995; Yoshio Hori, Murata, and Takahashi, 1989; Yoshio Hori, Takahashi, Koga, and Hoshi, 2002; Hussain, Skúlason, and Jónsson, 2015). It has been found that CO_2 reduces to one of the products mentioned previously through CO and CO derived intermediates. The electrolytes that are employed, strongly determines the product distribution. But, Cu has few drawbacks to its credit before being considered as an ideal catalyst for CO_2 reduction: (1) large overpotential of value greater than 1V is required for the reaction with Cu catalyst, (2) it is unstable for a longer period, and (3) has very poor selectivity (Lu et al., 2015; Wang et al., 2016). If the copper surface is implemented in nanoscale range, then HER reaction dominates over CO_2 reduction (Lu et al., 2015). From the previous study by Reske et al. selectivity of H_2 increases with a reduction in the size of Cu nanoparticles (NPs) as shown in Figure 5.4. Hence, low selectivity, less stability, high cost, and less active towards CO make these metals less efficient catalysts towards CO_2 reduction (Vasileff et al., 2017). As per the above discussions, the development of non-metal nano-structured electrocatalysts is highly required for ECR, which can be implemented in practical device applications. In this context, carbon-based electrocatalysts should be discussed first, which are cost-effective, and their activities towards ECR are equal to or better than the precious metals. These catalysts are basically of two types, i.e., metal-free and metal-carbon composites (Vasileff et al., 2017). Although carbon catalysts are preferred for catalyzing the competing HER compared to ECR, active sites for ECR can be formed by doping electronegative heteroatoms dopants (S, P, N, etc.) into the carbon networks leading to change in charge and spin densities of carbon atoms. As a result, an active site is induced, where adsorption of reaction intermediates occurs to run the electrocatalysis efficiently. Since carbon-based nanostructures have the high surface area and high electrical conductivity, these catalysts provide a high density of active sites and provide better catalytic activity.

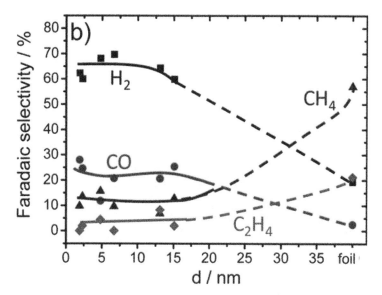

FIGURE 5.4   Selectivity of reaction products during CO_2 reduction reaction using Cu nanoparticles.
Source: Reprinted with permission from Reske et al. (2014). © American Chemical Society.

In summary, the development of catalysts with lower overpotential, low cost, higher selectivity, and higher activity is essential for an efficient ECR. With the help of theoretical calculations, the reaction mechanisms of ECR for various catalysts can be well understood and the prediction of better catalysts for enhancing the reduction can be done. Density functional theory (DFT) (Hohenberg, Kohn, and Sham, 1990; Hohenberg and Kohn, 1964; Kohn and Sham, 1965) calculations have become an authoritative tool for the theoretical study of catalytic processes and also give the mechanistic insights which corroborate the experimental results. In this chapter, we will discuss the reaction pathways for various products obtained from CO_2 reduction in the presence of various catalysts with the help of computational study. The stability of surface intermediates in ECR can be clearly understood on the basis of electronic structure calculations.

5.4   THEORETICAL APPROACH

While the electrochemical processes are in progress, it's difficult to study the details on an atomic scale. Interpretation of experiments occurring at the atomic level is very much important for the selection, design, and application of catalysts. Due to complexity in the reaction processes and very short life span of corresponding reaction intermediates, the selectivity mechanism seems unclear. Experimentally, it is very difficult to predict the active sites and reaction pathways (Chai and Guo, 2016). Catalytic activities can be accurately studied and described through theoretical approaches. From the last few decades, first-principles calculations based on DFT has made significant contributions in providing significant insights regarding the molecular level rearrangements and transfer of electrons and ions from one reactants to another during the electrolysis process (Alfonso, Tafen, and Kauffmann, 2018). Catalytic interfacial properties and their influences in the presence of solvents can be described through the relevant theoretical model (Letchworth-Weaver and Arias, 2012). Electrochemical potential, geometric structures, electronic structures, reaction pathways, and free energies can be computed from the outputs of computational calculations. Also, the electronic structure analyses provide specific idea for the development of optimal catalysts for efficient electrochemical reductions (She et al., 2017). Therefore, computational calculations have gained considerable acceptance over the few years, for studying the reaction mechanisms of various electrocatlaysts and designing the new possible catalysts, which will make the whole process more streamlined. In the following sections, the concept behind the theoretical modeling for electrochemistry will be reviewed. First, we will discuss the computational hydrogen electrode (CHE) method based on free energy calculations proposed by Norskov and his coworkers, which is used widely for study of electrochemical reaction theoretically (Nørskov et al., 2004; Nørskov, Bligaard, Rossmeisl, and Christensen, 2009). Theoretical study of ECR in the presence of various nano-electrodes with reaction pathways and mechanisms will be discussed in subsequent sections.

5.4.1   COMPUTATIONAL HYDROGEN ELECTRODE (CHE) METHOD

The reaction pathway contains a number of reaction intermediate steps in the electrochemical reduction of CO_2 to a final product. The electrochemical electron-proton transfer in each elementary step leads to the change in free energy which depends on the applied electrode potential. To study the change in free energy during the reaction with respect to applied potential, Norskov, and his co-workers have developed a linear free energy method, which is incorporated with first principles as the CHE method (Nørskov et al., 2004). Without calculating explicit free energies of electron-proton pair, their chemical potential can be calculated through this model with the help of energy values calculated from density functional study (Luo, Nie, Janik, and Asthagiri, 2016). From the thermochemical point of view, numerous materials and active sites can be screened through this method (Rendón-Calle, Builes, and Calle-Vallejo, 2018).

The change in free energy is defined as follows:

$$\Delta G = \Delta H - T\Delta S \tag{5.8}$$

which is the core of the reaction mechanism. But, for each elementary step, for individual reaction intermediates, we can rewrite the above equation with component 'i' for a specific step as:

$$G_i = H_i - TS_i \tag{5.9}$$

The enthalpy can be calculated as the sum of electronic energy, solvation energy, zero-point energy, thermal energy, and PV terms of individual step (Alfonso et al., 2018).

$$H_i = E_{el} + E_{sol} + E_{ZPE} + E_{thermal} + PV \tag{5.10}$$

where, electronic energy can be calculated from DFT method. The solvation energy term is due to the interaction of species with solvent, which can be evaluated from the solvent model in which constant dielectric continuum is considered (Mathew, Sundararaman, Letchworth-Weaver, Arias, and Hennig, 2014). The zero-point energy terms include vibrational frequencies, which can be evaluated from DFT by taking harmonic oscillator approximation into account with diagonalization of the Hessian matrix (Alfonso et al., 2018). The thermal energy term and entropy term both consist of translational, rotational, and vibrational components.

For calculation of enthalpy of the gaseous molecule, solvation energy term can be neglected and in place of PV, we can write RT assuming the ideal gas case. When we consider solvated adsorbed species, the PV term can be excluded. Vibrational contributions of thermal energy term and entropy will be considered for the calculation of free energy.

Thus, the free energy for the adsorbate is:

$$G_{ads} = E_{el} + E_{sol} + E_{ZPE} + E_{vib} - TS_{vib} \qquad (5.11)$$

The key feature of this CHE approach is that explicit treatment of solvated protons and electrons is avoided by taking their equilibrium with hydrogen present in the solution (Peterson, Abild-Pedersen, Studt, Rossmeisl, and Nørskov, 2010; Rendón-Calle et al., 2018). This technique is based on the equality in the chemical potential of electron-proton pair with hydrogen molecule in the gaseous phase on reversible hydrogen electrode (RHE). At any pH value and temperature, the following reaction is in equilibrium at 0V-RHE;

$$H^+ + e^- = \frac{1}{2} H_2 \qquad (5.12)$$

Hence, the chemical potential of the electron-proton pair is equal to the half of the chemical potential of hydrogen gas molecule at 0V potential from the following equation:

$$\mu(H^+) + \mu(e^-) = \frac{1}{2}\mu(H_2) \qquad (5.13)$$

Free energies of all the species which are involved in the electrochemical reduction process can be calculated consistently by this method. Hence, from free energy of H_2 molecule, the chemical potential of the electron-proton pair can be calculated. Free energy of proton-electron pair can be calculated with the help of a linear relation between chemical and electrical potential as follows:

$$\Delta G = -eU \qquad (5.14)$$

where, e is the fundamental unit of charge and U is the electrode potential or applied bias on RHE scale. The pH correction is not included, since RHE is defined at 0V and all pH and temperature values. Therefore, the net chemical potential of electron-proton pair at all temperatures and pH values can be written as:

$$\mu(H^+) + \mu(e^-) = \frac{1}{2}\mu(H_2) - eU \qquad (5.15)$$

As an example, let us consider an electro reduction reaction that can be written as:

$$A^* + H^+ + e^- \leftrightarrow AH^* \tag{5.16}$$

Where, the adsorbed species, A*, changes to another adsorbed species, AH*. Here, '*' represents the species in a bound state on the electrode surface. The change in free energy for the above reaction (Eqn. (5.16)) can be calculated from the following equation:

$$\Delta G = G(AH^*) - G(A^*) - \left[\frac{1}{2}G(H_2) - eU\right] \tag{5.17}$$

The potential (U) is therefore explicitly contained in the change in free energy of each intermediate steps of the electrochemical reduction reaction, which is described in the CHE model. Using DFT, the chemical potential of each adsorbed species can be calculated.

5.5 CARBON-BASED NANOSTRUCTURES AS ELECTROCATALYSTS FOR ECR

Due to the ease of their availability and cost-effectiveness, carbon-based nanostructures are considered to be the most suitable catalysts for electrochemical reduction of carbon dioxide. In addition, these carbon-based materials have certain advantages over metal surfaces as electrocatalysts for ECR, because of their chemical inertness at negative potential ranges in both aqueous and non-aqueous media and high overpotential values for the competing HER (Siahrostami et al., 2017; Yang, Waldvogel, and Jiang, 2016). Longer life span, efficient, and tunable surface chemistry by making stronger bonds between carbon and different surface modifiers and better electrochemical activity for a variety of compounds makes them potential electrodes for the electrochemical reduction of CO_2 (Yang, Swain, and Jiang, 2016).

5.5.1 NITROGEN-DOPED GRAPHENE

Graphene, a widely known two-dimensional allotrope of carbon has attracted enormous research interest due to its extraordinary chemical, electronic, electrical, and mechanical properties originating from unique linear dispersive band structure (Geim and Novoselov, 2007; Novoselov, Mishchenko, Carvalho, and Castro Neto, 2016). Its high surface area, high electrical and thermal conductivity, ease of functionalization and scalable production methods makes graphene an ideal candidate for wide range of applications such as nanoelectronics, nanophotonics, sensors, and green energy technologies (Luo et al., 2011; Wang, Shao, Matson, Li, and Lin, 2010). Being a zero-gap semiconductor, graphene shows metallic conductivity even at zero carrier concentration. The absence of bandgap is the only limitation of graphene for direct implementation in practical devices. Therefore, tailoring, and improvement of its electronic properties is highly necessary, which can be achieved by various methods such as doping with heteroatoms, chemical functionalization, confining the geometry into reduced dimensions, i.e., nanoribbons or quantum dots, and placing the monolayer (ML) graphene on various metallic or non-metallic substrates to induce interactions between them (Gao et al., 2012). In general, chemical doping with heteroatoms into the graphene is the most feasible and promising way which modifies the intrinsic material properties, tailors the electronic structure and changes the surface chemistry which are very much essential from the technological point of view (Shao, Sui, Yin, and Gao, 2008; Wang et al., 2010). Also, doping with heteroatoms leads to an increase in the chemical activity of graphene, which may allow further chemical functionalization or modification (Gao et al., 2012). Being situated in the nearest neighborhood of carbon (C) in the periodic table, boron (B) and nitrogen (N) are the most studied dopants for carbon-based structures, acting as p-type and n-type dopant, respectively. Among various dopants, nitrogen is considered to be an excellent element, at least for carbon-based structures, because it has comparable atomic size with carbon and also has five valence electrons that help create strong bonding with carbon. Also, electronegativity of nitrogen is more compared to carbon atoms, and hybridization between the lone pair electrons of N atoms and π-orbitals of graphene makes the C-N bond stronger and tailors the electronic properties of graphene, while preserving its excellent conducting properties (Lherbier, Blase, Niquet, Triozon, and Roche, 2008; Luo et al., 2011). In summary, N-doped graphene may be considered as novel nanomaterials with potential applications in various fields.

In this section, we will systematically discuss the catalytic activities of N-doped graphene for ECR, the mechanism behind it and the reaction pathways towards different end products, with the help of DFT. It has been theoretically studied that a high catalytic activity is shown by the N-doped graphene towards ORRs (Wang et al., 2012; Zhang and Xia, 2011). Doping of N atoms into the graphene matrix induces high spin and charge densities, which enhances the catalytic activities of the doped material. Hence, these materials would be considered as the alternatives for precious metal catalysts such as Pt, for fuel cells (Liu, Zhao, and Cai, 2016). Since this chapter will provide a brief review about the electrochemical reduction of CO_2, we will first focus on the different types of N-doped graphene composites depending upon the doping sites and the amount of nitrogen atoms in the ML graphene sheet. Then, mechanism of various N-doped graphene composites as catalysts for ECR will be overlooked and the possible products obtained at the end of the reaction will be analyzed with the help of theoretical study, which will show a calculative methodology for the successive improvement of non-metal, and carbon-based catalysts with high selectivity towards CO_2 reduction reactions.

Depending on the number and position of nitrogen atoms doped in the graphene lattice, N-doped graphene is broadly classified into three categories such as graphitic N (graphN), pyridinic N (pyriN) and pyrrolic N (pyroN) (Jing and Zhou, 2015; Liu et al., 2016; Luo et al., 2011; Ma, Shao, and Cao, 2012; Yu, 2013). When one carbon atom is replaced by one N atom in the graphene sheet, it is termed as graphitic N (graphN1) as shown in Figure 5.5a. Similarly, double graphitic N (graphN2) can be considered when two N atoms replace two C atoms in the graphene sheet. Since N atom has more electronegativity than C atom, there is a slight change in the bond length in graphN1. In this case, C-N bond length (1.411 Å) is slightly less than that of C-C bond (1.42 Å) in the graphene sheet. When there is a vacancy of three carbon atoms in the graphene sheet, as shown in Figure 5.5 (c) and (d), pyridinic, and pyrrolic configurations can be obtained by placing the N atoms. In the tri N pyridinic (pyriN3) structure, three C atoms of three conjugate hexagons are replaced by three N atoms (Figure 5.5c). But in the pyrrolic N (pyroN3) structure, one N atom is placed in the pentagon instead of the hexagon (Figure 5.5d). The bond length (1.405Å) between N and C present in the pentagon is more than the N-C bond length (1.322Å and 1.340Å) present in the hexagon. Another type of pyridinic structure consisting of four N atoms in the vacancies called tetra N pyridinic (pyriN4) is also taken into account, where N-C bond lengths are 1.329Å and 1.350Å, respectively (Jing and Zhou, 2015).

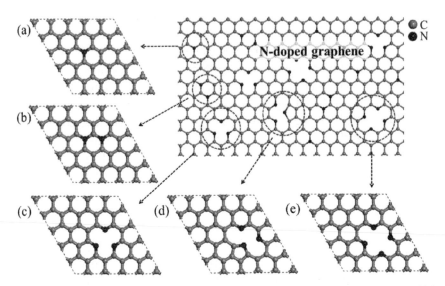

FIGURE 5.5   Different types of N-doped graphenes. (a) graphitic N (graphN1), (b) graphitic N (graphN2), (c) Pyridinic N (pyriN3), (d) Pyrrolic N (PyroN3), (e) Pyridinic N (PyriN4). *Source:* Reprinted with permission from Jing and Zhou (2015).

Now we will consider the above N-doped graphene structure for the study of catalytic activities for ECR. Any material that is to be considered as a catalyst should have good electrochemical stability. Stabilities of the above-described structures can be determined from the calculation of formation energy by the following equation:

$$E_f = E_{NG} - n\mu_C - m\mu_N \tag{5.18}$$

where, E_{NG} is the total energy of N-doped graphene. Here, n and m represent the number of carbon and nitrogen atoms present in the N-doped graphene structure, respectively. The chemical potentials of C and N are denoted by μ_C and μ_N, which can be calculated from pristine graphene and nitrogen in gas phase with the help of DFT.

From the theoretical study done by Liu et al. it is found that the formation energy of graphN1 structure is- 0.83 eV, which is less than that of the other N-doped graphene structures (Liu et al., 2016). Hence, graphN1 structure can be prepared easily than other four configurations. They have taken into consideration the following reactions, where two electrons and two protons steps are involved with reaction pathways shown in Figure 5.6.

$$CO_2 + 2H^+ + 2e^- \rightarrow HCOOH \qquad (5.19)$$

$$CO_2 + 2H^+ + 2e^- \rightarrow CO + H_2O \qquad (5.20)$$

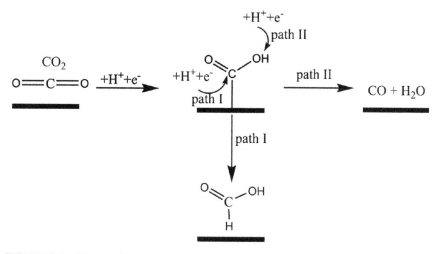

FIGURE 5.6 The reaction pathways of ECR with presence of N-doped graphene producing HCOOH in pathway I and CO in pathway II.
Source: Reprinted with permission from Liu, Zhao, and Cai (2016). © Royal Society of Chemistry.

The first step of the reaction pathway consists of the transfer of one proton and one electron from the solution to the adsorbed CO_2 molecule resulting in the formation of O-H bond to generate intermediate COOH* species attached to the surface of N-doped graphene. The COOH* species obtained will undergo for further hydrogenation through addition of a second proton/electron. There are two possible sites in COOH* species for addition of next proton-electron pair. If the pair ($H^+ + e^-$) is attached to C atom of COOH*, then the reaction will follow the path I and yield HCOOH as the final product. If the pair is attached to the O atom (OH) in COOH*, then the reaction will follow the path II to give CO as the end product.

Here, in the presence of N-doped graphene catalysts, CO_2 reduces to CO or HCOOH. Calculating the adsorption energy of reaction intermediates like HCOOH, COOH, and CO with the N-doped graphene composites at various adsorption sites, catalytic activities of catalysts can be determined. As shown in Figure 5.7, adsorption energy of COOH species with different N-doped graphene (ranging from −1.27 to −3.24 eV) are much higher than the adsorption energy of COOH with pristine graphene (−0.18 eV). Also, the adsorption energies of CO and HCOOH species with the catalysts are −0.14 eV and −0.26 eV, respectively, indicating the weak interactions between product species and N-doped graphene. It clearly explains the capability of holding intermediate COOH very tightly, whereas CO and HCOOH can loose from the surface easily at the same time. So, before going for further reduction by successive electrons, CO, and HCOOH can detach from the surfaces. Hence, doping with N atom in graphene sheet enhances the catalytic activity of graphene in ECR.

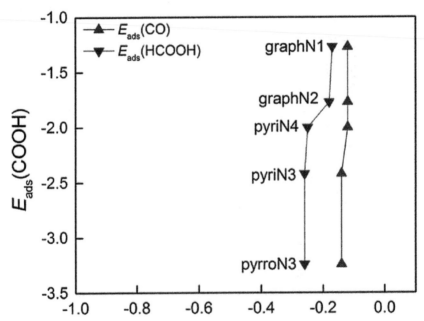

FIGURE 5.7   Adsorption energies of COOH, CO, and HCOOH with different N-doped graphene.
Source: Reprinted with permission from Liu, Zhao, and Cai (2016). © Royal Society of Chemistry.

The first step of the reduction reaction, i.e., hydrogenation of CO_2 to COOH*(CO_2 (g) +H⁺+e⁻ → COOH*) is an uphill process considering the required energies which are of the order of 1.73, 1.03, 0.60, and 1.01 eV in the free energy profile (zero electrode potential, U=0) for graphN1, graphN2, PyriN3, and PyriN4, respectively. But, in the case of pyrrolic N, the reaction is a downhill process, where the required energy is only 0.21 eV. In the second step of the reaction, the reduction reaction is a downhill process requiring energies of 1.50, 0.80, 0.37, and 0.78 eV for path I (COOH* species reduced to HCOOH) for, respectively, while for path II, these values are 1.22, 0.52, 0.09 and 0.50 V respectively for graphN1, graphN2, pyriN3, and pyriN4. In contrast, the second step reduction reaction for pyrrolic N requires 0.44 eV for path I and 0.72 eV for path II, as shown in Figure 5.8. As per the above values of free energies obtained from the theoretical studies done by Liu et al. the first step in the ECR is the rate-limiting step for graphN1, graphN2, pyriN3, and pyriN4, whereas the second step in the ECR is the rate-limiting step for pyrrolic N due to higher free energy profile (Liu et al., 2016). To eliminate the energy barriers of rate-limiting steps, the least negative electrode potentials applied are -1.73, -1.03, -0.60, -1.01, and -0.44 V for graphN1, graphN2, pyriN3, pyriN4 and PyrroN3, respectively. Since equilibrium potential for the reduction of CO_2 to HCOOH is taken as -0.20 V at pH $=0$, then the required overpotential for ECR is 1.53, 0.83, 0.40, 0.81 and 0.24 V for graphN1, graphN2, pyriN3, pyriN4, and pyroN3,

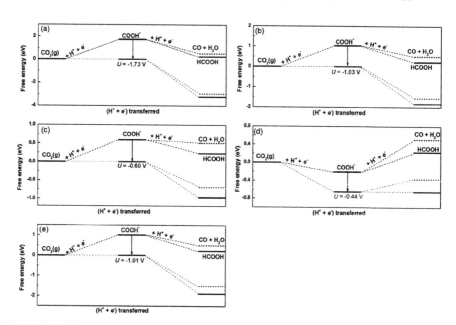

FIGURE 5.8 Free energy diagrams for the electrochemical reduction of CO_2 to HCOOH and CO on different N-doped graphene surfaces: (a) graphN1, (b) graph N2, (c) PyriN3, (d) PyrroN3 and (e) PyriN4. Red and black lines indicate the reaction with and without the applied potential.

Source: Reprinted with permission from Liu, Zhao, and Cai (2016). © Royal Society of Chemistry.

respectively. Hence, pyrrolic N exhibits high catalytic activity than other N-doped graphene surfaces due to low overpotential. Also, at U= −0.44 V, for pyroN3, reduction of COOH* to CO requires 0.28 eV of energy suggesting that path I dominates over path II. For the other four N-doped graphene surfaces, COOH* reduction to both CO and HCOOH are downhill processes indicating that both CO and HCOOH are the final products at the end of ECR.

In summary, due to low overpotential, high adsorption energy towards COOH* and low adsorption energy towards the end product CO and HCOOH, pyrrolic N shows high catalytic activity among all N-doped graphene composites. Since HCOOH is the only product of ECR on pyrrolic N, this indicates the high selectivity of the pyrrolic N towards the production of HCOOH. Theoretical studies on ECR by N-doped graphene suggest that pyrrolic N is the most efficient metal-free catalyst for the ECR reaction with high selectivity.

5.5.2   METAL DIMER DOPED GRAPHENE

Band structure, carrier concentrations, or local magnetic moments of graphene can be modified by doping it with impurities like metal atoms as surface adsorbents, through substitution, or interstitial arrangements. When the transition metal (TM) atoms are adsorbed on the surface of the pristine graphene, then ad-atoms can move freely on the surface. As a result, stable, and precise configurations cannot be attained which leads to a decrease in the efficiency of the adsorbed ad-atoms in nanoelectronics and magnetic applications. Hence, substitutional or interstitial doping of TM ad-atoms on the graphene surface is a suitable method for potential applications (He et al., 2014). He et al. showed that, introducing Fe dimer on graphene vacancy provides the possibilities of tuning the carrier states, which enhances the efficiency of the composites for various applications (He et al., 2014).

Motivated by this, Li et al. have systematically studied the catalytic activities of TM dimers (Cu_2, CuMn, and CuNi) supported on graphene with adjacent vacancies through DFT calculations and showed better activities of the composites for electroreduction of CO_2 with enhanced current density (Li, Su, Chan, and Sun, 2015). They choose graphene with two adjacent single vacancies as the suitable substrate for metal dimers to form composite systems to be used as electrocatalysts for CO_2 reduction. The dual vacancies created on the graphene surface were replaced with TM atoms such as Cu, Mn, and Ni. When same type of metal dopant is introduced on the graphene surface, it is termed as M_2@2SV, whereas the structure containing two different TM atoms is termed as MN@2SV. Among the various possible structures, adsorption energies of important reaction intermediates like CO, COOH, CHO, H, O, and OH are taken into account, for choosing efficient electroctalysts for CO_2 reduction due to the following reasons: (1) since the final products of ECR are mainly CO, CH_3OH and CH_4 from various theoretical investigations, adsorption energy of CO determines whether it will remain as a final product or undergo further reduction to produce CH_3OH or CH_4 (Back, Yeom, and Jung, 2015; Peterson and Nørskov, 2012a), (2) The overpotential of CO formation and its next possible reduction can be determined by relative adsorption strength of COOH and CHO (Back et al., 2015; Li et al., 2015), (3) The selectivity of CH_3OH and CH_4 can be determined from the adsorption strength of O (Back et al., 2015; Li et al., 2015), (4) The adsorption strengths of H and OH determine whether the competing HER dominates over ECR or not under the operating potential range (Hirunsit, Soodsawang, and Limtrakul, 2015; Li et al., 2015).

Figure 5.9 shows the adsorption energies of various intermediates with respect to CO and OH. Li et al. have determined the binding energy of CO to be- 0.98 eV at which CO will undergo further reduction process. From Figure 5.9, it can be observed that there is no linear relationship between the adsorption strengths of H, CHO, and CHO with that of CO on M2@2SV and MN@2SV. In this case, an increase in the adsorption of COOH and

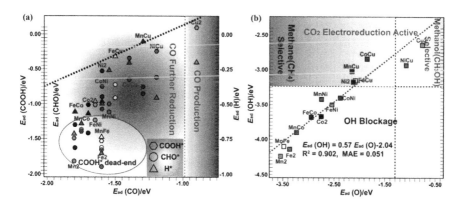

FIGURE 5.9   Adsorption of reaction intermediates with respect to (a) CO and (b) O species on different MN@2SV structure. The triangles, circles, and hexagonal indicate the adsorption of H, CHO, and COOH w.r.t CO respectively.

Source: Reprinted with permission from Li et al. (2015). © American Chemical Society.

CHO with respect to CO leads to a decrease in the overpotential required for the protonation of CO_2. On the other hand, there is a clear linear relationship between the adsorption strength of O and OH. In the case of few M2@2SV or MN@2SV, strong OH adsorption prevents CO_2 reduction, as catalytically active sites are blocked by the OH species. Among the various dopants on adjacent single graphene vacancy investigated by Li et al. they found Cu2@2SV to be the only potential catalyst for reduction of CO_2 to CO, whereas further reduction of CO occurs only on specific catalysts like Ni2@2SV or MCu@2SV.

For further reduction of CO on Ni2@2sv and MCu@2SV, the reaction obeys the following mentioned path through a number of proton-electron pair additions. In the first step, CO reduces to CHO after the addition of proton followed by an electron. As shown in the free energy profile diagram in Figure 5.10, in the successive addition of the electron-proton pair, the reduction reaction follows the path, $CO \rightarrow CHO \rightarrow CH_2O \rightarrow OCH_3$, which is thermodynamically favorable. The reduction from CO to CHO is an uphill process which is considered as the potential determining step. The free energy barrier in this process is 0.99 eV for Cu (111) surfaces, whereas the value is slightly higher than 0.80 eV in the presence of Ni2@2SV, FeCu@2SV, and CoCu@2SV due to stronger CO adsorption as shown in Figure 5.10. In contrast, when Ni or Mn atoms are incorporated into the graphene vacancy, the free energy barrier gets lowered with values 0.70 eV and 0.64 eV for NiCu@2SV and MnCu@2SV structure, respectively. Hence, a notable decrease in the free energy barrier (approximately 0.3 V) is observed leading to the reduction in applied overpotential to make the whole reaction step an exothermic process. Thus, Ni-Cu, and Mn-Cu dimers incorporated in graphene surface act as potential electrocatalyst for electroreduction of CO_2 and CO.

In summary, among various possible configurations of graphene supported TM dimer structures, Cu_2@2SV is the most catalytically active composite for CO production from CO_2, same as in the case of gold electrode. When one Cu atom in the dopant pair is replaced by Mn or Ni atom, the required overpotential drops to approximately 0.2–0.3 V lower than that for Cu electrode, and also allows further CO reduction to provide high selectivity. Mn-Cu and Ni-Cu dopant pairs supported on graphene are more selective towards CH_4 and CH_3OH production, respectively. Thus, compared to pure metal catalysts, graphene supported metal dopant pairs have certain prominent features when used as electrocatalysts for CO_2 reduction. Robust CO and CH_4 production occur with low overpotential and high current density unlike in the case of bulk Au or Cu electrodes. Generation of liquid fuel like methanol is also thermodynamically feasible. Finally, the important fact is that the higher binding energy between metal dopant and defective graphene prevents the metal atoms to form cluster, which leads to high spatial dispersion and high surface to volume ratio with less metal consumption which are not observed in the case of metal bulk or surface electrodes. Embedding metal dimer pair in defective graphene substrate, therefore, acts as possible effective catalysts for the reduction of CO_2 with low overpotential and promotes selectivity towards liquid fuel production (Li et al., 2015).

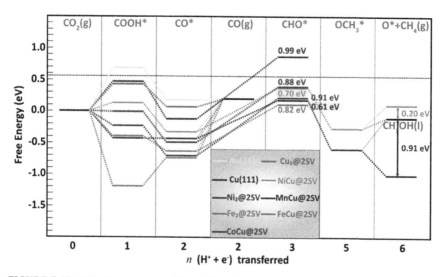

FIGURE 5.10 The free energy profile of ECR of CO_2 on various TM dimer doped graphene. *Source:* Reprinted with permission from Li et al. (2015). © American Chemical Society.

5.5.3 CARBON NANOTUBES (CNTS)

As discussed in the earlier sections, the search for cost-effective and metal-free electrocatalysts to replace the precious metals for electrochemical reduction of carbon dioxide process has gained a remarkable attention in the research community. In this context, carbon-based nanostructures such as carbon nanotube is the most emerging catalysts for various electrochemical processes due to their unique physical and electronic properties (Dresselhaus, Dresselhaus, and Saito, 1995). Besides that, compositional, and structural fine-tuning can be done by doping hetero-atoms on their surfaces. Strong structural dependent properties such as high electrical and thermal conductivity, large surface area and high tensile strength and economically low price make them potential electrocatalysts candidate (Yang et al., 2017). Due to the absence of metals, these are also environmentally friendly (Vasileff et al., 2017). The curvature present in carbon nanotube (CNT) has a great influence on the decreased overpotential for producing specific desired products from ECR, which makes it promising electrocatalyst for reduction process than any other planar structures (Chai and Guo, 2016). Comparing the catalytic activities of graphene and CNT, the selectivity of desire product and limiting potential is highly dependent on the curvature effect. For a same product, limiting potential can be controlled by tuning the degree of curvature. Since there is a mixture of SP^2 and SP^3 hybridizations in graphitic lattice due to curvature in nanotube, intermediated species are bound more strongly and yields higher order hydrocarbons (Vasileff et al., 2017).

5.5.3.1   NITROGEN-DOPED CARBON NANOTUBES (NCNT)

Recently, a great interest is observed in research community for the study of development and control of properties of CNTs through various functionalization methods by putting structural defects or by doping heteroatoms and makes them suitable for a number of technological applications (Ayala, Arenal, Rümmeli, Rubio, and Pichler, 2010). The heteroatoms to be doped should have the size approximately equal to that of carbon atom so that the crystalline structure of CNT would not be affected much. In other words, the heteroatom position should not be exclusively substitutional in the nanotube network. In this context, boron or nitrogen atoms have proved to be better dopants than any other atoms, and conserve the structural properties of nanotube. N atom contains one electron more than the C atom which may modify the electronic properties of CNT if N atom directly replaces C atom in the CNT which could create an n-type material. But, possibility of direct replacement is very less, because size of the N atom is comparatively larger than that of the C atom, which will create defects in the CNT structure in the immediate neighborhood due to the breaking of SP^2 hybridization between carbon atoms. As a result, atomic rearrangement around the locality of doping takes place in the tube. Hence, the overall electronic behavior of CNT depends on the combination of defect, substitution, and atomic rearrangements (Ayala et al., 2010). It has been studied before that doping N in carbon nanotube controls the growth mechanism and increases the metallic behavior due to additional single pair of electron from the nitrogen atom (Jang, Lee, Lyu, Lee, and Lee, 2004; Panchakarla, Govindaraj, and Rao, 2007). Hence, to make the CNT electrochemically active, surface modification or functionalization is required. In this context, incorporation of N atom in CNT removes the chemical inertness of nanotube by creating small defects on the surface, where active sites for various hydrocarbons or organic molecules can be created. The defects in hexagonal graphitic lattice due to nitrogen can be the origin of catalytic activities. N-doped carbon nanotubes are basically classified into three types: pyridinic N, pyrrolic N, and graphitic N based on the position and amount of N in nanotube structure, same as in the case of N-doped graphene. In this section, we will discuss the catalytic activities of these NCNTs and the underlying mechanism for ECR with the help of theoretical study.

Wu et al. have found that N-doped carbon nanotube acts as a highly efficient catalyst for CO_2 reduction reaction to give CO as the final product without any hydrocarbons or alcohols, indicating its high selectivity (Sharma et al., 2015; Wu et al., 2015). With the help of CHE and DFT, they showed that pyridinic N is the most efficient configuration with lower overpotential. On pure CNT or NCNT, the reaction mechanism of electroreduction of CO_2 to CO involves two electron-proton transfer steps. In the first step, the adsorbed CO_2 molecule gets attached with the electron-proton pair, which is transferred from the solution, and a COOH* intermediate species is formed. Then, after the addition of the second electron-proton pair, the COOH* species reduces to form CO*. The final product, CO, is obtained after the desorption of CO*. The reaction mechanisms of reduction of CO_2 to CO are expressed in the following equations:

$$CO_2(g) + {}^* + H^+(aq) + e^- \rightarrow COOH^* \tag{5.21}$$

$$COOH^* + H^+(aq) + e^- \rightarrow CO + H_2O \tag{5.22}$$

$$CO^* \rightarrow CO(g) + {}^* \tag{5.23}$$

The adsorbed site is denoted by '*.' Since the free energies of the reactions control the overpotential, the corresponding free energy diagram of the electrochemical reactions is shown in Figure 5.11. As shown in the figure, it is clear that the first electrochemical step is an uphill process at 0 V vs. RHE, which is maximum in the case of pure CNT (1.9 eV), and minimum for pyridinic N (0.30 eV). But, the second electron-proton pair addition to COOH* to form CO is a downhill process. Thus, the first step, where adsorbed COOH* species is formed, is a rate-limiting step and require an onset potential of −0.30 V vs. RHE to make the reduction downhill for pyridinic N. This value suggests better catalytic activities of pyridinic N than the theoretically calculated values for Ag (−0.79 V), Au (−0.63 V), and Cu (−0.31 V) (Peterson et al., 2010; Peterson and Nørskov, 2012b).

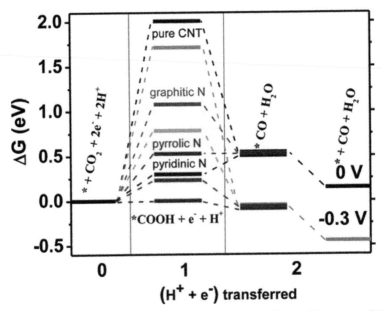

FIGURE 5.11   Free energy diagram for electroreduction of CO_2 to CO on pure CNT and NCNTs.
Source: Reprinted with permission from Wu et al. (2015).

In order to explain the high selectivity of NCNTs towards CO production, Wu et al. have calculated the adsorption energies of COOH and CO with CNT and NCNTs as shown in Figure 5.12. For reduction of CO_2 to CO, a high selective catalyst should have the capability to hold the intermediate COOH* species tightly and loosening the final product CO simultaneously. That means, COOH should be strongly bound to the catalyst, whereas weak binding energy will be obtained for CO. From the Figure 5.12, it is clear that pyridinic N binds COOH more strongly than any other NCNTs, whereas binding energy of COOH for pristine CNT gives positive value indicating a weak bound. In contrast, low binding energy of CO is observed for all the nanotube structures. All the NCNTs show higher COOH binding and lower CO binding than the CO-selective metal catalysts like Cu, Ag, and Au. Hence, due to very low binding energy, CO is easily escaped from the surface before going for further reduction. Thus, CO is the only product obtained during ECR on NCNTs without the production of any hydrocarbon and/or alcohol indicating high selectivity. Thus, for CO_2 reduction to CO, NCNT can be used as a potential, and low cost alternative to expensive metal catalysts.

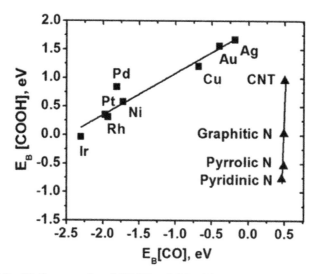

FIGURE 5.12   Binding energies of COOH and CO with CNT and NCNTs. The values are compared with various metals.
Source: Reprinted with permission from Wu et al. (2015). © American Chemical Society.

5.5.3.2   COVALENTLY CONNECTED CARBON NANOTUBES (CNTS)

The three dimensional (3D) carbon solids with novel properties can be obtained by interconnecting carbon nanotubes (CNTs) by means of chemical functionalization (Ozden et al., 2015). These type of carbon-based functional materials can be synthesized by making covalent C-C bond through the Suzuki cross-coupling reaction (Chemler, Trauner, and Danishefsky, 2001), and make them potentially suitable for electronic and energy storage/ harvest device applications (Pal et al., 2017). As a result, the highly porous 3D solid nanotube networks are obtained due to the cross-linking of CNTs.

In addition to this, multi-terminal junctions are observed in these new 3D structures. Therefore, Suzuki coupling is one of the most efficient methods to interconnect carbon-based nanostructures such as nanotubes through controllable interface chemistry, where multifunctional properties can be easily tuned for a wide range of applications (Ozden et al., 2015). In this section, we will briefly describe the catalytic activities of covalently coupled carbon nanotubes (CCNT) for ECR, through electronic structure methods.

Previously, it has been reported that in the presence of boron-doped graphene, reduction of CO_2 leads to production of high-yield formate (Sreekanth, Nazrulla, Vineesh, Sailaja, and Phani, 2015). Here, the role of boron doping is the same as in the case of nitrogen-doped graphene composites, as an electrocatalyst for ECR. As per another theoretical investigations done by Chai et al. the nitrogen doping and curvature in graphene/CNT system play a crucial role in deciding limiting potential and product selectivity such as CH_3OH and CO (Chai and Guo, 2016). Motivated by this, Pal et al. have designed a system containing dual CNTs, covalently connected by a molecular (C_6H_4) junction through Suzuki coupling to study the ECR (Pal et al., 2018). They considered two single-walled carbon nanotubes (SWCNTs) with chirality (5, 5), which are coupled through a linker, C_6H_4, yielding a CCNT structure as shown in Figure 13c. Here, the plane of the linker is taken to be perpendicular to the axis of nanotube, which is energetically the most stable structure than any possible orientations of the linker plane with respect to the axis of nanotube.

Pal et al. calculated the band structure and density of states (DOS) of single nanotube, nanotube with the functional COOH, coupled nanotubes, and coupled nanotubes with COOH through DFT, for comparison of electron mobility near the Fermi level (Pal et al., 2018). Form the charge transfer analysis, it is found that CNT surface gains a charge of +0.04e, whereas C_6H_4 acquires a negative charge of −0.08e resulting in a Fermi level shift downward in the band structure. As shown in Figure 5.13a, there is finite value in DOS at Fermi level indicating the metallic nature of CNT (5, 5), whereas in the case of couple nanotube, the band structure gets distorted due to the breaking in symmetry and bands with very less dispersion is observed near Fermi level (Figure 5.13c), which indicates the higher effective mass of electron with less mobility. But, when COOH functional group is attached with coupled nanotube, then bands near the Fermi level get higher dispersion (Figure 5.13d) implying the higher electron mobility and facilitate the reduction of CO_2.

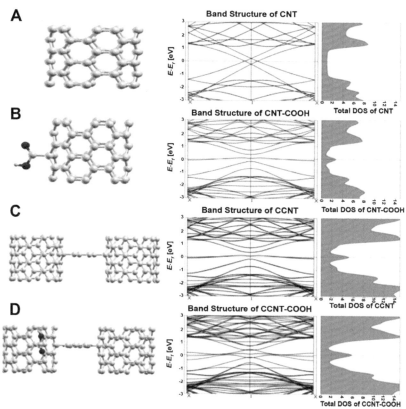

FIGURE 5.13 Relaxed geometrical structures with band structures and DOS of (A) CNT, (B) CNT with COOH, (C) coupled CNTs and (D) coupled CNT with COOH.
Source: Reprinted with permission Pal et al. (2018). © American Chemical Society.

To describe the product selectivity of ECR on CCNT structures, adsorption energies of reaction intermediate COOH* with CNT and CCNT surfaces are calculated and found to be −0.88 eV and −1.01 eV, respectively. These values indicate that stronger binding of intermediates occurs on Suzuki coupled nanotube surfaces than pristine nanotube. In addition to this, the adsorption energies of HCOOH on pristine and couple CNT surfaces are found to be −0.77eV and −0.53 eV, respectively implying easier desorption of HCOOH from CCNT surfaces as a final product after the reduction of CO_2. Thus, compared to bare CNT, CCNT acts as a promising electrocatalyst for ECR to achieve liquid product formate, where molecular origin plays a vital role in the charge transfer and enhances the CO_2 reduction process.

5.6   CONCLUSION

In summary, product selectivity of electroreduction of CO_2 can be tuned by engineering catalysts through chemical modification, which is a quite promising method (Pal et al., 2018). Understanding the reaction mechanism and pathways are highly important to design an efficient, highly active, and cost-effective catalyst with high selectivity. Thus, the theoretical studies enlighten the path for various possible modifications of carbon-based catalysts through chemical functionalization, for the reduction of carbon dioxide to get desired products.

KEYWORDS

- **carbon-nanostructures**
- **computational hydrogen electrode model**
- **covalently coupled carbon nanotubes**
- **density functional theory**
- **electrochemical CO_2 reduction**
- **standard hydrogen electrode**

REFERENCES

Alfonso, D. R., Tafen, D. N., & Kauffmann, D. R., (2018). First-principles modeling in heterogeneous electrocatalysis. *Catalysts, 8*, 424.

Alivisatos, A. P., (1996). Semiconductor clusters, nanocrystals, and quantum dots. *Science, 271*, 933–937.

Ayala, P., Arenal, R., Rümmeli, M., Rubio, A., & Pichler, T., (2010). The doping of carbon nanotubes with nitrogen and their potential applications. *Carbon, 48*, 575–586.

Back, S., Yeom, M. S., & Jung, Y., (2015). Active Sites of Au and Ag nanoparticle catalysts for CO_2 electroreduction to CO. *ACS Catalysis, 5*, 5089–5096.

Balogun, M. S., Huang, Y., Qiu, W., Yang, H., Ji, H., & Tong, Y., (2017). Updates on the development of nanostructured transition metal nitrides for electrochemical energy storage and water splitting. *Materials Today, 20*, 425–451.

Balogun, M. S., Qiu, W., Luo, Y., Meng, H., Mai, W., Onasanya, A., et al., (2016). A review of the development of full cell lithium-ion batteries: The impact of nanostructured anode materials. *Nano Research, 9*, 2823–2851.

Benson, E. E., Kubiak, C. P., Sathrum, A. J., & Smieja, J. M., (2009). Electro catalytic and homogeneous approaches to conversion of CO_2 to liquid fuels. *Chemical Society Reviews, 38*, 89–99.

Centi, G., & Perathoner, S., (2009). Opportunities and prospects in the chemical recycling of carbon dioxide to fuels. *Catalysis Today, 148*, 191–205.

Chai, G. L., & Guo, Z. X., (2016). Highly effective sites and selectivity of nitrogen-doped graphene/CNT catalysts for CO_2 electrochemical reduction. *Chemical Science, 7*, 1268–1275.

Chemler, S. R., Trauner, D., & Danishefsky, S. J., (2001). The B-alkyl Suzuki-Miyaura cross-coupling reaction: Development, mechanistic study, and applications in natural product synthesis. *Angewandte Chemie-International Edition, 40*, 4544–4568.

Dresselhaus, M. S., Dresselhaus, G., & Saito, R., (1995). Physics of carbon nanotubes. *Carbon, 33*, 883–891.

Endo, M., Strano, M. S., & Ajayan, P. M., (2007). *Potential Applications of Carbon Nanotubes* (pp. 13–62). Springer, Berlin, Heidelberg. https://doi.org/10.1007/978-3-540-72865-8_2 (accessed on 16 May 2020).

Etacheri, V., Marom, R., Elazari, R., Salitra, G., & Aurbach, D., (2011). Challenges in the development of advanced Li-ion batteries: A review. *Energy and Environmental Science, 4*, 3243.

Gao, D., Cai, F., Wang, G., & Bao, X., (2017). Nano structured heterogeneous catalysts for electrochemical reduction of CO_2, *Curr. Opin. Green Sustain. Chem., 3*, 39–44.

Gao, H., Song, L., Guo, W., Huang, L., Yang, D., Wang, F., Zuo, Y., et al., (2012). A simple method to synthesize continuous large area nitrogen-doped graphene. *Carbon, 50*, 4476–4482.

Gasteiger, H. A., Kocha, S. S., Sompalli, B., & Wagner, F. T., (2005). Activity benchmarks and requirements for Pt, Pt-alloy, and non-Pt oxygen reduction catalysts for PEMFCs. *Applied Catalysis B: Environmental, 56*, 9–35.

Geim, A. K., & Novoselov, K. S., (2007). The rise of graphene. *Nature Materials, 6*, 183–191.

Guo, Y. G., Hu, J. S., & Wan, L. J., (2008). Nanostructured materials for electrochemical energy conversion and storage devices. *Advanced Materials, 20*, 2877–2887.

He, Z., He, K., Robertson, A. W., Kirkland, A. I., Kim, D., Ihm, J., et al., (2014). Atomic structure and dynamics of metal dopant pairs in graphene. *Nano Letters, 14*, 3766–3772.

Hirunsit, P., Soodsawang, W., & Limtrakul, J., (2015). CO_2 electrochemical reduction to methane and methanol on copper-based alloys: Theoretical insight. *Journal of Physical Chemistry C, 119*, 8238–8249.

Hohenberg, P. C., Kohn, W., & Sham, L. J., (1990). The beginnings and some thoughts on the future. In: *Advances in Quantum Chemistry* (Vol. 21, pp. 7–26).

Hohenberg, P., & Kohn, W., (1964). Inhomogeneous electron gas. *Physical Review, 136*, B864.

Hori, Y., Koga, O., Yamazaki, H., & Matsuo, T., (1995). Infrared spectroscopy of adsorbed CO and intermediate species in electrochemical reduction of CO_2 to hydrocarbons on a Cu electrode, *Electrochimica Acta, 40*, 2617–2622.

Hori, Y., Konishi, H., Futamura, T., Murata, A., Koga, O., Sakurai, H., & Oguma, K., (2005). "deactivation of copper electrode" in electrochemical reduction of CO_2. *Electrochimica Acta, 50*, 5354–5369.

Hori, Y., Murata, A., & Takahashi, R., (1989). Formation of hydrocarbons in the electrochemical reduction of carbon dioxide at a copper electrode in aqueous solution. *Journal of the Chemical Society, Faraday Transactions 1: Physical Chemistry in Condensed Phases, 85*, 2309–2326.

Hori, Y., Takahashi, I., Koga, O., & Hoshi, N., (2002). Selective formation of C_2 compounds from electrochemical reduction of CO_2 at a series of copper single crystal electrodes. *Journal of Physical Chemistry B, 106*, 15–17.

Hussain, J., Skúlason, E., & Jónsson, H., (2015). Computational study of electrochemical CO_2 reduction at transition metal electrodes. *Procedia Computer Science, 51*, 1865–1871.

Jang, J. W., Lee, C. E., Lyu, S. C., Lee, T. J., & Lee, C. J., (2004). Structural study of nitrogen-doping effects in bamboo-shaped multiwalled carbon nanotubes. *Applied Physics Letters, 84*, 2877–2879.

Jiao, Y., Zheng, Y., Jaroniec, M., & Qiao, S. Z., (2015). Design of electro catalysts for oxygen- and hydrogen-involving energy conversion reactions. *Chemical Society Reviews, 44*, 2060–2086.

Jin, H., Guo, C., Liu, X., Liu, J., Vasileff, A., Jiao, Y., Zheng, Y., & Qiao, S. Z., (2018). Emerging two-dimensional nanomaterials for electro catalysis. *Chemical Reviews, 118*, 6337–6408.

Jing, Y., & Zhou, Z., (2015). Computational insights into oxygen reduction reaction and initial Li_2O_2 nucleation on pristine and N-doped graphene in Li-O_2 batteries. *ACS Catalysis, 5*, 4309–4317.

Jones, J. P., Prakash, G. K. S., & Olah, G. A., (2014). Electrochemical CO_2 reduction: Recent advances and current trends. *Israel Journal of Chemistry, 54*, 1451–1466.

Kohn, W., & Sham, L. J., (1965). Self-consistent equations including exchange and correlation effects. *Physical Review, 140*, A1133.

Kondratenko, E. V., Mul, G., Baltrusaitis, J., Larrazábal, G. O., & Pérez-Ramírez, J., (2013). Status and perspectives of CO_2 conversion into fuels and chemicals by catalytic, photocatalytic and electrocatalytic processes. *Energy and Environmental Science, 6*, 3112–3135.

Kuhl, K. P., Hatsukade, T., Cave, E. R., Abram, D. N., Kibsgaard, J., & Jaramillo, T. F., (2014). Electrocatalytic conversion of carbon dioxide to methane and methanol on transition metal surfaces. *Journal of the American Chemical Society, 136*, 14107–14113.

Leite, E. R., (2009). *Nanostructured Materials for Electrochemical Energy Production and Storage*. Springer. https://doi.org/10.1007/978-0-387-49323-7 (accessed on 16 May 2020).

Letchworth-Weaver, K., & Arias, T. A., (2012). Joint density functional theory of the electrode-electrolyte interface: Application to fixed electrode potentials, interfacial capacitances, and potentials of zero charge. *Physical Review B-Condensed Matter and Materials Physics, 86*, 75140.

Lherbier, A., Blase, X., Niquet, Y. M., Triozon, F., & Roche, S., (2008). Charge transport in chemically doped 2D graphene. *Physical Review Letters, 101*, 036808.

Li, H., & Oloman, C., (2005). The electro-reduction of carbon dioxide in a continuous reactor. *Journal of Applied Electrochemistry, 35*, 955–965.

Li, Y., Su, H., Chan, S. H., & Sun, Q., (2015). CO_2 electro reduction performance of transition metal dimers supported on graphene: A theoretical study. *ACS Catalysis, 5*, 6658–6664.

Liu, G., Bai, H., Zhang, B., & Peng, H., (2018). Role of organic components in electrocatalysis for renewable energy storage. *Chemistry-A European Journal, 24*, 18271–18292.

Liu, N., Li, W., Pasta, M., & Cui, Y., (2014). Nanomaterials for electrochemical energy storage. *Frontiers of Physics, 9*, 323–350.

Liu, Y., Zhao, J., & Cai, Q., (2016). Pyrrolic-nitrogen doped graphene: A metal-free electro catalyst with high efficiency and selectivity for the reduction of carbon dioxide to formic acid: A computational study. *Physical Chemistry Chemical Physics, 18*, 5491–5498.

Lu, Q., & Jiao, F., (2016). Electrochemical CO_2 reduction: Electro catalyst, reaction mechanism, and process engineering. *Nano Energy, 29*, 439–456.

Lu, Q., Rosen, J., & Jiao, F., (2015). Nanostructured metallic electrocatalysts for carbon dioxide reduction. *Chem. Cat. Chem., 7*, 38–47.

Luo, W., Nie, X., Janik, M. J., & Asthagiri, A., (2016). Facet dependence of CO_2 reduction paths on Cu electrodes. *ACS Catalysis, 6*, 219–229.

Luo, Z., Lim, S., Tian, Z., Shang, J., Lai, L., MacDonald, B., Fu, C., Shen, Z., Yu, T., & Lin, J., (2011). Pyridinic N doped graphene: Synthesis, electronic structure, and electrocatalytic property. *Journal of Materials Chemistry, 21*, 8038–8044.

Ma, C., Shao, X., & Cao, D., (2012). Nitrogen-doped graphene nanosheets as anode materials for lithium ion batteries: A first-principles study. *Journal of Materials Chemistry, 22*, 8911.

Ma, J., Sun, N., Zhang, X., Zhao, N., Xiao, F., Wei, W., & Sun, Y., (2009). A short review of catalysis for CO_2 conversion. *Catalysis Today, 148*, 221–231.

Mathew, K., Sundararaman, R., Letchworth-Weaver, K., Arias, T. A., & Hennig, R. G., (2014). Implicit solvation model for density-functional study of nanocrystal surfaces and reaction pathways. *Journal of Chemical Physics, 140*, 084106.

Nørskov, J. K., Bligaard, T., Rossmeisl, J., & Christensen, C. H., (2009). Towards the computational design of solid catalysts. *Nature Chemistry, 1*, 37–46.

Nørskov, J. K., Rossmeisl, J., Logadottir, A., Lindqvist, L., Kitchin, J. R., Bligaard, T., & Jónsson, H., (2004). Origin of the over potential for oxygen reduction at a fuel-cell cathode. *Journal of Physical Chemistry B, 108*, 17886–17892.

Novoselov, K. S., Mishchenko, A., Carvalho, A., & Castro, N. A. H., (2016). 2D materials and Van Der Waals heterostructures. *Science, 353*, 9439.

Ozden, S., Narayanan, T. N., Tiwary, C. S., Dong, P., Hart, A. H. C., Vajtai, R., & Ajayan, P. M., (2015). 3D macroporous solids from chemically cross-linked carbon nanotubes. *Small, 11*, 688–693.

Pal, S., Narayanaru, S., Kundu, B., Sahoo, M., Bawari, S., Rao, D. K., Nayak, S. K., Pal, A. J., & Narayanan, T. N., (2018). Mechanistic insight into formate production via CO_2 reduction in C-C coupled carbon nanotube molecular junctions. *The Journal of Physical Chemistry C, 122*, 23385–23392.

Pal, S., Sahoo, M., Veettil, V. T., Tadi, K. K., Ghosh, A., Satyam, P., Biroju, R. K., Ajayan, P. M., & Nayak, S. K., & Narayanan, T. N., (2017). Covalently connected carbon nanotubes as electrocatalysts for hydrogen evolution reaction through band engineering. *ACS Catalysis, 7*, 2676–2684.

Panchakarla, L. S., Govindaraj, A., & Rao, C. N. R., (2007). Nitrogen-and boron-doped double-walled carbon nanotubes. *ACS Nano, 1*, 494–500.

Peterson, A. A., & Nørskov, J. K., (2012a). Activity descriptors for CO_2 electroreduction to methane on transition-metal catalysts. *Journal of Physical Chemistry Letters, 3*, 251–258.

Peterson, A. A., & Nørskov, J. K., (2012b). Activity descriptors for CO_2 electroreduction to methane on transition-metal catalysts. *Journal of Physical Chemistry Letters, 3*, 251–258.

Peterson, A. A., Abild-Pedersen, F., Studt, F., Rossmeisl, J., & Nørskov, J. K., (2010). How copper catalyzes the electroreduction of carbon dioxide into hydrocarbon fuels. *Energy and Environmental Science, 3*, 1311–1315.

Qiao, J., Liu, Y., Hong, F., & Zhang, J., (2014). A review of catalysts for the electro reduction of carbon dioxide to produce low-carbon fuels. *Chemical Society Reviews, 43*, 631–675.

Rendón-Calle, A., Builes, S., & Calle-Vallejo, F., (2018). A brief review of the computational modeling of CO_2 electroreduction on Cu electrodes. *Current Opinion in Electrochemistry, 9*, 158–165.

Reske, R., Mistry, H., Behafarid, F., Cuenya, B. R., & Strasser, P., (2014). Particle size effects in the Catalytic electro reduction of CO_2 on Cu nano particles, *J. Am. Chem. Soc., 136*, 6978–6986.

Shao, Y., Sui, J., Yin, G., & Gao, Y., (2008). Nitrogen-doped carbon nanostructures and their composites as catalytic materials for proton exchange membrane fuel cell. *Applied Catalysis B: Environmental, 79*, 89–99.

Sharma, P. P., Wu, J., Yadav, R. M., Liu, M., Wright, C. J., Tiwary, C. S., Yakonson, B. I., Lou, J., Ajayan, P. M., & Zhou, X. D., (2015). Nitrogen-doped carbon nanotube arrays for high-efficiency electrochemical reduction of CO_2: On the understanding of defects, defect density, and selectivity. *Angewandte Chemie-International Edition, 54*, 13701–13705.

She, Z. W., Kibsgaard, J., Dickens, C. F., Chorkendorff, I., Nørskov, J. K., & Jaramillo, T. F., (2017). Combining theory and experiment in electro catalysis: Insights into materials design. *Science, 355*, 4998.

Siahrostami, S., Jiang, K., Karamad, M., Chan, K., Wang, H., & Nørskov, J., (2017). Theoretical investigations into defected graphene for electrochemical reduction of CO_2. *ACS Sustainable Chemistry and Engineering, 5*, 11080–11085.

Simakov, D. S. A., (2017). Electro catalytic reduction of CO_2. In: *Renewable Synthetic Fuels and Chemicals from Carbon Dioxide* (pp. 27–42), Springer, Cham. https://doi.org/10.1007/978-3-319-61112-9_2 (accessed on 16 May 2020).

Slater, M. D., Kim, D., Lee, E., & Johnson, C. S., (2013). Sodium-ion batteries. *Advanced Functional Materials, 23*, 947–958.

Sreekanth, N., Nazrulla, M. A., Vineesh, T. V., Sailaja, K., & Phani, K. L., (2015). Metal-free boron-doped graphene for selective electro reduction of carbon dioxide to formic acid/formate. *Chemical Communications, 51*, 16061–16064.

Vasileff, A., Zheng, Y., & Qiao, S. Z., (2017). Carbon solving carbon's problems: Recent progress of nanostructured carbon-based catalysts for the electrochemical reduction of CO_2. *Advanced Energy Materials, 7*, 1700759.

Wang, S., Zhang, L., Xia, Z., Roy, A., Chang, D. W., Baek, J. B., & Dai, L., (2012). BCN graphene as efficient metal-free electro catalyst for the oxygen reduction reaction. *Angewandte Chemie-International Edition, 51*, 4209–4212.

Wang, W., Wang, S., Ma, X., & Gong, J., (2011). Recent advances in catalytic hydrogenation of carbon dioxide. *Chemical Society Reviews, 40*, 3703–3727.

Wang, Y., Shao, Y., Matson, D. W., Li, J., & Lin, Y., (2010). Nitrogen-doped graphene and its application in electrochemical biosensing. *ACS Nano, 4*, 1790–1798.

Wang, Z. L., Li, C., & Yamauchi, Y., (2016). Nanostructured nonprecious metal catalysts for electrochemical reduction of carbon dioxide. *Nano Today, 11*, 373–391.

Wu, J., Yadav, R. M., Liu, M., Sharma, P. P., Tiwary, C. S., Ma, L., Zou, S., et al., (2015). Achieving highly efficient, selective, and stable CO_2 reduction on nitrogen-doped carbon nanotubes. *ACS Nano, 9*, 5364–5371.

Xu, Y., Kraft, M., & Xu, R., (2016). Metal-free carbonaceous electro catalysts and photo catalysts for water splitting. *Chemical Society Reviews, 45*, 3039–3052.

Yabuuchi, N., Kubota, K., Dahbi, M., & Komaba, S., (2014). Research development on sodium-ion batteries. *Chemical Reviews, 114*, 11636–11682.

Yang, N., Swain, G. M., & Jiang, X., (2016). Nanocarbon electrochemistry and electro analysis: Current status and future perspectives. *Electroanalysis, 28*, 27–34.

Yang, N., Waldvogel, S. R., & Jiang, X., (2016). Electrochemistry of carbon dioxide on carbon electrodes. *ACS Applied Materials and Interfaces, 8*, 28357–28371.

Yang, Z., Liu, Z., Zhang, H., Yu, B., Zhao, Y., Wang, H., Ji, G., Chen, Y., Liu, X., & Liu, Z., (2017). N-Doped porous carbon nanotubes: Synthesis and application in catalysis. *Chemical Communications, 53*, 929–932.

Yu, Y. X., (2013). Can all nitrogen-doped defects improve the performance of graphene anode materials for lithium-ion batteries? *Physical Chemistry Chemical Physics, 15*, 16819–16827.

Zhang, L., & Xia, Z., (2011). Mechanisms of oxygen reduction reaction on nitrogen-doped graphene for fuel cells. *Journal of Physical Chemistry C, 115*, 11170–11176.

Zhang, Q., Uchaker, E., Candelaria, S. L., & Cao, G., (2013). Nanomaterials for energy conversion and storage. *Chemical Society Reviews, 42*, 3127–3171.

Zheng, Y., Liu, J., Liang, J., Jaroniec, M., & Qiao, S. Z., (2012). Graphitic carbon nitride materials: Controllable synthesis and applications in fuel cells and photo catalysis. *Energy and Environmental Science, 5*, 6717–6731.

Zhu, C., Yang, P., Chao, D., Mai, W., & Fan, H. J., (2015). Heterogeneous nanostructures for sodium ion batteries and supercapacitors. *Chem. Nano Mat., 1*, 458–476.

Zou, X., & Zhang, Y., (2015). Noble metal-free hydrogen evolution catalysts for water splitting. *Chemical Society Reviews, 44*, 5148–5180.

CHAPTER 6

Catalyst for Hydrogen Oxidation Reaction and Its Application Towards Energy Storage

AVIJIT BISWAL and AVINNA MISHRA

College of Engineering and Technology, Ghatikia, Bhubaneswar – 751029, Odisha, India

ABSTRACT

With an increasing demand for an alternative to fossil fuels, hydrogen energy is found to be attracted significant attention as a green, clean, and sustainable fuel from the last few years. Hydrogen can be used in fuel cells which converts chemical energy to electrical energy directly. The best suitable method to produce hydrogen is by splitting water electrochemically or photo electrochemically in the presence of an electrolyzer or a photoelectrolyzer. The whole phenomena of the production of hydrogen to its application in fuel cells are the result of two major reactions, i.e., hydrogen oxidation reaction (HOR) and hydrogen evolution reaction (HER). However, these two reactions are kinetically not feasible without a catalyst. Various methods have been developed in order to replace Pt which is the best catalyst so far for HOR reaction. In this chapter, we mainly focus on HOR, its mechanism along with various measurement activities, its catalytic activities, and their dependent controlling parameters. A detailed discussion also presented on different electrocatalyst like noble metals nanoparticles, graphene-based nanocomposites, non-precious electrocatalyst, etc. Some of the recent advances made on the catalytic performances, fuel cell efficiency will be included in the conclusion section.

6.1   INTRODUCTION

Energy is an inevitable need for modern civilization beyond which, moving a step ahead seems to be impossible. Day-by-day human demand for energy is increasing with the increasing growth in population. The total global use of primary energy based on fossil fuels. However, due to the continuous consumption of fossil fuel, its limited reserves and the day-by-day increasing environmental issues of these non-renewable energy sources motivate the researchers to the hunt for renewable or alternative energy sources. Due to the increasing demand of alternatives for fossil fuels, various research groups throughout the globe are trying to explore new sustainable energy sources and technologies. Among the various alternatives present till date, electrochemical energy conversion technology is found to be the most viable solutions for the accounted problem. This may be due to the fact of its environmental friendliness, high power density affordability, and stability. In this respect, fuel cells are considered to be one of the favorable technologies for the generation of sustainable and green energy.

Hydrogen energy is found to be attracted significant attention as a green, clean, and sustainable fuel from the last few years. Regardless of the abundance of Hydrogen in our environment and being the simplest and a unique energy carrier, as it can be easily extracted continuously by water splitting either by electrochemically or photoelectrochemically with the help of renewable energy sources such as solar and wind (Strmcnik et al., 2010; Turner, 2004; Dresselhaus and Thomas, 2001). Most importantly during oxidation of hydrogen, water is formed as the only by-product due to combustion and hence has no warning for the environment in the context of green chemistry. Basically, we use electrolyzer or photoelectrolyzer for water splitting purposes. In general, we found a variety of electrochemical reactions involved in this particular energy conversion technology such as hydrogen oxidation reaction (HOR), hydrogen evolution reaction (HER), oxygen evolution reaction (OER), etc. For the successful occurrence of these reactions, we need efficient electrocatalyst which acts a major candidate for the calculation of conversion efficiency of electrochemical systems. (Bashyam and Zelenay, 2006; Zhuang et al., 2016; Tang et al., 2010, 2013; Gasteiger and Marković, 2009; Wang et al., 2011). The only disadvantages associated with fuel cell are its high cost. The major issues for the commercialization and scaling up fuel cell application involves many factors such as high cost of the electrocatalyst, activity reduction in catalyst during long-term operation, and a more clear interface kinetic mechanism. (Yoo and Sung, 2010). Hence, Hydrogen fuel cells are one of the best choices when switching over to sustainable energy systems.

Among the various types of fuel cells, such as alkaline fuel cells (AFC), proton exchange membrane (PEM) fuel cells, phosphoric acid fuel cells (PAFC), molten carbonate fuel cells (MCFC), solid oxide fuel cells (SOFC), direct alcohol fuel cells (DAFCs), polymer electrolyte membrane (PEM) fuel cells dominant over other fuel cells due to its high power density and an added advantages of low weight and volume compared with other fuel cells. These are also called PEM fuel cell in which a solid polymer is used as an electrolyte with porous carbon electrodes containing a platinum or platinum alloy catalyst. These fuel cells use air as the source of hydrogen and oxygen. Pure hydrogen supplied to these cells is stored in storage tanks or reformers. In recent years, there is a shift towards the development of advanced anion-exchange membrane fuel cells (AEMFCs) from normal fuel cell to make the fuel cells economically viable. However, these fuel cells are still based on platinum group metals (PGM) anode and cathode electrocatalysts. Significant efforts have been made in the development of PGM-free catalysts for the oxygen reduction reaction (ORR) and the HOR in acidic and basic medium. In comparison, recently much emphasis is given to HOR (Davydova et al., 2018).

However, due to the lethargic kinetics of the process, various electro-catalysts are used for hydrogen oxidation/HER. Considerable efforts have been made to resolve the problems and attempts are taken for the partial or complete replacement of Pt electrocatalyst in PEM fuel cell cathodes such as Co- or Fe-based catalysts (Lefèvre et al., 2005) non-precious metal catalysts (Wood et al., 2008), carbonized catalysts (Maruyama and Abe, 2005), and pyrolyzed macrocyclic compounds (Zhang et al., 2006). However, these catalysts couldn't meet the requirements like platinum-based catalysts. Tungsten carbide (WC) was used as the electrocatalyst for various HOR reactions (Palanker et al., 1977; Okamoto et al., 1987; Barnett et al., 1997; Meng and Shen, 2005; Morishita et al., 2007; Bronoel et al., 1991; Lee et al., 2004; Rees et al., 2009; Nagai et al., 2007; Izhar et al., 2009; Li et al., 2009; Obradović et al., 2012; Krischer and Savinova, 2008; Cong and Song, 2018; Sheng et al., 2010; Durst et al., 2014; Zheng et al., 2016; Elbert et al., 2015). Later, synthesis of foreign metal (such as Mo, Ni, Fe, Mn, Co) containing WC were carried out (Bronoel et al., 1991; Lee et al., 2004; Rees et al., 2009; Nagai et al., 2007; Izhar et al., 2009).

The above investigations revealed that WC could not achieve a significant shift in HOR due to its lower surface area and the presence of Ni, Mn, and Fe decreases the surface area after their addition. However, Co-WC showed better catalytic activity in comparison to WC. Recently Iridium and Iridium based nanocatalyst (Li et al., 2009) and WC/Pt nanocatalysts (Obradović et al., 2012) have been reported to have excellent catalytic activity for HOR reaction.

6.2 GENERAL MECHANISM OF HOR

Fuel cell is very similar to a redox cell, i.e., battery which contains two electrodes with a suitable electrolyte which facilitates the free passage of ions. The fuel cell needs hydrogen (H_2) and oxygen (O_2) as the fuel to work. Hydrogen enters at the anode of the fuel cell and oxygen at cathode. At anode, hydrogen undergoes oxidation to form H^+. This oxidation half-reaction is called HOR as hydrogen gets oxidized to hydrogen ion at anode releasing an electron. The released electron at anode traveled through external circuits providing current, which are taken by the oxygen at cathode completing the circuit and powers the device. The hydrogen ions pass from cathode to anode side through the electrolyte and combine with oxygen to form water. The detailed half-cell reaction is given below. The water thus formed is removed as waste from the fuel cell. Series of fuel cells can be used for higher power generation.

The detailed mechanism is shown schematically in Figure 6.1.

$$\text{At Anode: } 2H_2 \rightarrow 4H^+ + 4e^- \tag{6.1}$$

$$\text{At Cathode: } O_2 + 4H^+ + 4e^- \rightarrow 2H_2O \tag{6.2}$$

$$\text{Net reaction: } 2H_2 + O_2 \rightarrow 2H_2O \tag{6.3}$$

It is very much essential to have a clear understanding on the overall mechanism of HOR in order to synthesis and design electrocatalyst in alkaline medium (Figure 6.2). Usually, the overall HOR mechanism can be understood by the following equation:

$$H_2 + 2OH^- \rightarrow 2H_2O + 2e^- \tag{6.4}$$

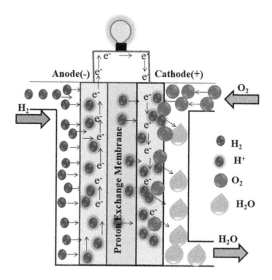

FIGURE 6.1   Schematic diagram of fuel cell showing hydrogen oxidation reaction (HOR) at anode.

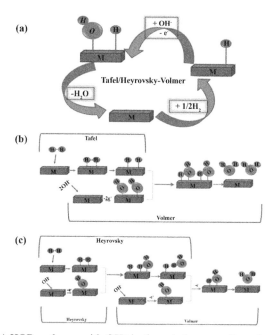

FIGURE 6.2   (a) HOR pathway with OH⁻ in the solution overall mechanism, (b) Possible alkaline HOR path way by Tafel-Volmer route and (c) Heyrovsky-Volmer route with OH⁻ in an electrolyte in detail.

But the whole HOR activity passes through a combination of three subsequent elementary steps in alkaline or basic medium, i.e., (a) Dissociative adsorption of molecular dihydrogen (H₂) without electron transfer (ET), which is known as the Tafel (T) reaction, (b) ET from molecular dihydrogen to the catalyst known as Heyrovsky (H) reaction, and (c) discharge of the adsorbed hydrogen atom known as Volmer (V) reaction (Figure 6.3). The elementary steps of HOR are shown in the following Eqns. (6.5)–(6.7).

$$H_2 + 2M \rightarrow 2M - H_{ad} \, (Tafel) \tag{6.5}$$

$$H_2 + OH^- \rightarrow M - H_{ad} + H_2O + e^- \, (Heyrovsky) \tag{6.6}$$

$$H_{ad} + OH^- \rightarrow H_2O + e^- + M \, (Volmer) \tag{6.7}$$

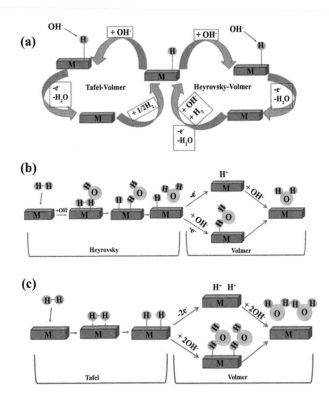

FIGURE 6.3 (a) HOR pathways with OH$_{ad}$ on the electrocatalysts surface overall mechanism, (b) Possible alkaline HOR pathways by Tafel-Volmer, and (c) Heyrovsky-Volmer with adsorbed (OH$_{ad}$) on the surface of an electrocatalyst in detail.

Here M denotes surface sites with a high affinity to H atoms and H_{ad} is the adsorbed hydrogen. Due to the involvement of OH$^-$ in the Heyrovsky and Volmer step, fundamental understanding of the HOR pathway in alkaline media remains under considerable discussion to date. According to Cong and Song (2018) there are two different types of ideas for HOR, i.e., (i) HOR pathway with OH$^-$ in the solution (Rheinlander et al., 2014) and (ii) HOR pathway with OH_{ad} on the electrocatalysts surface (formed by $OH^- \rightarrow OH_{ad}$ + e$^-$) (Strmcnik et al., 2013; Alesker et al., 2016; Alia et al., 2013).

6.3 TYPES OF ELECTROCATALYSTS FOR HOR

Electrocatalysts significantly affect the HOR activity, hence wide varieties of electrocatalysts are reported for HOR individually, or in combined form have their unique advantages and disadvantages. In-order to make an easy understanding, all the electrocatalysts reported to be used for HOR till date are categorized and shown in Figure 6.4.

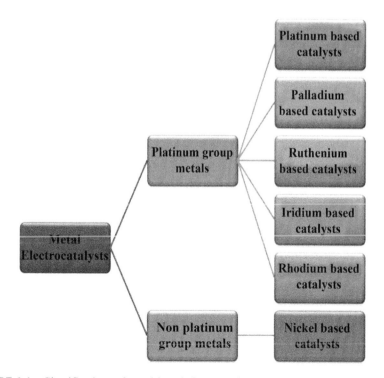

FIGURE 6.4   Classifications of metal-based electrocatalysts.

6.3.1 PLATINUM-BASED CATALYSTS

Among the other reported catalysts, platinum is the most active catalyst for HOR due to its exchange current density and mass activity at low overpotential (Davydova et al., 2018). Different forms of Pt-based catalysts were investigated for HOR in alkaline electrolyte, including single-crystal Pt (Schmidt et al., 2002) bulk/polycrystalline Pt (Harrison and Khan, 1971), carbon-supported Pt (Durst et al., 2014; Zheng et al., 2016; Mahoney et al., 2014) Pt-Metal alloys (Wang et al., 2015; Scofield et al., 2016), Pt modified gold (Mahoney et al., 2014) and core-shell structures (Alia et al., 2013; Elbert et al., 2015). Although numerous investigations were carried out on Pt and Pt-based electrocatalysts, however, the detailed mechanism of HOR is still needs to explore more clearly along with the insights into the nature of rate-determining step. Later the kinetic data for HOR was summarized and a conclusion given by Conway et al. (2002) that the HOR rate on Pt is determined by the dissociative adsorption of H_2. However, their report has excluded pH effect on HOR kinetics which was later investigated by Bagotzky et al. (1974) and found a similar conclusion. In the gas phase, the rate of HOR was found to be close to the rate of hydrogen dissociation (Ross and Stonehart, 1974). This suggests the fact that the electrolyte having pH below 4 does not affect the behavior of potential energy along the hydrogen-platinum reaction coordinate. An important conclusion also drawn from above that, the mechanism of HOR on smooth platinum surface is very similar to a sluggish dual-site dissociation and adsorption of Hydrogen molecule followed by fast electrochemical oxidation of adsorbed hydrogen into hydronium ions. The kinetics of HOR on Pt(pc) using 1 M KOH electrolyte shows similar discharge mechanism to the overall reaction $(H_2 + 2OH^- \leftrightarrow 2H_2O + 2e^-)$ (Harrison and Khan, 1971).

Researches were carried out in investigating the properties of Pt and Pt supported carbon on HOR (Ross and Stonehart, 1974). According to Poltorak, the rate-determining step in HOR is hydrogen dissociation and it takes place only on Pt surface of Pt supported carbon materials. Therefore, the presence of carbon in these composites does not affect the HOR activity. From the literature, it was also confirmed that the basically HOR follows the mechanism of dissociative chemisorption of hydrogen on Pt surface. This is according to Tafel mechanism. Attempts were taken by modifying Pt catalyst with other transition metals such as nanocatalysts of Ru with monolayers (MLs) of Pt (Elbert et al., 2015). This modified catalyst showed enhanced surface-specific activity in comparison to unmodified nanocatalyst of Pt in alkaline solution. However, the results are different when experimented in acidic solution in comparison with Pt/C catalyst. It was due to the increase in energy of activation value from acidic to alkaline medium which indicates the fact that the rate-determining step switching from Tafel to Volmer reaction. Further investigations on HOR in alkaline medium were carried out on modifying Pt catalyst by core-shell nanocatalyst containing Au in the core with layers of Pt (Friebel et al., 2012). However, the Pt layered Au cored nanocatalyst showed similar behavior to that of bulk Pt and/or Pt-supported carbon catalyst and significantly enhanced behavior than pure disc-like Au catalyst. Many attempts were also made to modify Pt catalyst by platinum coated copper nanowires (NWs), Cu template Pt nanotubes, and Pt with 5% monolayered Cu as HOR catalysts (Alia et al., 2013) in alkaline medium. Comparing the results of the above catalysts, it was informed that the specific activity of HOR enhanced due the insertion of Cu in Pt catalyst. Cu provides compressive strain on Pt and supports hydroxyl adsorption thereby improving the HOR activity.

6.3.2   PALLADIUM BASED CATALYSTS

Various investigations are reported on Pd-based catalysts. Experiments carried out using a rotating disc electrode using 10% Pd modified catalyst. It has been observed the exchange current density (by Butler-Volmer fitting and micro polarization region) is much lower in alkaline medium compared to acidic medium (Durst et al., 2014). On Pd/C catalyst, the Volmer step as the rate-limiting step. Similar results of exchange current density were reported for 20%Pd modified catalysts by rotating disc electrode method (Zheng et al., 2016). Hydrogen absorption into bulk Pd may create complexity in the HOR activity, hence the problem can be avoided by the thin layer coating of Pd metal on bulk substrates of Pt, Rh, and Au metals through electrodeposition (Henning et al., 2015; Zheng et al., 2016). Effect of particle size of Pd/C catalysts have also been investigated (Zheng et al., 2016). The results revealed that there is an increase in specific exchange current density for HOR with an increase in particle size of Pd from 3 to 19 nm, after that, it shows a similar activity as like bulk Pd. This might be due to the increased ratio of the sites having weaker hydrogen binding energy (HBE). Summarizing the reports of the kinetic data on HOR in alkaline media, it can be concluded that the Pd layered Au catalysts shows the lowest exchange current density and the highest value was achieved for PdIr/C and for 1.5% Pd/Ni. It was also concluded that Kinetics of HOR is pH-dependent as reported for Pt catalysts. Materials containing crystalline interconnected phases of Ni and Pd were deposited by Alesker et al. (2016). This material when used as anode catalyst for HOR shows higher activity exhibiting the best ever high performance for a Platinum-free alkaline membrane fuel cell which was recorded to 0.4 A cm^{-2} at 0.6V vs. RHE. The higher HOR activity of this Pd/Ni electrocatalyst may be due to the better contact between Pd and Ni nanoarchitecture. This observation was also supported by the negative shift in HOR onset potential (about 200 mV) in comparison to pure Pd-catalyst.

6.3.3   RUTHENIUM BASED CATALYSTS

Handful literatures are available on Ru as an electrocatalyst for HOR reaction in alkaline medium. The Ru/C catalyst shows size-dependent HOR activity and maximum performance has been observed for particles of 3 nm (Ohyama et al., 2013). This was quite different observation to Ir and Pd-based catalysts, as in the case of Ir (Zheng et al., 2015) and Pd (Zheng et al., 2016) based catalysts, the exchange current density is associated with decrease in particle size. 3 nm Ru/C catalyst showed HOR activity higher than the commercially available carbon-supported platinum nanoparticles (PtNPs). As the surface-specific activity of Ru NPs changes its behavior with respect to its size, i.e., amorphous structure (below 3 nm size) to well defined crystalline structure (above 3 nm size) and nanocrystallite rough surface (at about 3 nm size), it was compared to show volcano-shaped specific activity. This was the key in achieving high HOR activity (Ohyama et al., 2013). But the pure Ru catalysts show reduced current density above 0.2 V. This may be due to the inhibition in the H$_2$ adsorption by oxygenated species.

6.3.4   IRIDIUM BASED CATALYSTS

A few investigations are carried out on the catalytic activity of the Iridium catalyst for HOR in alkaline medium. Zheng et al. (2015) investigated taking Ir/C catalyst and found that with an increase in particle size (from 3–12 nm) results in an increase in the fraction of Ir sites which have low metal-HBE. The HOR activities associated with the surface area of the lowest HBE site does not depend on the total electrochemical surface area. This observation suggests the fact that sites with low HBE are the active sites for HOR. It was further suggested the fact that modified Ir nanostructures such as Ir nanowires, Ir nanotubes having a high fraction of low index planes, and higher HOR activities than small Ir-NPs. Investigations made by Montero et al. (2016) suggested that the HOR follows the Tafel-Volmer route when it takes place on Ir electrodes and was confirmed in both alkaline and acidic solutions (Montero et al., 2013). However, the values are different for the Heyrovsky-Volmer route in alkaline and acidic solutions. A model was proposed by Montero et al. (2016) explaining the difference in both the route in acidic and alkaline solutions. According to the model in, for an acidic solution, incorporation of proton takes place in the water network with the transfer of an electron to the metallic substrate. However, in an alkaline solution, the transfer of electron occurs in parallel to the disappearance of hydroxyl ions. The report suggests that; here transfer of Hydrogen bonds occurs through the network of water. Hence, it was difficult for OH^- ion to acquire a suitable spatial configuration for ET. Due to this activation energy of Volmer, route increases in alkaline solution resulting decrease of the corresponding equilibrium reaction rates. In spite of significant catalytic activity, the high cost (like Pt) limits its commercial application.

6.3.5 NICKEL-BASED CATALYSTS

Being the only non-platinum group metals, Ni-based catalysts was extensively investigated as an efficient electrocatalyst for HOR with enhanced surface activity in alkaline solution. Ni-based catalysts used in HOR are in the form of Raney nickel catalyst with doped metals (Ewe et al., 1974, 1975; Chatterjee et al., 2004; Mund et al., 1977; Lee et al., 1998; Kenjo, 1988; Jenseit et al., 1990; Raj and Vasu, 1990; Al-Saleh et al., 1996, 1998; Sleem et al., 1997; Kiros and Schwartz, 2000; Kiros et al., 2003; Linnekoski et al., 2007), binary, and ternary metal catalysts as Ni one component, (Sheng et al., 2014), supported Ni catalysts (Lu et al., 2008) and single-crystalline Ni facet (Floner et al., 1990), etc. Among which, commonly used Ni-based catalysts are Raney Nickel catalysts due to their flexibility in liquid electrolyte or gas diffusion electrodes in alkaline medium. But the activity of pure Raney Ni is very poor (Kiros and Schwartz, 2000). Hence extensive studies were carried out to improve the catalytic activity of Ni by doping other transition metals such as Cr (Izhar and Nagai, 2009), Fe (Chatterjee et al., 2004), Ti (Ewe et al., 1974; Mund et al., 1977), Co,(Chatterjee et al., 2004; Lee et al., 1998; Jenseit et al., 1990), Cu (Al-Saleh et al., 1996), Mo (Ewe et al., 1974) into the Ni-Al alloy before its extraction by alkali KOH. Another problem associated with Ni is that, it works in high alkaline media of about 6M KOH hence was unable to work in AEMFCs which generally use an alkaline media having 0.1–1M KOH (Lu et al., 2008) (Figure 6.5).

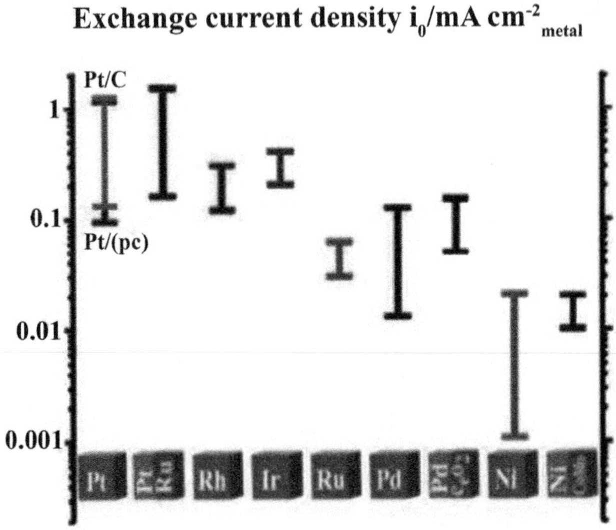

FIGURE 6.5 Ranges of experimental values of exchange current density in HOR for monometallic surfaces (green cubes) and for bifunctional catalysts (green-orange cubes). *Source:* Reprinted with permission from Davydova et al. (2018). © American Chemical Society.

Ternary Ni alloys were also investigated for HOR in alkaline media. Electrochemically deposited CoNiMo catalyst found to exhibit significantly higher activity than Ni for HOR (Sheng et al., 2014). Various analyses such as density functional theory (DFT) and temperature-programmed desorption showed that CoNiMo catalyst shows similar metal-HBE to that of Pt and significantly lower metal-HBE of Ni. This was the consequence of a multimetallic bond which changes the HBE of Ni and ultimately became the key factor for the superior HOR activity (Sheng et al., 2014). Zhuang et al. (2016) has achieved the highest HOR activity (till date) of 9.3 Ag^{-1} at 0.05V in alkaline KOH media at room temperature. They have used composite catalyst of Ni NPs with support of nitrogen-doped carbon nanotubes. It is interesting to note that nitrogen supported carbon nanotubes themself are very poor HOR catalysts, but they surprisingly enhance the catalytic activity of Ni-nano particles. A comparison of some selected polarization curves recorded by RDE method was given below in Figure 6.3. From which it is easily predicted the activity of the existing catalysts: The most active and the most expensive catalyst is Pt, along with Rh and Ir. Ru is showing competitive current densities among the PGM. Among the non-PGM catalysts, Ni is considered to be the best with little modification and doping with other metal ions. This suggests the fact that certain non-PGM catalysts can replace PGM catalysts in the future with further modification. Figure 6.5 shows the result analysis of the surface-specific exchange current density. The result reveals that the catalytic activity of monometallic systems can be enhanced by several ways such as i) modification of catalysts by doping into bulk or modifying the surface in several MLs, by the help of ligands using bi-functional mechanism ii) by changing the strain of the particles with the use of a substrate or by iii) modification of crystal structure by thermal treatment.

6.3.6 TRANSITION METAL PHOSPHIDES (TMPS)

Recently transition metal phosphides (TMPs) have gathered significant attention due to their unique electrochemical properties and showed some impressive catalytic performances in various fields. It shows high activity, better durability, and nearly cent percent Faradic efficiency in strong acid, strong base as well as the neutral medium. In general, a good catalytic material is a combination of transition metal elements and non-metallic elements. The commonly used transition metals are Co, Fe, Ni, Mo, etc., whereas non-metal elements are C, N, O, S, P, etc. A wide variety of phosphides can be synthesized using these transition metal elements. In general, the TMPs can be categorized as (a) single TMPs and (b) multiple TMPs. Typical single TMPs mainly include CoP, Co_2P, Ni_2P, Ni_5P_4, FeP, Fe_2P, etc. Multiple TMPs are mainly $NiCo_2P_x$, Co-Fe-P, NiFe-P, etc. There are some other element doped TMPs, such as Fe-doped CoP, Fe-doped Ni_2P, etc. (Izhar and Nagai, 2009; Pei et al., 2018). Single TMPs are the most common class of phosphides. Among them, the Co and Ni-based TMPs are broadly explored for HOR by various research groups. Many interesting morphologies like nanocubes, cuboids, nanosheets, nanoflowers, quantum dots have been synthesized by several groups following different synthetic methodologies including hydrothermal and deposition techniques. Compared to single TMPs, bimetallic phosphides presently conquer a very imperative position. Many reports on bimetallic phosphides have been established which shows different synthesis routes of preparation. Some of these phosphides found to exhibit excellent catalytic performance which is comparable to those of precious metal-based catalysts.

A series of TMPs like WP, CoP, NiP, Ni-WP, Co-MoP have been investigated by Izhar and Nagai group (2009). The HOR activity was examined by the chronoamperometric method through the current time plot where mass activity indicates the HOR activity. Among the TMPs series, Co-WP showed the highest mass activity (0.087 $\mu A \mu g^{-1}$) which was reconfirmed by slow-scan voltammetry measurements. The exchange current density (J_0) for all phosphides was calculated using Butler-Volmer rate law and all the results were in a well-agreement with results obtained from chronoamperometric. The bimetallic phosphide Co-WP showed the highest exchange current density of 3.03 $mA Cm^{-2}$ among all (Popczyk et al., 2005; Budniok and Kupka, 1989). It was also found that all the bimetallic phosphide samples showed an increased electrocatalytic activity than tungsten phosphide. The activity of metal phosphides realizes on metal promoter which ultimately enhances the HOR activity which showed a probability for transition metals to be used as electrocatalyst for fuel cell anodes (Zhou et al., 2018; Chang et al., 2018) (Figure 6.6).

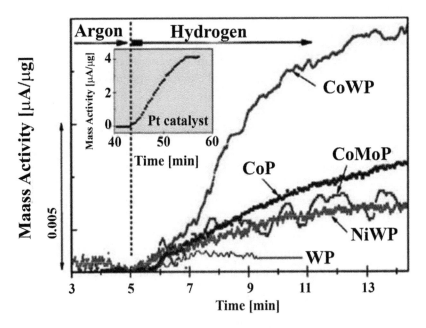

FIGURE 6.6 Mass activity plot for the series of transition metal phosphides (TMPs) like WP, CoP, NiP, Ni-WP, Co-MoP, and CoWP. Inset shows a similar variation for Platinum. *Source:* Reprinted with permission from Izhar and Nagai (2009). © Elsevier.

6.4 CONCLUSION

In this chapter, initially, we have discussed about the general mechanism of HOR in fuel cell and different pathways in alkaline medium. In the second section, we approached to discuss the performance, nature, efficiency, etc., of a wide variety of HOR electrocatalyst based on metal and non-metal. A wide comparative explanation regarding the platinum group with non-platinum group electrocatalysts is provided in this context.

KEYWORDS

- **electrocatalyst**
- **fuel cell**
- **hydrogen oxidation reaction**
- **nanoparticles**
- **oxygen evolution reaction**
- **proton exchange membrane**

REFERENCES

Al-Saleh, M. A., Al-Zakri, A. S., & Gultekin, S. (1997). Preparation of Raney-Ni gas diffusion electrode by filtration method for alkaline fuel cells. *J. Appl. Electrochem., 27*, 215–220

Al-Saleh, M. A., Gultekin, S., Al-Zakri, A. S., & Khan, A. A. A., (1996). Steady state performance of copper impregnated Ni/PTFE gas diffusion electrode in alkaline fuel cell. *Int. J. Hydrogen Energy., 21*, 657–661.

Al-Saleh, M. A., Sleem-Ur-Rahman, Kareemuddin, S. M. M., & Al-Zakri, A. S., (1998). Novel methods of stabilization of Raney-nickel catalyst for fuel-cell electrodes. *J. Power Sources, 72*, 159–164.

Alesker, M., Page, M., Shviro, M., Paska, Y., Gershinsky, G., Dekel, D. R., & Zitoun, D., (2016). Palladium/nickel bifunctional electrocatalyst for hydrogen oxidation reaction in alkaline membrane fuel cell. *J. Power Sources, 304*, 332–339.

Alia, S. M., Pivovar, B. S., & Yan, Y., (2013). Platinum-coated copper nanowires with high activity for hydrogen oxidation reaction in base. *J. Am. Chem. Soc., 135*, 13473–13478.

Bagotzky, V. S., & Osetrova, N. V., (1973). Investigations of hydrogen ionization on platinum with the help of micro-electrodes. *J. Electroanal. Chem. Interfacial Electrochem., 43*, 233–249.

Barnett, C. J., Burstein, G. T., Kucernak, A. R. J., & Williams, K. R., (1997). Electrocatalytic activity of some carburized nickel, tungsten, and molybdenum compounds. *Electrochim. Acta, 42*, 2381–2388.

Bashyam, R., & Zelenay, P., (2006). A class of non-precious metal composite catalysts for fuel cells. *Nature, 443*, 63.

Bronoel, G., Museux, E., Leclercq, G., Leclercq, L., & Tassin, N., (1991). Study of hydrogen oxidation on carbides. *Electrochim. Acta, 36*, 1543–1547.

Budniok, A., & Kupka, J., (1989). The evolution of oxygen on amorphous Ni-Co-P alloys. *Electrochim. Acta, 34*, 871–873.

Chang, J., Li, K., Wu, Z., Ge, J., Liu, C., & Xing, W., (2018). Sulfur-doped nickel phosphide nanoplates arrays: A monolithic electrocatalyst for efficient hydrogen evolution reactions. *ACS Appl. Mater. Interfaces, 10*, 26303–26311.

Chatterjee, A. K., Banerjee, R., & Sharon, M., (2004). Enhancement of hydrogen oxidation activity at a nickel-coated carbon beads electrode by cobalt and iron. *J. Power Sources, 137*, 216–221.

Cong, Y., Yi, B., & Song, Y., (2018). Hydrogen oxidation reaction in alkaline media: From mechanism to recent electrocatalysts. *Nano Energy, 44*, 288–303.

Conway, B. E., & Tilak, B. V., (2002). Interfacial processes involving electrocatalytic evolution and oxidation of H$_2$, and the role of chemisorbed H. *Electrochim. Acta, 47*, 3571–3594.

Davydova, E. S., Mukerjee, S., Jaouen, F., & Dekel, D. R., (2018). Electrocatalysts for hydrogen oxidation reaction in alkaline electrolytes. *ACS Catal., 8*, 6665–6690.

Dresselhaus, M. S., & Thomas, I. L., (2001). Alternative energy technologies. *Nature, 414*, 332.

Durst, J., Siebel, A., Simon, C., Hasché, F., Herranz, J., & Gasteiger, H. A., (2014). New insights into the electrochemical hydrogen oxidation and evolution reaction mechanism. *Energy Environ. Sci., 7*, 2255–2260.

Elbert, K., Hu, J., Ma, Z., Zhang, Y., Chen, G., An, W., Liu, P., et al., (2015). Elucidating hydrogen oxidation/evolution kinetics in base and acid by enhanced activities at the optimized Pt shell thickness on the Ru core. *ACS Catal., 5*, 6764–6772.

Ewe, H., Justi, E., & Schmitt, A., (1974). Ewe, H., Justi, E., & Schmitt, A., (1974). Structure and properties of Raney nickel catalysts with alloys for alkaline fuel cells. Electrochim. *Acta, 19*, 799–808.

Ewe, H., Justi, E., & Schmitt, A., (1975). Ewe, H., Justi, E., & Schmitt, A., (1975). Studies on the preservation of Raney nickel mixed catalysts. *Energy Convers., 14*, 35–41.

Floner, D., Lamy, C., & Leger, J. M., (1990). Electrocatalytic oxidation of hydrogen on polycrystal and single-crystal nickel electrodes. *Surf. Sci., 234*, 87–97.

Friebel, D., Viswanathan, V., Miller, D. J., Anniyev, T., Ogasawara, H., Larsen, A. H., O'Grady, C. P., et al., (2012). Balance of nanostructure and bimetallic interactions in Pt model fuel cell catalysts: *In situ* XAS and DFT Study. *J. Am. Chem. Soc., 134*, 9664–9671.

Gasteiger, H. A., & Marković, N. M., (2009). Chemistry. Just a dream-or future reality? *Science, 324*, 48–49.

Harrison, J. A., & Khan, Z. A., (1971). The oxidation of hydrogen. *J. Electroanal. Chem. Interfacial Electrochem., 30*, 327–330.

Henning, S., Herranz, J., & Gasteiger, H. A., (2015). Bulk-palladium and palladium-on-gold electrocatalysts for the oxidation of hydrogen in alkaline electrolyte. *J. Electrochem. Soc., 162*, F178–F189.

Izhar, S., & Nagai, M., (2009). Transition metal phosphide catalysts for hydrogen oxidation reaction. *Catal. Today, 146*, 172–176.

Izhar, S., Yoshida, M., & Nagai, M., (2009). Characterization and performances of cobalt-tungsten and molybdenum-tungsten carbides as anode catalyst for PEFC. *Electrochim. Acta, 54*, 1255–1262.

Jenseit, W., Khalil, A., & Wendt, H., (1990). Material properties and processing in the production of fuel cell components: I. hydrogen anodes from Raney nickel for lightweight alkaline fuel cells. *J. Appl. Electrochem., 20*, 893–900.

Kenjo, T., (1988). Doping effects of transition metals on the polarization characteristics in Raney nickel hydrogen electrodes. *Electrochim. Acta, 33*, 41–46.

Kiros, Y., & Schwartz, S., (2000). Long-term hydrogen oxidation catalysts in alkaline fuel cells. *J. Power Sources, 87*, 101–105.

Kiros, Y., Majari, M. A., & Nissinen, T., (2003). Effect and characterization of dopants to Raney nickel for hydrogen oxidation. *J. Alloys Compd., 360*, 279.

Krischer, K., & Savinova, E. R., (2008). Fundamentals of electro catalysis. In: *Handbook of Heterogeneous Catalysis* (pp. 1873–1905). American Cancer Society.

Lee, H. K., Jung, E. E., & Lee, J. S., (1998). Enhancement of catalytic activity of Raney Nickel by cobalt addition. *Mater. Chem. Phys., 55*, 89–93.

Lee, K., Ishihara, A., Mitsushima, S., Kamiya, N., & Ota, K., (2004). Stability and electrocatalytic activity for oxygen reduction in WC + Ta catalyst. *Electrochim. Acta, 49*, 3479–3485.

Lefèvre, M., Dodelet, J. P., & Bertrand, P., (2005). Molecular oxygen reduction in PEM fuel cell conditions: ToF-SIMS analysis of co-based electrocatalysts. *J. Phys. Chem. B, 109*, 16718–16724.

Li, B., Qiao, J., Yang, D., Zheng, J., Ma, J., Zhang, J., & Wang, H., (2009). Synthesis of a highly active carbon-supported Ir–V/C catalyst for the hydrogen oxidation reaction in PEMFC. *Electrochim. Acta, 54*, 5614–5620.

Linnekoski, J. A., Krause, A. O. I., Keskinen, J., Lamminen, J., & Anttila, T., (2007). Processing of Raney-nickel catalysts for alkaline fuel cell applications. *J. Fuel Cell Sci. Technol., 4*, 45–48.

Lu, S., Pan, J., Huang, A., Zhuang, L., & Lu, J., (2008). Alkaline polymer electrolyte fuel cells completely free from noble metal catalysts. *Proc. Natl. Acad. Sci., 105*, 20611–20614.

Mahoney, E. G., Sheng, W., Yan, Y., & Chen, J. G., (2014). Platinum-modified gold electrocatalysts for the hydrogen oxidation reaction in alkaline electrolytes. *Chem. Electro. Chem., 1*, 2058–2063.

Maruyama, J., & Abe, I., (2005). Formation of platinum-free fuel cell cathode catalyst with highly developed nanospace by carbonizing catalase. *Chem. Mater., 17*, 4660–4667.

Meng, H., & Shen, P. K., (2005). Tungsten carbide nanocrystal promoted pt/c electro catalysts for oxygen reduction. *J. Phys. Chem. B, 109*, 22705–22709.

Montero, M. A., De Chialvo, M. R. G., & Chialvo, A. C., (2016). Evaluation of the kinetic parameters of the hydrogen oxidation reaction on nanostructured iridium electrodes in alkaline solution. *J. Electroanal. Chem., 767*, 153–159.

Montero, M. A., Fernández, J. L., Gennero, D. C. M. R., & Chialvo, A. C., (2013). Kinetic study of the hydrogen oxidation reaction on nanostructured iridium electrodes in acid solutions. *J. Phys. Chem. C, 117*, 25269–25275.

Morishita, T., Soneda, Y., Hatori, H., & Inagaki, M., (2007). Carbon-coated tungsten and molybdenum carbides for electrode of electrochemical capacitor. *Electrochim. Acta, 52*, 2478–2484.

Mund, K., Richter, G., & Von, S. F., (1977). Titanium-containing raney nickel catalyst for hydrogen electrodes in alkaline fuel cell systems. *J. Electrochem. Soc., 124*, 1–6.

Nagai, M., Yoshida, M., & Tominaga, H., (2007). Tungsten and nickel tungsten carbides as anode electrocatalysts. *Electrochim. Acta, 52*, 5430–5436.

Obradović, M. D., Gojković, S. L., Elezović, N. R., Ercius, P., Radmilović, V. R., et al., (2012). The kinetics of the hydrogen oxidation reaction on WC/Pt catalyst with low content of Pt nano-particles. *J. Electroanal. Chem., 671*, 24–32.

Ohyama, J., Sato, T., Yamamoto, Y., Arai, S., & Satsuma, A., (2013). Size specifically high activity of Ru nanoparticles for hydrogen oxidation reaction in alkaline electrolyte. *J. Am. Chem. Soc., 135*, 8016–8021.

Okamoto, H., Kawamura, G., Ishikawa, A., & Kudo, T., (1987). Characterization of oxygen in WC catalysts and its role in electrocatalytic activity for methanol oxidation. *J. Electrochem. Soc., 134*, 1645–1649.

Palanker, V. S., Gajyev, R. A., & Sokolsky, D. V., (1977). On adsorption and electro-oxidation of some compounds on tungsten carbide; their effect on hydrogen electro-oxidation. *Electrochim. Acta, 22*, 133–136.

Pei, Y., Cheng, Y., Chen, J., Smith, W., Dong, P., Ajayan, P. M., et al., (2018). Recent developments of transition metal phosphides as catalysts in the energy conversion field. *J. Mater. Chem. A, 6*, 23220–23243.

Popczyk, M., Budniok, A., & Lasia, A., (2005). Electrochemical properties of Ni-P electrode materials modified with nickel oxide and metallic cobalt powders. *Int. J. Hydrogen Energy, 30*, 265–271.

Raj, I. A., & Vasu, K. I., (1990). Transition metal-based hydrogen electrodes in alkaline solution electrocatalysis on nickel based binary alloy coatings. *J. Appl. Electrochem., 20*, 32–38.

Rees, E. J., Essaki, K., Brady, C. D. A., & Burstein, G. T., (2009). Hydrogen electrocatalysts from microwave-synthesized nanoparticulate carbides. *J. Power Sources, 188*, 75–81.

Rheinlander, P. J., Herranz, J., Durst, J., & Gasteiger, H. A., (2014). Kinetics of the hydrogen oxidation/evolution reaction on polycrystalline platinum in alkaline electrolyte reaction order with respect to hydrogen pressure. *J. Electrochem. Soc., 161*, F1448–F1457.

Ross, P. N., & Stonehart, P., (1974). Correlations between electrochemical activity and heterogeneous catalysis for hydrogen dissociation on platinum. *J. Res. Inst. Catal. Hokkaido Univ., 22*, 22–41.

Schmidt, T. J., Ross, P. N., & Markovic, N. M., (2002). Temperature dependent surface electrochemistry on Pt single crystals in alkaline electrolytes: Part 2. The hydrogen evolution/oxidation reaction. *J. Electroanal. Chem., 524,525*, 252–260.

Scofield, M. E., Zhou, Y., Yue, S., Wang, L., Su, D., Tong, X., Vukmirovic, M. B., et al., (2016). Role of chemical composition in the enhanced catalytic activity of pt-based alloyed ultrathin nanowires for the hydrogen oxidation reaction under alkaline conditions. *ACS Catal., 6*, 3895–3908.

Sheng, W., Bivens, A. P., Myint, M., Zhuang, Z., Forest, R. V., Fang, Q., et al., (2014). Non-precious metal electrocatalysts with high activity for hydrogen oxidation reaction in alkaline electrolytes. *Energy Environ. Sci., 7*, 1719–1724.

Sheng, W., Gasteiger, H. A., & Shao-Horn, Y., (2010). Hydrogen oxidation and evolution reaction kinetics on platinum: Acid vs. alkaline electrolytes. *J. Electrochem. Soc., 157*, B1529–B1536.

Strmcnik, D., Escudero-Escribano, M., Kodama, K., Stamenkovic, V. R., Cuesta, A., & Markovic, N. M., (2010). Enhanced electrocatalysis of the oxygen reduction reaction based on patterning of platinum surfaces with cyanide. *Nat. Chem., 2*, 880–885.

Strmcnik, D., Uchimura, M., Wang, C., Subbaraman, R., Danilovic, N., Van, D. V. D., Paulikas, A. P., et al., (2013). Improving the hydrogen oxidation reaction rate by promotion of hydroxyl adsorption. *Nat. Chem., 5*, 300–306.

Tang, D., Pan, J., Lu, S., Zhuang, L., & Lu, J., (2010). Alkaline polymer electrolyte fuel cells: Principle, challenges, and recent progress. *Sci. China Chem., 53*, 357–364.

Tang, H., Yin, H., Wang, J., Yang, N., Wang, D., & Tang, Z., (2013). Molecular architecture of cobalt porphyrin multilayers on reduced graphene oxide sheets for high-performance oxygen reduction reaction. *Angew. Chemie Int. Ed., 52*, 5585–5589.

Turner, J. A., (2004). Sustainable hydrogen production. *Science, 80, 305*, 972–974.

Wang, H., Liang, Y., Li, Y., & Dai, H. (2011). $Co_{1-x}S$–graphene hybrid: A high-performance metal chalcogenide electrocatalyst for oxygen reduction. *Angew. Chemie Int. Ed., 50*, 10969–10972.

Wang, Y., Wang, G., Li, G., Huang, B., Pan, J., Liu, Q., Han, J., Xiao, L., Lu, J., & Zhuang, L., (2015). Pt-Ru catalyzed hydrogen oxidation in alkaline media: Oxophilic effect or electronic effect? *Energy Environ. Sci., 8*, 177–181.

Wood, T. E., Tan, Z., Schmoeckel, A. K., O'Neill, D., & Atanasoski, R., (2008). Non-precious metal oxygen reduction catalyst for PEM fuel cells based on nitroaniline precursor. *J. Power Sources, 178*, 510–516.

Yoo, S. J., & Sung, Y. E., (2010). Design of palladium-based alloy electrocatalysts for hydrogen oxidation reaction in fuel cells. In: *Fuel Cell Science* (pp. 111–146). John Wiley & Sons, Ltd.

Zhang, L., Zhang, J., Wilkinson, D. P., & Wang, H., (2006). Progress in preparation of non-noble electrocatalysts for PEM fuel cell reactions. *J. Power Sources, 156*, 171–182.

Zheng, J., Sheng, W., Zhuang, Z., Xu, B., & Yan, Y., (2016a). Universal dependence of hydrogen oxidation and evolution reaction activity of platinum-group metals on pH and hydrogen binding energy. *Sci. Adv., 2*.

Zheng, J., Zhou, S., Gu, S., Xu, B., & Yan, Y., (2016b). Size-dependent hydrogen oxidation and evolution activities on supported palladium nanoparticles in acid and base. *J. Electrochem. Soc., 163*, F499–F506.

Zheng, J., Zhuang, Z., Xu, B., & Yan, Y., (2015). Correlating hydrogen oxidation/evolution reaction activity with the minority weak hydrogen-binding sites on Ir/C catalysts. *ACS Catal., 5*, 4449–4455.

Zhou, Q., Shen, Z., Zhu, C., Li, J., Ding, Z., Wang, P., Pan, F., et al., (2018). Nitrogen-doped CoP electrocatalysts for coupled hydrogen evolution and sulfur generation with low energy consumption. *Adv. Mater., 30*, 1800140.

Zhuang, Z., Giles, S. A., Zheng, J., Jenness, G. R., Caratzoulas, S., Vlachos, D. G., & Yan, Y., (2016). Nickel supported on nitrogen-doped carbon nanotubes as hydrogen oxidation reaction catalyst in alkaline electrolyte. *Nat. Commun., 7*, 10141.

CHAPTER 7

Theoretical and Computational Investigations of Li-Ion Battery Materials and Electrolytes

ANOOP KUMAR KUSHWAHA

School of Basic Sciences, Indian Institute of Technology, Bhubaneswar, Odisha – 752050, India

ABSTRACT

In the last two decades, very much efforts have been made to develop next-generation energy storage devices, which could be implemented in electric vehicles. Therefore, in recent years, the development of cheap and sustainable electrochemical energy storage systems (e.g., lithium, sodium, and potassium, etc., ions batteries) is a hot topic for research. The alkali metals ion batteries, e.g., Li-ion battery, Na-ion battery, and K-ion battery, have been widely studied as the power sources for a wide range application from portable electronic devices to electric vehicles. However, due to having high capacity and long cycle life, Li-ion battery has been found as a suitable candidate for such purpose. Since this field is advancing rapidly and attracting an increasing number of researchers, therefore it is required to summaries the current progress in the development of electrode materials and electrolytes as well as key scientific challenges in Li-ion battery from a theoretical and computational point of view. First-principles density functional theory (DFT) and molecular dynamics based computational methodology usually considered for designing the electrode/electrolytes materials and further used to understand the chemical behavior. In this chapter, we have discussed the key aspect of Li-ion battery from a theoretical perspective, e.g., working principles of Li-ion batteries, cathode, and anode materials and electrolyte solution, as well as future research direction based on computational, predicted and designed electrode materials and electrolyte.

7.1 INTRODUCTION

7.1.1 ENERGY STORAGE

Energy is the basic requirement of our daily life and is directly associated with the economy and productivity of a country. From ancient times till now, many of the energy sources, e.g., combustion of wood, coal, and nuclear power have been discovered/invented to fulfill the requirement. Nowadays, the energy is harvested either by the natural resources or by the transformation of such sources to other kinds of energy (energy carriers). For instance, energy can directly be extracted by the combustion of natural gas, coal, petrol, diesel, etc., while in other cases; these fuels are used to generate electricity which is then used to carry out effective work. Currently, fossil fuels (petroleum, coal, natural gas) are the main source of energy, which can be realized in Figure 7.1. The global energy consumption data of 2017 (Figure 7.1) shows that the fossil fuel share stands at ~85% (~ 34% by petroleum, ~ 28% by coal, and ~ 23% by natural gas) of total available primary energy sources ("BP Statistical Review on World Energy," 2017). Non-fossil sources including hydroelectric (7%), nuclear (4%), and others like solar/wind and bio (4%) share the rest of ~15% of total available energy sources. In 2017, the world's primary energy consumption increased up to 2.2% (1.6% in 2016), which is found to be highest since 2013 (Homewood, 2018). The growing energy consumption would lead to serious issues in the future because of the limited stock of fossil fuels. Furthermore, the use of fossil fuels has been found as the major cause of environmental pollutions. For example, fossil fuels release the greenhouse gases (e.g., methane, carbon dioxide, etc.) and harmful gases (e.g., sulfur dioxide, carbon monoxide (CO), nitrous oxide, etc.) during combustion, which lead to several serious problems such as damage to the ozone layer, acid rain, and health-related issues. To resolve the aforementioned serious problems, clean, and renewable energy sources must be adopted.

The renewable energy, extracted from the natural resources such as sun, wind, geothermal, biomass, rain, tides of the ocean, etc., can be replenished continuously. According to the International Renewables Energy Agency (IRENA), the renewable energy produced 24.5% of the world's total electricity in 2016, which was 18.5% in 2008 ("Renewable Energy Highlights," 2018). The increment in global renewable energy capacity data shows that the renewable energy could be a better alternative for the fossil fuels. According to the REN21's report, the renewable energy contributed 19.3%

(8.9% by traditional biomass, 4.2% by geothermal and traditional biomass, 3.9% by hydroelectricity, and 2.2% by wind and solar) of total global energy consumptions in 2017 ("REN21's 2017 Annual Report," 2017). Efficient renewable energy sources provide a positive impact on the energy security, economic benefits, and climate changes. One more benefit is that renewable energy resources are widespread, while the fossil fuels have limited availability around the globe ("Executive Summary," 2012). Apart from several benefits, the renewable energy sources have few limitations such as dependency on the time and season. For example, the solar system can't be used during the night and hydroelectricity can't be generated at droughty places. This problem can be only solved by using the power storage system. Therefore, it is essential to develop efficient energy storage devices, which could store the energy (produced by renewable energy sources), for a certain period of time.

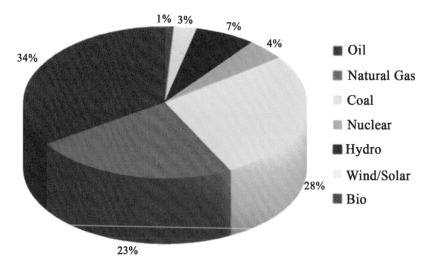

FIGURE 7.1   Global primary energy consumption in 2017.
Source: Reprinted with permission from BP Statistical Review of World Energy (2018).

Energy storage devices are used to store the energy produced by the energy resources and can release it according to the requirement. As the renewable energy, production cycles have a large fluctuations owing primarily to the dependency on the day and seasons, energy storage devices could be better option for the energy management. In last few years, the energy consumption and production by the renewable energy sources drastically increased only due to the development of efficient energy storage technologies. This increment can be realized by a recent report of market research firm IHS, which mentioned that energy storage market is growing exponentially. According to the IHS report, the annual installation size was ~0.34 gigawatts in 2013, which then increased to 6 gigawatts in 2017, and there are further plan to enhance it up to 40 gigawatts by 2022 ("Energy storage Association," n.d.). In spite of the increasing market range and improved storage capacity of energy storage system, the energy devices are still not sufficient to fulfill the criteria. Therefore, further invention and research is required to improve the efficiency and scaled commercialization of storage devices. On the basis of the form of energy stored (for example, chemical, electrical, thermal, and mechanical) in the storage devices, the energy storage systems can be distinguished into several categories. However, on the basis of direct and indirect storages of energy, the energy storage is classified, which has been depicted in Figure 7.2.

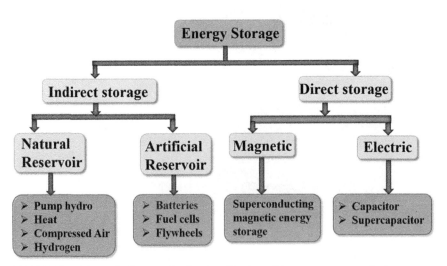

FIGURE 7.2   Schematic diagram for the classification of the energy storage devices.
Source: Adapted from Javed (2017).

In the case of direct energy storage, energy is stored and released in the same form without any conversion. Superconducting magnetic energy storage (SMES) and capacitor/supercapacitor are the examples of the direct storage, although they work through different principles. For example, SMES system stores the energy in the magnetic field, which is produced by the flow of direct current in the superconducting coil. This coil is cryogenically cooled to a temperature below the superconducting critical temperature. On the other hand, capacitor, a passive two-terminal electronic component, stores electrical energy in the presence of an applied external electrical field. The capacitors store electrical energy as electrostatic charge and the stored energy can be kept for a long period with negligible loss. In case of indirect energy storage, the energy is stored in one form while released in another form. Indirect energy storage can be further classified into the natural and artificial reservoirs. Pumped hydroelectric energy storage (PHES) and compressed air energy storage (CAES) are examples of the natural reservoir. The PHES consists of two reservoirs and turbine/generator connected via waterway. In this system, energy is stored in the form of gravitational potential energy of water which is pumped from a lower reservoir to the higher one. Currently, this storage method is the most popular and widely used, generating over 90 gigawatts at global level. In the case of CAES, the ambient air is compressed and stored under pressure in an underground cavern. During the requirement of electricity, the pressurized air is heated and expands in an expansion turbine driving a generator for power production. Generally, PHES and CAES methods are used for high power storage application. In the case of an artificial reservoir, the energy is stored either in the form of kinetic energy (flywheel), or electrochemical energy (battery in fuel cell). The energy storage devices based on the electrochemical energy are getting much attention due to their wide application, which have been discussed in details in the following section.

7.1.2   BATTERY SYSTEM

In general, battery is an energy storage device, which stores chemical energy, which can be converted to electric energy as and when required. It consists of one or more than one cells which are connected externally. Generally, battery consists of three major components, i.e., anode, cathode, and electrolytes along with separator which have been discussed in the next section. Batteries can be classified into three major categories; electrochemical cell, physical cell, and biological cell, which are represented in Figure 7.3. The biological battery is also known as bio-battery is an energy storage device, which is found in the living cells of both plants and animals (Ivanov et al., 2010; Kannan et al., 2009). The biological battery produces electricity through the carbohydrates (sugar) by utilizing an enzyme as a catalyst (Prabhulkar et al., 2012). The bio-battery consists of an anode (with the utilization of sugar-digesting enzyme and mediator), a cathode (comprises oxygen-reducing enzymes and mediator), and a separator. Both anode and cathode are separated through a cellophane separator. At the anode, the electrons and hydrogen ions are extracted from the glucose and the hydrogen ions so formed move towards the cathode through the separator. At the cathode, in the presence of oxygen, the hydrogen ions and electrons are absorbed to produce water. During this electrochemical reaction, the electrons pass through the outer circuit and generate electricity (Kannan et al., 2009). In the case of a physical battery, the electrical energy is directly produced and stored through the sunlight and heat. Solar battery and thermal batteries are an example of this hierarchy.

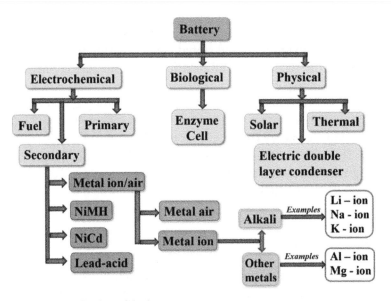

FIGURE 7.3   Classification of the battery system.

Electrochemical batteries are commonly used in several applications such as portable devices, transportation, and industries. These batteries consist of three segments, e.g., anode, cathode, and electrolyte. In this case, chemical energy is converted into electrical energy through a reduction-oxidation (redox) reaction. The electrochemical batteries are of three categories; fuel, primary, and secondary. In the case of fuel cells, the chemical potential energy, supposed to be stored in the molecular bonds, is converted to electrical energy. Among the many available fuel cells, such as proton exchange membrane (PEM) fuel cell, direct methanol fuel cell, alkaline fuel cell (AFC), molten carbonate fuel cell (MCFC), etc., PEM fuel cell has been implemented in large commercial applications due to comparatively lower weight and sufficient power in comparison to the other fuel cells (Barbir and Gomez, 1997; Gamburzev and Appleby, 2002; Wee, 2007). The PEM fuel cell consists of a polymer electrolyte (a material with very high molecular weight composed of a chain of many small molecules with relatively low molecular weights. Furthermore, hydrogen and oxygen are used as the inputs in PEM fuel cell, which provide pure water and electricity as the output products. Although the PEM fuel cell has wide applications, it is limited due to the storage of hydrogen gas, which needs to be stored as compressed gas in a pressurized tank for the use as input fuel cell (Ticianelli et al., 1988). The primary cell (also known as a galvanic cell) is also a kind of the electrochemical energy storage devices as depicted in Figure 7.3. In the primary cell, the chemical oxidation takes place at the anode while reduction occurs at the cathode. The electrochemical reaction, occur inside the cell, are not reversible which lead to non-rechargeable characteristics and the cell is discarded after it is fully discharged.

The secondary cell, also known as rechargeable battery, can be charged and discharged many times in comparison to the primary cell. Similar to the primary cell and fuel cell, the secondary cell consists of anode and cathode separated by electrolyte along with separator and electrolyte additives. On the basis of the electrolyte appearance, the secondary cell could be separated into two categories; aqueous and non-aqueous (Kiehne, 2003; Root, 2011). Lead-acid battery, nickel-cadmium (Ni-Cd) battery, and Nickel-metal hydride (Ni-MH), etc., are some of the examples of aqueous electrolyte based battery. Metal-ion battery (e.g., lithium-ion battery, sodium-ion battery, potassium-ion battery, aluminum-ion battery, etc.), metal-air battery (e.g., Li-air battery, Na-air battery, etc.), sodium-sulfur battery, etc., are the examples of non-aqueous electrolyte based secondary battery systems. The portable electronic devices such as, laptops, cell phones, etc., require smaller and lighter batteries which could be fully charged quickly as well as store the energy for a long time. Therefore, the specific goal of the development of an efficient battery system is related to the improvement in the energy density and power density and reduction/minimization of volumetric and mass constraints. In Figure 7.4, the specific energy density and volumetric energy density has been illustrated and correlated to the various rechargeable battery systems for getting the comparative analysis to pick out the most efficient battery system. Among all the rechargeable battery system, alkali-metal ion battery, especially Li-ion battery, has been found to be the most efficient, which can be realized from Figure 7.4. Apart from being lightweight and small, the Li-ion battery possesses high energy density, high cycle stability, and high energy efficiency, which popularized their use in numerous portable electronic devices (Van Schalkwijk and Scrosati, 2002). Although Li-ion battery is widely used in electronic devices, it is limited by several problems, which requires the development of other alkali metal-ion battery such as Na-ion battery and K-ion battery. In the following section, we have discussed the Li-ion battery in detail.

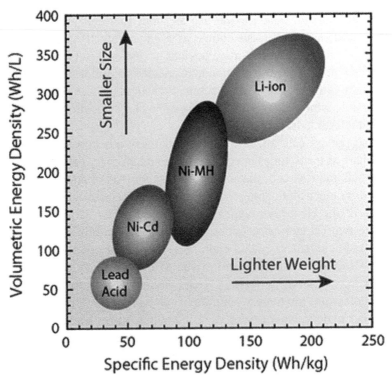

FIGURE 7.4 Diagram comparing the rechargeable battery technologies as a function of volumetric and specific energy densities. The arrows indicate the direction of development to reduce battery size and weight.

Source: Reprinted with permission from Kiehne (2003). © CRC Press.

7.1.3   LI-ION BATTERY: COMPUTATIONAL APPROACH

With the increasing demand of small battery systems in portable electronic devices and large scale battery systems in electric vehicles, research, and development of efficient battery system need significant push from the scientific community. As previously discussed, Li-ion battery has been one of the most efficient rechargeable batteries; its application therefore, is rapidly increasing in small electronic devices and electric vehicles (Lu et al., 2013). However, Li-ion battery is limited by several factors such as cost, aging, safety, etc. The limited resources and subsequent difficulty in extraction of lithium from its mineral, increases the cost of lithium, and leads to increment in the total cost of the battery. Another major disadvantage of Li-ion battery is the burn/fire due to highly reactive nature of the lithium. In order to develop the next generation battery, more effort is required to make batteries of other metal ions such as Na-ion, K-ion, Mg-ion, etc. Recently sodium-ion battery and potassium ion battery have gathered much attention as the low-cost alternatives of lithium-ion battery (LIB) due to abundance of Na and K rich minerals and ores. However, in comparison to the Li-ion battery, finding suitable electrode materials and electrolytes for Na-ion and K-ion battery is difficult due to the larger ionic size of Na and K, which leads to complications in the thermodynamical and kinetic properties. There are several electrodes and electrolyte materials developed, and require further research to facilitate the battery optimization process. Therefore, in this chapter, we have focused on the already commercialized battery systems and also the future scope of the development of electrode and electrolyte materials in the case of Li-ion based batteries.

In the last few decades, several electrode/electrolytes materials for Li-ion batteries have been investigated with various traditional experimental techniques. However, these traditional trial-and-error methods are very time consuming as well as economically challenging. Therefore, it is necessary to develop some useful techniques which could provide an efficient prediction of the electrode/electrolyte materials without performing the experiment. The combination of high-performance computer and efficient computational methodologies has been found to provide good predictions for the electrode/electrolyte materials. With the implementation of efficient algorithm and theoretical model, several computational methods/tools have been developed for the investigation and explanation of the material's properties as well as to provide the productive information of their geometrical structure along with the rational composition. As a result, the design of electrode/electrolyte materials are getting much attentions and requires a parallel development of the computational techniques, from the atomic scale to mesoscale and continuum scale. The strategy of the computational technique is to use it as a predictive tool to solve the problems associated with electrons, atoms, clusters, particles, porous electrodes, cells, and the whole pack. There are several computational methods such as first-principles calculations, molecular dynamics simulation, and Monte Carlo simulations, etc., that have been developed for the investigation of alkali metal ion batteries at different scales, and vary from atomic to continuum scale.

The first-principles density functional theory (DFT) deals with the understanding of fundamental physics/chemistry that occurs in the materials. The first-principles calculations describe the movement of the electrons with the mathematical equation in the many-body quantum mechanics without using any empirical parameters in the calculation. This method provides ultra-fast and accurate solutions to the Schrödinger equation. Currently, the first-principles calculation has become the core simulation technique at the atomic scale. Further, with the combination of molecular dynamics and Monte Carlo techniques, most of the physical and chemical problems of the materials at the atomic scale could easily be solved. In the case of the electrode/electrolyte materials of alkali metal-ion battery, this method describe the ionic interaction at atomic level, charge distribution and charge transfer, metal cation movement and trajectories as well as the kinetics and thermodynamics, which are important parameters that provide the deep understanding of the physical and chemical behavior of electrode and electrolyte materials. With the basic information computed by the first-principles calculations, we can understand the fundamentals of physical and chemical characteristic of electrode/electrolyte materials and can find a way to modify and design the material with superior performance for alkali metal-ion battery. In this chapter, we have discussed the first-principles investigation, understanding, and prediction of the electrode/electrolyte materials of Li-ion battery. Before proceeding to the main context of the Li-ion battery, it is essential to discuss the parameters which decide the quality of the electrode materials.

7.1.4   ELECTRODE MATERIALS PARAMETERS

To enhance the performance of the alkali metal ion battery system, various electrode/electrolyte materials have been developed. These materials could be proposed on the basis of the few parameters by the computational tools and experimental methods. In this section of the chapter, we have discussed these parameters which provide the qualitative information about the electrode materials.

7.1.4.1 CELL VOLTAGE

Cell voltage could be defined either with the open-circuit voltage or closed-circuit voltage. The voltage between two terminals without external load is represented by the open-circuit voltage, while the voltage between two terminals connected with external circuit gives the closed-circuit voltage. Open-circuit voltage (V_{OC}) of the alkali metal ion battery is computed with the chemical potential difference of the negative and positive electrodes as given;

$$V_{OC} = \frac{\mu^N - \mu^P}{F}$$

where, μ^N and μ^P are the chemical potential of negative and positive electrode, respectively. The term, F, represents the Faraday constant. Above equation, shows that the high cell voltage could be achieved by the high value of positive electrode and low value of the negative electrode. Furthermore, thermodynamic stability of the electrode could be obtained by choosing the electrolyte material which should have the bandgap within the electrode material.

7.1.4.2 CONDUCTIVITY

The conductivity of the electrolyte is one transport property that helps to determine how fast a cell can be charged or discharged. The movement of the charge carrier ions between the electrodes through the electrolytic medium is usually investigated by the calculation of ionic conductivity. This phenomenon is the characteristic of the liquid as well as solid-state electrolytes, which could provide initial information for the prediction of electrolyte/electrode materials. Using the first-principles calculation, the ionic conductivity can be computed and further investigated in the context of the alkali metal ion battery system. The mathematical expression for the calculation of specific conductivity could be defined as;

$$\sigma = qnb$$

where, q, n, and b are the charge carrier, concentration, and mobility of the ion, respectively. In the case of ionic conductivity, the charge carriers are encountered due to the other ionic movement, therefore, the electron mobility shows a higher value than that of the ionic mobility. The ionic conductivity could be defined in the term of the specific conductivity as,

$$\lambda_i = \frac{\sigma_i}{|z_i|c_i} = Fb_i$$

where, c_i, and z_i are the molar concentration and charge number respectively. F represents the Faraday's constant. Total ionic conductivity can be expressed as,

$$\sigma = \sum q_i n_i b_i = \sum |z_i| c_i \lambda_i$$

7.1.4.3   SPECIFIC CAPACITY

The capacity represents the maximum amount of energy that can be extracted from the battery under certain specific conditions. The specific capacity, capacity per gram of active material could be calculated as,

$$\text{Specific capacity} = \frac{F \times \Delta X}{molecular\ weight}\ Ah/g$$

where, the term, F, represents the Faraday constant, ΔX shows the amount of reversible charge carriers. For the superior performance of the battery system, the specific capacity should be high. The higher specific capacity of electrode materials could be obtain with selecting low molecular weight and highly reactive charge carriers.

7.1.4.4   POWER DENSITY

Power density is defined as the amount of battery's power per unit volume (or mass). It is also known as volume power density or volume-specific capacity. The power of the battery depends on the cell impedance, charge carrier diffusion through electrolytes, etc. In many cases, the charge carries diffusion in the electrode materials found as the limiting factor and determines the power of the battery. The mathematical expression for the power density of the battery could be expressed as;

Power density = Current (A/kg or A/liter) × Voltage (V) = Energy density/time

7.1.4.5   ENERGY DENSITY

Energy density represents the amount of energy that is stored in a given system or the region of space per unit volume (or weight). The energy per unit weight is defined as gravimetric energy density and energy per unit volume defined as volumetric energy density. The expression is given as:

Gravimetric energy density = (Specific capacity/kg) × Cell voltage

Volumetric energy density = (Specific capacity/liter) × Cell voltage

For the rechargeable battery system, the energy density must be high so that more energy could be stored per a fixed volume or weight. Details of the energy density have been discussed in the next section.

7.2   LITHIUM-ION BATTERY (LIB)

Lithium is the lightest of all metals and has the greatest electrochemical potential which leads to its use as a potential candidate in the battery system. Being a highly reactive element, lithium can store lot of energy in their atomic bonds which imparts high energy density to the battery system. The pioneer research work on the lithium battery began in 1912 by G. N. Lewis, although first-generation non-rechargeable lithium batteries were commercialized in 1970. From 70's till now, LIBs are the fastest-growing energy storage devices due to various transformations including non-rechargeable to rechargeable lithium battery. In 1991, Sony Corporation, for the first time, commercialized rechargeable Li-ion battery which was comparatively safer and lighter than earlier developed lithium based batteries. For example, the energy density of the Li-ion battery has been found twice than energy density of standard Ni-Cd battery. Therefore, the high cell voltage at 3.6 volts allows battery pack design with a single cell but a nickel-based pack requires three 1.2 volt cells to be connected in series. Thus, LIBs have become an ideal energy storage device and have a wide application due to long cycle life and high energy density (Nitta et al., 2015). Among the available battery systems, Li-ion battery exhibits the highest specific energy and energy density as well as the highest individual cell potential over others. One great advantage of Li-ion battery is their low self-discharge rate (~5% per month) which is comparatively lower than other rechargeable batteries.

A typical Li-ion battery has been shown in Figure 7.5, illustrating the major components, i.e., anode, cathode, and electrolyte. During the charging of the Li-ion battery, Li ions move from the positive electrode to negative electrode through the electrolytic medium. The electrons flow from the positive electrode to the negative electrode via an outer circuit. Both Li-ion and electrons are combined at the negative electrode and form neutral lithium atom. When the battery is fully charged, no more ions move between the electrodes. During the discharging of Li-ion battery, the Li cations travel from the negative electrode to the positive electrode through the electrolytic solvent, however, electron flow back through the external circuit along with providing power to external energy-consuming devices. When all the ions and electrons move back then the battery becomes fully discharged and requires recharging. The Li-ion battery comprises of a graphite anode, a lithium salt-based electrolyte dissolved in a non-aqueous organic solvent and a transition metal oxide cathode (Fergus, 2010; Nitta et al., 2015), which is discussed in details in next section. In addition to the electrodes, the two other constituents for a battery are electrolyte additives and separators (Zhang,

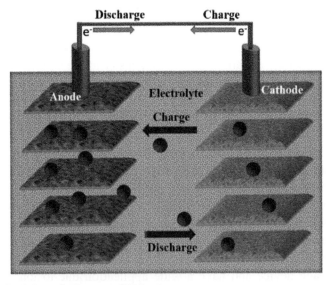

FIGURE 7.5   Schematic diagram of Li-ion battery.

2006, 2007). The electrolyte is an ionic conducting material, and to further improve its conductivity, a small amount of additional components, known as electrolytic additives, are incorporated (Zhang, 2006). While the separator is a membrane which prevents the direct, contact between anode and cathode and allows the movement of ions through the electrolytes (Zhang, 2007). Electrolyte additives and separator improve the performance and other properties, e.g., safety of the Li-ion battery. The major components of the Li-ion battery have been discussed below;

7.2.1 POSITIVE ELECTRODE (CATHODE) MATERIALS

The major role of the cathode in Li-ion battery is related to the intercalation and storage of the Li^+ ion. Those electrode materials which show a potential more than 2.25 V with respect to lithium metal could be used as positive electrode materials. Initially, metal chalcogenides such as TiS_3, $NbSe_3$, and $LiTiS_2$ were used as the cathode materials in the Li-ion battery (Murphy and Christian, 1979). However, low operating voltage and low energy storage capability of metal chalcogenides motivated the development of other intercalation cathode materials such as transition metal oxides. The transition metal oxides are generally used in Li-ion batteries due to their specific capacity (100–300 mAh/g) and higher operating voltage (3–5 V vs. Li/Li^+) in comparison to other compounds. On the basis of the structural arrangement, the transition metal oxides can be divided into several categories, e.g., spinel, layered, olivine, and tavorite, the details of which have been discussed below;

7.2.1.1 LAYERED TRANSITION METAL OXIDES

Layered di-chalcogenide was the first layered compound which was initially investigated for the intercalation of the Li-ion in the 70's (Dines, 1974). Furthermore, metal selenide and metal sulfide have also been investigated for their application as intercalation electrode materials (Lerf and Schöllhorn, 1977). However, the cell voltage of the battery using such intercalated materials exhibit comparatively lower output voltages, for example, ~ 2.0 V found for the $LiTiS_2$ with respect to lithium cell. Lower value of cell voltage motivated the development of other layered transition metal compound which could be used for the intercalation of Li-ion with maintaining higher cell voltage. The reason for the superior performance of the layered transition metal oxide over selenide and sulfide is due to their comparatively higher value of total free energy. The free energy for the transition metal oxide compounds could be calculated with the chemical reaction; ($xLi +$ $MX_n <=> Li_x MX_n$; X=oxygen, M= transition metal, x = number of atom). Due to having higher free energy, the layered transition metal oxides have been detailed studied through computational methods for the application as cathode material in Li-ion battery.

Several layered transition metal oxides, i.e., $LiCoO_2$, $LiNiO_2$, $LiMnO_2$, Li_2MnO_3, etc., are generally used as cathode materials in Li-ion battery (Etacheri et al., 2011). The compound, $LiCoO_2$, developed by Goodenough and commercialized by Sony, is found to be the most commercially applied cathode material (Mizushima et al., 1980). Starting from the initial proposal of layered $LiCoO_2$ as a cathode material, this material still remains as the most used cathode material in commercial Li-ion batteries. The primary reason behind such performance is high stability during electrochemical process, superior capacity, and ease of the preparation in bulk quantities. The crystal structure of the $LiCoO_2$ (M=Co) has been depicted in Figure 7.6, where lithium-ion is intercalated between the layered structure of cobalt oxide (Daniel et al., 2014). The layer arrangement provides the 2D channel for the lithium-ion diffusion. Therefore, the lithium-ion could be reversibly inserted into/from the layered structure with the concurrent reduction/oxidation of cobalt ion. With the removal of all 1Li+/Co, the ideal specific capacity of the $LiCoO_2$ could be ~280 mAh/g. However, in practice, the specific capacity shows the value ~140 mAh/g due to removal of only 0.5 Li in the potential window of 3–4.2 volt. The charging the Li-ion battery at higher voltage is limited by oxygen evolution on the cathode and leads to safety issues. Furthermore, due to toxicity and expensiveness of cobalt, it is required to investigate and develop other transition metal-based cathode materials.

As discussed earlier, $LiCoO_2$ is a good candidate for the cathode material; however, $LiCoO_2$ is limited by several factors such as high cost, low thermal stability, and fast fading of capacity at deep cycling process. Due to having similar crystal structure and theoretical specific capacity (275 mAh/g), $LiNiO_2$ could be a suitable candidate for the replacement of $LiCoO_2$ as the cathode materials (Kraytsberg and Ein-Eli, 2012). The higher energy density and lower cost of Ni-based oxide materials in comparison to Co-based material suggests the implementation of $LiNiO_2$ as cathode material for the intercalation and storage of Li ion (Kalyani and Kalaiselvi, 2005). The major drawback of the use of $LiNiO_2$ is associated with the blocking of the Li diffusion pathway during synthesis and de-lithiation, because of high tendency of the substitution of Li^+ by the Ni^{2+} ion. Furthermore, higher thermal instability than $LiCoO_2$ due to more readily reduction of Ni^{3+} than Co^{3+} shows that $LiNiO_2$ could not be favored for use as a cathode material. However, the thermal stability and electrochemical performance of $LiNiO_2$ could be improved by Mg doping as well as addition of small amount of aluminum. Therefore, the doped compound, $LiNi_{0.8}Co_{0.15}Al_{0.05}O_2$, has been found to be a high-performance cathode material which is why their commercial application is growing drastically (Zeng et al., 2014). However, at elevated temperatures (40–70C), the capacity fades due to solid electrolyte interface (SEI) formation and micro-crack formation at the grain boundaries (Daniel et al., 2014).

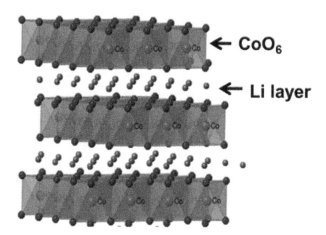

FIGURE 7.6   Crystal structure of layered transition metal oxides $LiMO_2$ (M=Co, Mn, Fe, Ni, V, Cr, etc.) exhibit the intercalated Li ion between the layered structure.
Source: Reprinted with permission from Daniel et al. (2014). © AIP Publishing.

Apart from $LiCoO_2$ and $LiNiO_2$, Mn-based compound, especially, $LiMnO_2$ could be used as cathode material due to comparatively much cheaper and less toxic nature of Mn than Co and Ni (Fergus, 2010). In the last decade, the performance of $LiMnO_2$ has improved with different methods, for example, by introducing impurities. Although there is an improvement in the performance, the cycling performance of $LiMnO_2$ is still not satisfactory due to its tendency to change the structural arrangement from layered structure to spinal structure during Li-ion extraction. The continuous research effort on the development of cathode materials, less expensive than $LiCoO_2$, resulted in the formulation of the Li $(Ni_{0.5}Mn_{0.5})$ O_2 cathode. Li $(Ni_{0.5}Mn_{0.5})$ O_2 could be an attractive material as it possesses similar energy density as that of $LiCoO_2$ as well as reduction in cost due to rich availability of transition metals (Santhanam and Rambabu, 2010). The presence of Ni allows higher Li extraction capacity to be achieved. However, cation mixing can cause low Li diffusivity and may result in unappealing rate capability. Furthermore, with the addition of Co in the Li $(Ni_{0.5}Mn_{0.5})$ O_2, the structural stability could be enhanced which can withstand elevated temperatures and wide potential range. The compound, $LiNi_xCoyMn_zO_2$ (NCM), has been found to possess similar or higher achievable specific capacity than $LiCoO_2$ and similar operating voltage while having lower cost due to reduction in the Co content. NCM compound provides a specific capacity of ~200 mAh/g (higher than $LiCoO_2$); however, major issue arises during the synthesis of material without cation mixing due to the presence of Ni^{2+} ion. $LiNi_{0.33}Co_{0.33}Mn_{0.33}O_2$ is the common form of NCM and is widely used in the battery market. Other transition metal oxides such as $LiCrO_2$ and $LiVO_2$ show the layered structure however does not perform the electrochemical activity because of spinal phase formation during cycling process (Chikkannanavar et al., 2014). The synthesis of iron and tin oxides ($LiFeO_2$ and $LiTiO_2$) at high temperature is very difficult; therefore, these materials could not be used as cathode material in Li-ion battery.

7.2.1.2   SPINEL TRANSITION METAL OXIDES

Spinel transition metal oxides could be represented by the general formula AB_2X_4 (A=Li, Na; B=transition metal such as Mn, Co, Ni, etc.; X=oxygen) (Whittingham, 2004). These structures are thermodynamically very stable than layered transition metal oxides, and therefore could be used as high voltage cathode materials. Among the other spinel metal oxides, $LiMn_2O_4$ have been extensively investigated for its application as cathode material in the LIB because of the lower cost, safety, and environmentally friendly nature of manganese (Kim et al., 2008). The crystal structure of $LiMn_2O_4$ in the spinel form with the cubic symmetry has been depicted in Figure 7.7, which shows the presence of 3D channel for the diffusion of lithium (Kumar et al., 2003; Yi et al., 2009). In the spinel $LiMn_2O_4$ structure, Li+ occupies the tetrahedral site (8a) while Mn^{3+}/Mn^{4+} occupy the octahedral sites (16d) in the cubic closed packed arrangement of oxygen anion. The edge-sharing arrangement of MnO_6 octahedral structure provides the well-ordered three-dimensional Mn_2O_4 array doe the lithium diffusion. The diffusion of lithium cation occurs through the vacant tetrahedral and octahedral structured interstitial sites in the 3D arrangement. More specifically, in the de-insertion process, the Li^+ diffuses from one site to other site through the octahedral site (16c). The diffusion of the Li^+ through the octahedral site (16c) is not favorable and requires very high energy. Therefore, the Li^+ de-insertion process occurs at the 4V potential region without changing the cubic symmetry of the spinal $LiMn_2O_4$. During discharging, this process is reversible and maintains a stable specific capacity of ~125 mAh/g.

In the spinel $LiMn_2O_4$, during discharging, the octahedral site (16c) is not fully occupied; hence, further insertion of the Li^+ ion in the structure could be possible. This process takes place at a lower potential region (3V) in comparison to 4V in earlier cases. Therefore, the spinel $LiMn_2O_4$ is capable to adopt/insert two lithium-ions inside the octahedral structures. Further increment in the number of intercalating lithium-ions in the octahedral structure leads to serious problems such as distortion of the octahedral structure. Therefore, at 3V, the cycling process transforms the spinel structure to tetragonal structure with huge changes in the unit cell volume. The cycling process at lower potential leads to reduction in the structural integrity, resulting enhancement in the capacity fadedness. This capacity fadedness could be reduced by cycling $LiMn_2O_4$ electrode material with adopting high potential region of 4V. The structural stability of spinel $LiMn_2O_4$ cathode during the intercalation of lithium could be further enhanced by doping it with metallic ions, e.g., Al^{3+}, Zn^{2+}, etc., or coating it with Al_2O_3. Apart from the doping, the structural stability and electrochemical performance could be further improved by few other techniques such as; substitution of manganese site with other transition metal ions, e.g., Co, Cr, and Ni, etc. The spinel electrode material, $LiNi_{0.5}Mn_{1.5}O_4$ is found to be free from Mn^{3+} and involves only Ni^{2+}/Ni^{4+} oxidation/reduction process. This spinel material is used as cathode in the high voltage Li-ion battery (4.7 V in comparison to 4.2 V for $LiMn_2O_4$ with respect to Li/Li+) due to stability at high voltage as well as providing high energy density. There are few other spinel electrode materials, e.g., $LiFe_2O_4$, $LiCo_2O_4$, and $LiCr_2O_4$ that have been proposed, although these materials could not be commercialized due to the difficulty in their chemical synthesis.

7.2.1.3 OLIVINE AS CATHODE MATERIAL

The layered and spinel transition metal oxide-based electrode materials were initially developed for enhancing the electrochemical performance of Li-ion battery. However, fading in the capacity after about 1000 cycles and instability motivated the development of other kinds of cathode materials (Zaghib et al., 2013). Polyanion compound has thus been proposed and investigated in last decades in this context. Polyanion could be represented by the general formula $(XO_4)^{3-}$ (X=P, S, As, W, Mo, etc.). These polyanion occupy the lattice positions and enhance the redox potential of cathode as well as increase the stability of the whole structure. Olivine ($LiFePO_4$) is one of the polyanion based electrode materials, which possesses high thermal stability and high power capacity (Zhang, 2011, 2012). Furthermore, long cyclability without capacity fadedness and high safety of this material make it one of the strong and unique electrode materials for the application in the Li-ion battery technology. The $LiFePO_4$ consists of the corner shared FeO_6 and edge shared LiO_6 octahedral structures connected through the PO_4 tetrahedral structure. In this olivine structure, Li^+ and Fe^{2+} ions generally occupy the octahedral sites while P is located in the tetrahedral site with slightly distorted hexagonal closed packed structure. As a cathode material in Li-ion battery, $LiFePO_4$ was first introduced by Padhi et al. which exhibit the reversible extraction of lithium from $LiFePO_4$ and insertion of lithium into $FePO_4$ (Padhi et al., 1997). Earlier studies have found that the specific capacity of the $LiFePO_4$ is ~170 mAh/g, which is close to the theoretical capacity (Chen and Dahn, 2002; Prosini et al., 2001). The lithium extraction and re-insertion process is a two-step process, during which olivine maintains the structural stability. Therefore, this cathode material is capable of being cycled for more than 1000 cycles without showing any reduction in the capacity. Although $FePO_4$ shows characteristics of a good electrode material, it is however limited by several factors such as lower average potential as well as lower value of ionic and electrical conductivity. In last few decades, these electrochemical performances have been improved by the reduction in the particle size, which could be obtained by carbon coating and metal cation doping. Apart from $LiFePO_4$, other olivine structures like $LiMnPO_4$, $LiCoPO_4$, and $LiNi_{0.5}Co_{0.5}PO_4$ have been developed and investigated for their possible application in Li-ion battery (Marom et al., 2011; Martha et al., 2009; Zhang, 2010). In one aspect, these olivines show better properties as compared to $LiFePO_4$, while in other aspect $LiFePO_4$ shows superior performance. For example, the olivine structure $LiMnPO_4$ exhibits higher average voltage in comparison to another olivine, $LiFePO_4$, though with lower conductivity. Furthermore, the olivine, $LiNiPO_4$ shows very high operating voltage (~ 5.2 V) which is much higher than that of $LiFePO_4$. Apart from olivines, other class of materials has been developed to enhance the electrochemical performance of Li-ion batteries, which have been discussed in the following section.

7.2.1.4   TAVORITE AS CATHODE MATERIAL

Recently developed tavorite-structures, e.g., $LiFeSO_4F$ are getting much attention for their application as cathode material in Li-ion battery due to having high cell voltage (~4 V), high specific capacity (~ 200 mAh/g), and higher electronic/ionic conductivity (Tripathi et al., 2011). Furthermore, these are economically cheaper than olivine due to the abundance of their composite elements. $LiFeSO_4F$ is composed of two slightly distorted $Fe^{2+}O_4F_2$ oxyfluoride octahedral structures connected by the fluorine vertices, forming a chain structures. These chains are then linked by the SO_4 tetrahedral structures and lithium is intercalated in the tunnel as shown in the. $LiFeSO_4F$ exhibits faster lithium diffusion as well as high specific capacity. The major drawback of this material is related to its low solubility in water and high instability at high temperatures. One more issue related to the $LiFeSO_4F$ is its synthesis, which makes its bulk production an uphill task. Other tavorite-structured materials such as fluorosulfates and fluorophosphates have been found as promising candidates for cathode material in high voltage Li-ion battery. These tavorite-structured materials exhibit low activation energy, which provides the charging and discharging of $Fe(SO_4)F$ and $V(PO_4)F$ at very high rates in comparison to the olivine structured materials such as $FePO_4$. The vanadium-containing tavorite materials show high capacity and high voltage along with a large number of charging-discharging cycles. However, the environmental toxicity of this material limits its application as cathode materials in Li-ion battery (Tripathi et al., 2011).

7.2.2   NEGATIVE ELECTRODE (ANODE) MATERIALS

The selection of an efficient anode material is another important task because it influences the electrochemical performance such as the charging rate and cyclability of the Li-ion battery. In the ideal case, lithium metal should be the best anode material for the Li-ion battery due to its low molecular weight and high specific capacity. However, in rechargeable LIB, the lithium metal anode creates serious problem such as safety and reversibility. For example, the metallic lithium anode forms dendritic growths at the interface of the anode-electrolyte, which eventually leads to the short circuiting of the battery. The major role of the anode materials are the intercalation and extraction of lithium cation during the charging and discharging process. There are large numbers of anode materials that have been proposed in the last few decades to enhance the electrochemical performance of the Li-ion battery. Generally, the anode materials are distinguished in different categories such as intercalation-based, conversion-reaction-based, and alloying-reaction based materials, on the basis of energy storage mechanism. The different energy storage mechanism as well as the anode material has been discussed in the following section.

7.2.2.1   INTERCALATION-BASED ANODE MATERIAL

In the case of intercalation based anode materials, Li^+ ions are electrochemically intercalated between the layered structures of the anode. Carbon-based anode material such as graphite is one of the good examples in this category. The graphite anode was commercialized before two decades and can still be found as a common element in the Li-ion battery. Currently, graphite is the most widely used anode material in commercially available Li-ion batteries due to its high electrical conductivity, high Li diffusivity, and low de-lithiation potential vs. Li. Further, low cost, and high abundance and small volume changes during lithiation/delithiation process, make graphite a unique candidate as the anode material. During the electrochemical activity, the lithium cations are intercalated between the layered graphite structures. One lithium cation could be stored by the six-carbon atom as described by the electrochemical reversible reaction;

$$xLi^+ + C_6 \text{ (in graphite)} + xe^- \leftrightarrow Li_xC_6$$

The lithium cations are intercalated in the vacant site of the graphite structure with the formation of lithiated carbon and de-intercalate from the lithiated carbon during the reverse process. The intercalation and accommodation of the lithium in the carbon structure depend on the microstructure, crystallinity, and the morphology of the layered graphite. The lithium cation is intercalated in the vacant place between layered graphite structures with the opening of the Van der Waals gap between the layers. The low capacity of the graphite anode is the major drawback, which motivated to develop anode material which shows better performance. The capacity could be enhanced with increment in the surface area (space) of the material that required for the accommodation of lithium cation between the layers (Cabana et al., 2010). Carbon nanotube, carbon nanofiber, and graphene are the other carbon-based materials which could be possible alternatives of the graphite anode. The aforementioned carbon-based anode materials exhibit larger surface area as well as higher electron conductivity than graphite-based anode materials, which make them suitable for the high rate charging/discharging (Cabana et al., 2010).

7.2.2.2   CONVERSION-REACTION-BASED ANODE MATERIAL

In order to get high-performance anode materials, several mechanisms have been proposed including conversion-reaction based material to explore the anode materials. Consider the case of conversion-reaction based anode materials, where the faradaic reaction is expressed as:

$$M_pX_q + (q,n)\,Li^+ + pe^- = pM + qLi_nX$$

where, M represents the transition metals (Ti, V, Mn, Fe, etc), X corresponds to the anion (O, N, S, P, etc.), n is the number of the negative charge on the X, and p and q are the positive integers. With the combination of M and X group, we can produce several anode materials. The theoretical capacity of this kind of material is higher in comparison to the graphite, which represents the unique character of these materials. However, larger volume expansion during lithiation/delithiation and relatively higher reaction potential (i.e., lower cell potential) in comparison to graphite have been the major drawback of these kinds of materials (Yu et al., 2016).

7.2.2.3   ALLOYING-REACTION-BASED ANODE MATERIAL

In the case of alloying reaction-based materials, the anode materials consist of alloy of lithium with metals (e.g., Si, Sn, Ge) (Liu et al., 2010; Obrovac and Chevrier, 2014). These materials have high theoretical capacity, for example, the theoretical specific capacity of Sn has been found ~995 mAh/g in case of $Li_{22}Sn_5$ (Wang et al., 2012). Major disadvantage of using such kind of the material related to the insertion/extraction of the Li+, which shows the drastically expansion in the volume (260%. larger than initial volume). The expanded volume leads to the inner stress and bring several negative consequences. In the case of the alloying Sn-based anode material, the Sn alloyed with both electrochemically inactive and active metals. The electro-chemically inactive metal is those metals which does not show lithiation with lithium-ion while the active metals show such characteristics and contribute in overall capacity of electrode. In the electrochemically inactive metals, consider the electrochemical reaction of $FeSn_2$, as shows below;

$$FeSn_2 + 8.8Li^+ + 8.8e^- \rightarrow 2Li_{4.4}Sn + Fe$$

$$Li_{4.4}Sn \leftrightarrow Sn + 4.4\,Li^+ + 4.4e^-$$

With the introduction of inactive metal in the Sn-based alloy in Li-ion battery, the conductivity is found to enhance and charge-discharge rate ability shows improvement. Although, the inactive metal treat as an ideal buffer agent for volume expansion of electrochemical active Sn (Zhang et al., 2008). Similar characteristic has been found for the Sn with elec-trochemically active metal; however, exhibit higher specific capacity and longer cycling life in comparison to the electrochemically inactive metal. Recently many groups are working to optimize the volume expansion by using various techniques such as by controlling size of Sn, with alloying modification and structural design of the Sn-alloy materials (Hassoun et al., 2008; Wang et al., 2010; Ying and Han, 2017; Zhang et al., 2008). Thus, the alloying reaction-based anode materials are not commercially used due to large volume changes during charging and discharging (Liu et al., 2010).

7.2.3 ELECTROLYTES

Electrolytes are one of the most important components of the Li-ion battery, which provide a conductive pathway for the movement of Li^+ cation between the electrodes. Since the electrolytes have physical contact with both of the electrodes, i.e., negative, and positive, it might be one of the reasons behind the dendrite formation, therefore it must be selected carefully. The electrolyte should not degrade or decompose within the voltage range of Li-ion battery. Furthermore, it must be inert and stable in the acceptable temperature range.

Electrolytes are distinguished in different categories such as; aqueous, non-aqueous, ionic liquids (ILs), polymer, and hybrid electrolytes. However, generally, liquid (aqueous and non-aqueous) electrolyte with lithium salts are used in the commercial Li-ion batteries, therefore our discussion has been focused on the liquid electrolytes. As previously discussed, the cathode and anode in the Li-ion battery possess oxidant and reductant respectively. The 'windows' of the electrolyte is defined as the energy difference of lowest unoccupied molecular orbital (LUMO) and highest occupied molecular orbital (HOMO) of the electrolytic solvent. For the thermodynamic stability, the electrochemical potential of the anode and cathode must be within the range of window of electrolytes.

In the case of the aqueous electrolyte, the concentrated saline solution is used as the electrolyte and provides a medium for the transportation of lithium-ions between the electrodes. Due to the water-based nature, the aqueous electrolytes are non-flammable. Furthermore, the aqueous electrolytes possess non-explosive and non-toxic nature, which are important for the safety and environmental concern of Li-ion battery. The major disadvantage associated with the aqueous electrolyte is the narrow electrochemical stability window (ESW) and small energy density. On the other hand, the non-aqueous electrolytes show comparatively higher ESW and larger energy density than the aqueous electrolyte, therefore generally non-aqueous electrolytes are used in commercial batteries. In the commercial Li-ion batteries, the non-aqueous electrolytes are composed of lithium salts, e.g., $LiPF_6$, $LiBF_4$, and $LiClO_4$ dissolved in polar organic solvents. These organic solvents include cyclic and acyclic carbonates such as, ethylene carbonate (EC), dimethyl carbonate (DMC), ethyl methyl carbonate (EMC), etc. The organic solvent with low viscosity and low melting point are required for the high ionic mobility within the operating voltage range. The conventional organic solvents are suitable for their uses in ~ 4V Li-ion battery. After the invention of high operating potential and high specific capacity cathode materials, it is required to develop a high voltage electrolyte (Armand and Tarascon, 2008). The conventional electrolytes are limited by several intrinsic problems such as high flammability (at high working potential values, >5 V) and high volatility (Goodenough and Kim, 2009). These safety issues demand for the urgent development of several electrolytes/additives which can remain stable at higher potential windows and have resistance toward thermal degradation. Even though there are significant reports on electrolytes/additives, (Balducci et al., 2011; Dalavi et al., 2010; Watanabe et al., 2008; Xu et al., 2011) fluorinated carbonates (FLC), owing to their desirable physical properties (low melting point, low flammability, and high electrochemical stability), have gained much attention recently (Achiha et al., 2009; Markevich et al., 2014; Smart et al., 2003; Wang et al., 2010; Zhang et al., 2013).

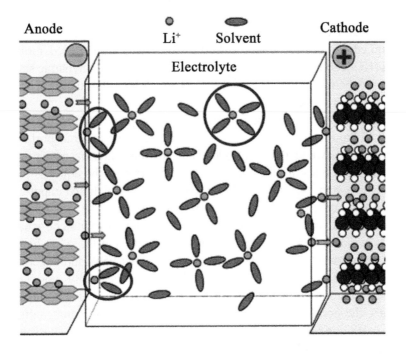

FIGURE 7.7   Schematic description of changes in the solvation number of Li⁺ during the movement between the electrodes.
Source: Reprinted with permission from Xu (2004).

Inside the Li-ion battery, the overall capacity directly depends on the electrodes while current density, time stability, and safety of the battery are associated with the electrolytes. Since the electrolytes have the interface with both the negative and positive electrodes, therefore it must be chemically stable. The chemical compatibilities could be ensured by the formation of the SEI (Aurbach et al., 2007). The physical properties and SEI formation are associated with the electrode, which suggests that electrolyte should not be decoupled from the electrode. Therefore, electrolyte requires careful investigation, especially during the charge transportation (Aurbach et al., 2004; Deng, 2015). During the movement of Li⁺ inside the electrolyte, Li⁺

is coordinated with the solvent molecules and form the solvation shell (Xu, 2004). The solvation number of Li^+ cation is found different in different regions, e.g., inside the electrolyte, solvation number is higher in comparison to near the electrode. Figure 7.7 shows the changes in the solvation number, especially near to electrode, which is selected from earlier report (Xu, 2004). Recently, Wei Cui et al. revisited the Li^+ ion solvation by EC in Li-ion battery electrolytes (Cui et al., 2016). In that model, Li^+ is coordinated by ethylene carbonates (EC_n; $n = 1-6$), because, usually solvation number of Li^+ changes during the transportation of Li^+ between two electrodes. For example, near to the anode/cathode single or double EC molecules interact with Li^+, while inside the electrolyte, Li^+ sometimes may interact with up to six ECs resulting in a most probable solvation number of four.

In the quantum chemical simulations, the electrolytic solvents are represented by the two solvation models; implicit solvent model and explicit solvent model (Bryantsev, 2012; Marenich et al., 2009; Rayne et al., 2010). The basis of the implicit continuum model is the sharp boundary between the solute and the bulk of the solvent, represented as a structure-less polarizable medium, characterized by its dielectric constant (Tomasi et al., 2005). In these models, the molecule/cluster under investigation is located inside a cavity surrounded by a homogeneous dielectric medium of the solvents such as; acetonitrile (MeCN, $\varepsilon = 35.6$), EC ($\varepsilon = 89.6$), propylene carbonate (PC, $\varepsilon = 64.0$), Diethylene carbonate (DEC, $\varepsilon = 2.40$), DMC ($\varepsilon = 7.15$) and Triethylene glycol dimethyl ether (Triglime, $\varepsilon = 7.94$). The implicit solvent model has been successfully applied in the investigation of the chemical reaction within the surrounding medium (Kushwaha et al., 2017, 2018; Kushwaha and Nayak, 2017), There are several continuum models has proposed for expressing the solvent media, out of which polarizable continuum model (PCM) is generally used to represent continuum dielectric medium. Mathematically, the PCM is expressed by the Poisson-Boltzmann equation which is an expansion term of Poisson's equation. Recently solvation model on density (SMD) is getting much attention for solvation model (Marenich et al., 2009). Similar to the PCM, the SMD model solve the Poisson-Boltzmann equation analytically, only difference is that SMD model used specific parameterize radii for the construction of cavity. The conductor boundary condition based COSMO solvation model is also an implicit solvent model (Klamt and Schüürmann, 1993). The computational cost calculation using the implicit model is lower in comparison to others while is does not maintain the high accuracy especially reaction mechanism. In the explicit solvation model, the molecular solvents form the solvation shell (changes during transfer between the electrodes) around the Li$^+$ ion, either in the electrolyte or at the electrode interfaces. For example, in the case of EC, the coordination number of Li$^+$ has been found to be four, which reduces at the interfaces (Bhatt et al., 2012; Bhatt and O'Dwyer, 2014; Cui et al., 2016). The explicit solvent model provides the more realistic picture of the solute-solvent interaction in comparison to implicit solvent model. Although the computational cost is the major drawback of the explicit solvent model.

7.3   CONCLUSION

In this chapter, we have discussed briefly about energy storage devices especially battery systems including alkali metal ion battery. Due to having the superior performance of Li-ion battery among all alkali metal ion battery, we have investigated and discussed their components. Earlier developed trial-and-error methods for the prediction of the electrode and electrolytic materials were time-consuming and economically challenging, therefore the computational methodologies has been discussed for the prediction of new materials. Various electrode/electrolyte materials have been discussed for their application as the high voltage Li-ion battery.

KEYWORDS

- electrode materials
- first-principles calculations
- implicit and explicit electrolytic solutions
- ions intercalations
- Li-ion battery
- superconducting magnetic energy storage

REFERENCES

Achiha, T., Nakajima, T., Ohzawa, Y., Koh, M., Yamauchi, A., Kagawa, M., & Aoyama, H., (2009). Electrochemical behavior of nonflammable organo-fluorine compounds for lithium-ion batteries. *J. Electrochem. Soc., 156*, A483–A488.
Armand, M., & Tarascon, J. M., (2008). Building better batteries. *Nature, 451*, 652.

Aurbach, D., Markovsky, B., Salitra, G., Markevich, E., Talyossef, Y., Koltypin, M., Nazar, L., Ellis, B., & Kovacheva, D., (2007). Review on electrode-electrolyte solution interactions, related to cathode materials for Li-ion batteries. *J. Power Sources, 165*, 491–499.

Aurbach, D., Talyosef, Y., Markovsky, B., Markevich, E., Zinigrad, E., Asraf, L., Gnanaraj, J. S., & Kim, H. J., (2004). Design of electrolyte solutions for Li and Li-ion batteries: A review. *Electrochim. Acta, 50*, 247–254.

Balducci, A., Jeong, S. S., Kim, G. T., Passerini, S., Winter, M., Schmuck, M., Appetecchi, G. B., et al., (2011). Development of safe, green, and high performance ionic liquids-based batteries (ILLIBATT project). *J. Power Sources, 196*, 9719–9730.

Barbir, F., & Gomez, T., (1997). Efficiency and economics of proton exchange membrane (PEM) fuel cells. *Int. J. Hydrogen Energy, 22*, 1027–1037.

Bhatt, M. D., & O'Dwyer, C., (2014). Density functional theory calculations for ethylene carbonate-based binary electrolyte mixtures in lithium-ion batteries. *Curr. Appl. Phys., 14*, 349–354.

Bhatt, M. D., Cho, M., & Cho, K., (2012). Conduction of Li^+ cations in ethylene carbonate (EC) and propylene carbonate (PC): Comparative studies using density functional theory. *J. Solid State Electrochem., 16*, 435–441.

Bryantsev, V. S., (2012). Calculation of solvation free energies of Li^+ and O_2-ions and neutral lithium-oxygen compounds in acetonitrile using mixed cluster/continuum models. *Theor. Chem. Acc., 131*, 1250.

Cabana, J., Monconduit, L., Larcher, D., & Palacin, M. R., (2010). Beyond intercalation-based Li-ion batteries: The state of the art and challenges of electrode materials reacting through conversion reactions. *Adv. Mater., 22*, E170–E192.

Chen, Z., & Dahn, J. R., (2002). Reducing carbon in $LiFePO_4$/C composite electrodes to maximize specific energy, volumetric energy, and tap density. *J. Electrochem. Soc., 149*, A1184–A1189.

Chikkannanavar, S. B., Bernardi, D. M., & Liu, L., (2014). A review of blended cathode materials for use in Li-ion batteries. *J. Power Sources, 248*, 91–100.

Cui, W., Lansac, Y., Lee, H., Hong, S. T., & Jang, Y. H., (2016). Lithium-ion solvation by ethylene carbonates in lithium-ion battery electrolytes, revisited by density functional theory with the hybrid solvation model and free energy correction in solution. *Phys. Chem. Chem. Phys., 18*, 23607–23612.

Dalavi, S., Xu, M., Ravdel, B., Zhou, L., & Lucht, B. L., (2010). Nonflammable electrolytes for lithium-ion batteries containing dimethyl methylphosphonate. *J. Electrochem. Soc., 157*, A1113–A1120.

Daniel, C., Mohanty, D., Li, J., & Wood, D. L., (2014). Cathode materials review. In: *AIP Conference Proceedings* (pp. 26–43). AIP.

Deng, D., (2015). Li-ion batteries: Basics, progress, and challenges. *Energy Sci. Eng., 3*, 385–418.

Dines, M. B., (1974). Intercalation in layered compounds. *J. Chem. Educ., 51*, 221.

Dudley, B. (2018). BP statistical review of world energy. *BP Statistical Review*, London, UK.

Energy Storage Association (WWW Document), (n.d.). URL: http://energystorage.org/energy-storage/facts-FIGUREs (accessed on 16 May 2020).

Etacheri, V., Marom, R., Elazari, R., Salitra, G., & Aurbach, D., (2011). Challenges in the development of advanced Li-ion batteries: A review. *Energy Environ. Sci., 4*, 3243–3262.

Executive Summary [WWW Document], (2012). *Energy Technol. Perspect.*

Fergus, J. W., (2010). Recent developments in cathode materials for lithium-ion batteries. *J. Power Sources, 195*, 939–954.

Gamburzev, S., & Appleby, A. J., (2002). Recent progress in performance improvement of the proton exchange membrane fuel cell (PEMFC). *J. Power Sources, 107*, 5–12.

Goodenough, J. B., & Kim, Y., (2009). Challenges for rechargeable Li batteries. *Chem. Mater., 22*, 587–603.

Hassoun, J., Derrien, G., Panero, S., & Scrosati, B., (2008). A nanostructured Sn-C composite lithium battery electrode with unique stability and high electrochemical performance. *Adv. Mater., 20*, 3169–3175.

Homewood, P., (2018). *BP Energy Review* [WWW Document]. https://notalotofpeopleknowthat. wordpress.com/2019/06/14/bp-energy-review-2018/ (accessed on 12 June 2020).

Ivanov, I., Vidaković-Koch, T., & Sundmacher, K., (2010). Recent advances in enzymatic fuel cells: Experiments and modeling. *Energies, 3*, 803–846.

Javed, A., (2017). Activated carbon fiber for energy storage. In: *Activated Carbon Fiber and Textiles* (pp. 281–303). Wood head Publishing.

Kalyani, P., & Kalaiselvi, N., (2005). Various aspects of $LiNiO_2$ chemistry: A review. *Sci. Technol. Adv. Mater., 6*, 689.

Kannan, A. M., Renugopalakrishnan, V., Filipek, S., Li, P., Audette, G. F., & Munukutla, L., (2009). Bio-batteries and bio-fuel cells: Leveraging on electronic charge transfer proteins. *J. Nanosci. Nanotechnol., 9*, 1665–1678.

Kiehne, H. A., (2003). *Battery Technology Handbook.* CRC Press.

Kim, D. K., Muralidharan, P., Lee, H. W., Ruffo, R., Yang, Y., Chan, C. K., Peng, H., Huggins, R. A., & Cui, Y., (2008). Spinel $LiMn_2O_4$ nanorods as lithium-ion battery cathodes. *Nano Lett., 8*, 3948–3952.

Klamt, A., & Schüürmann, G., (1993). COSMO: A new approach to dielectric screening in solvents with explicit expressions for the screening energy and its gradient. *J. Chem. Soc. Perkin Trans., 2*, 799–805.

Kraytsberg, A., & Ein, E. Y., (2012). Higher, stronger, better a review of 5 volt cathode materials for advanced lithium-ion batteries. *Adv. Energy Mater, 2*, 922–939.

Kumar, V. G., Gnanaraj, J. S., Ben-David, S., Pickup, D. M., Van-Eck, E. R. H., Gedanken, A., & Aurbach, D., (2003). An aqueous reduction method to synthesize spinel-$LiMn_2O_4$ nanoparticles as a cathode material for rechargeable lithium-ion batteries. *Chem. Mater., 15*, 4211–4216.

Kushwaha, A. K., & Nayak, S. K., (2017). Wobbled electronic properties of lithium clusters: Deterministic approach through first-principles. *Phys. E Low-Dimensional Syst. Nanostructures, 97*, 368–374.

Kushwaha, A. K., Sahoo, M. R., & Nayak, S., (2018). Probing potential Li-ion battery electrolyte through first-principles simulation of atomic clusters. In: *AIP Conference Proceedings* (p. 20014). AIP Publishing.

Kushwaha, A. K., Sahoo, M. R., Nanda, J., & Nayak, S. K., (2017). Engineering redox potential of lithium clusters for electrode material in lithium-ion batteries. *J. Clust. Sci., 28*, 2779–2793.

Landi, B. J., Ganter, M. J., Cress, C. D., DiLeo, R. A., & Raffaelle, R. P., (2009). Carbon nanotubes for lithium-ion batteries. *Energy Environ. Sci., 2*, 638–654.

Lerf, A., & Schöllhorn, R., (1977). Solvation reactions of layered ternary sulfides $AxTiS_2$, $AxNbS_2$, and $AxTaS_2$. *Inorg. Chem., 16*, 2950–2956.

Liu, C., Li, F., Ma, L., & Cheng, H., (2010). Advanced materials for energy storage. *Adv. Mater., 22*, E28–E62.

Lu, L., Han, X., Li, J., Hua, J., & Ouyang, M., (2013). A review on the key issues for lithium-ion battery management in electric vehicles. *J. Power Sources, 226*, 272–288.

Marenich, A. V., Cramer, C. J., & Truhlar, D. G., (2009). Universal solvation model based on solute electron density and on a continuum model of the solvent defined by the bulk dielectric constant and atomic surface tensions. *J. Phys. Chem. B, 113*, 6378–6396.

Markevich, E., Salitra, G., Fridman, K., Sharabi, R., Gershinsky, G., Garsuch, A., Semrau, G., et al., (2014). Fluoroethylene carbonate as an important component in electrolyte solutions for high-voltage lithium batteries: Role of surface chemistry on the cathode. *Langmuir, 30*, 7414–7424.

Marom, R., Amalraj, S. F., Leifer, N., Jacob, D., & Aurbach, D., (2011). A review of advanced and practical lithium battery materials. *J. Mater. Chem., 21*, 9938–9954.

Martha, S. K., Grinblat, J., Haik, O., Zinigrad, E., Drezen, T., Miners, J. H., Exnar, I., et al., (2009). $LiMn_{0.8}Fe_{0.2}PO_4$: An advanced cathode material for rechargeable lithium batteries. *Angew. Chemie. Int. Ed., 48*, 8559–8563.

Mizushima, K., Jones, P. C., Wiseman, P. J., & Goodenough, J. B., (1980). $LixCoO_2$ (0< x<-1): A new cathode material for batteries of high energy density. *Mater Res. Bull., 15*, 783–789.

Murphy, D. W., & Christian, P. A., (1979). Solid state electrodes for high energy batteries. *Science, 205*, 651–656.

Nitta, N., Wu, F., Lee, J. T., & Yushin, G., (2015). Li-ion battery materials: Present and future. *Mater Today, 18*, 252–264.

Obrovac, M. N., & Chevrier, V. L., (2014). Alloy negative electrodes for Li-ion batteries. *Chem. Rev., 114*, 11444–11502.

Padhi, A. K., Nanjundaswamy, K. S., & Goodenough, J. B., (1997). Phospho-olivines as positive-electrode materials for rechargeable lithium batteries. *J. Electrochem. Soc., 144*, 1188–1194.

Prabhulkar, S., Tian, H., Wang, X., Zhu, J. J., & Li, C. Z., (2012). Engineered proteins: Redox properties and their applications. *Antioxid. Redox Signal, 17*, 1796–1822.

Prosini, P. P., Zane, D., & Pasquali, M., (2001). Improved electrochemical performance of a $LiFePO_4$-based composite cathode. *Electrochim. Acta, 46*, 3517–3523.

Rayne, S., Rayne, S., & Forest, K., (2010). Accuracy of computational solvation free energies for neutral and ionic compounds: Dependence on level of theory and solvent model. *Nat. Preced.*, pp. 1–22.

Reddy, T. B., & Hossain, S., (2002). Rechargeable lithium batteries (ambient temperature). *Handb. Batter., 3*, 31–34.

Reddy, T. B., (2011). *Linden's Handbook of Batteries*. McGraw-hill New York.

REN21's 2017 Anual Report [WWW Document], (2017). *Renew. Energy Policy Netw. 21st Century.*

Renewable Energy Highlights [WWW Document], (2018). *Int. Renew. Energy Agency.*

Root, M., (2011). *The TAB™ Battery Book: An In-Depth Guide to Construction, Design, and Use.* New York: McGraw-Hill/TAB Electronics.

Santhanam, R., & Rambabu, B., (2010). Research progress in high voltage spinel $LiNi_{0.5}Mn_{1.5}O_4$ material. *J. Power Sources, 195*, 5442–5451.

Smart, M. C., Ratnakumar, B. V., Ryan-Mowrey, V. S., Surampudi, S., Prakash, G. K. S., Hu, J., & Cheung, I., (2003). Improved performance of lithium-ion cells with the use of fluorinated carbonate-based electrolytes. *J. Power Sources, 119*, 359–367.

Ticianelli, E. A., Derouin, C. R., Redondo, A., & Srinivasan, S., (1988). Methods to advance technology of proton exchange membrane fuel cells. *J. Electrochem. Soc., 135*, 2209–2214.

Tomasi, J., Mennucci, B., & Cammi, R., (2005). Quantum mechanical continuum solvation models. *Chem. Rev., 105*, 2999–3094.

Tripathi, R., Gardiner, G. R., Islam, M. S., & Nazar, L. F., (2011). Alkali-ion conduction paths in LiFeSO$_4$F and NaFeSO$_4$F favorite-type cathode materials. *Chem. Mater., 23*, 2278–2284.

Van, S. W., & Scrosati, B., (2002). Advances in lithium-ion batteries introduction. In: *Advances in Lithium-Ion Batteries* (pp. 1–5). Springer.

Wang, B., Luo, B., Li, X., & Zhi, L., (2012). The dimensionality of Sn anodes in Li-ion batteries. *Mater. Today, 15*, 544–552.

Wang, X. J., Lee, H. S., Li, H., Yang, X. Q., & Huang, X. J., (2010). The effects of substituting groups in cyclic carbonates for stable SEI formation on graphite anode of lithium batteries. *Electrochem. Commun., 12*, 386–389.

Wang, X. L., Han, W. Q., Chen, J., & Graetz, J., (2010). Single-crystal intermetallic M-Sn (M = Fe, Cu, Co, Ni) nanospheres as negative electrodes for lithium-ion batteries. *ACS Appl. Mater. Interfaces, 2*, 1548–1551.

Watanabe, Y., Kinoshita, S., Wada, S., Hoshino, K., Morimoto, H., & Tobishima, S., (2008). Electrochemical properties and lithium-ion solvation behavior of sulfone-ester mixed electrolytes for high-voltage rechargeable lithium cells. *J. Power Sources, 179*, 770–779.

Wee, J. H., (2007). Applications of proton exchange membrane fuel cell systems. *Renew. Sustain. Energy Rev., 11*, 1720–1738.

Whittingham, M. S., (2004). Lithium batteries and cathode materials. *Chem. Rev., 104*, 4271–4302.

Xu, K., (2004). Nonaqueous liquid electrolytes for lithium-based rechargeable batteries. *Chem. Rev., 104*, 4303–4418.

Xu, M., Zhou, L., Hao, L., Xing, L., Li, W., & Lucht, B. L., (2011). Investigation and application of lithium difluoro (oxalate) borate (LiDFOB) as additive to improve the thermal stability of electrolyte for lithium-ion batteries. *J. Power Sources, 196*, 6794–6801.

Yi, T. F., Zhu, Y. R., Zhu, X. D., Shu, J., Yue, C. B., & Zhou, A. N., (2009). A review of recent developments in the surface modification of LiMn$_2$O$_4$ as cathode material of power lithium-ion battery. *Ionics (Kiel), 15*, 779.

Ying, H., & Han, W., (2017). Metallic Sn-based anode materials: Application in high-performance lithium-ion and sodium-ion batteries. *Adv. Sci., 4,* 1700298.

Yu, S., Lee, S. H., Lee, D. J., Sung, Y., & Hyeon, T., (2016). Conversion reaction-based oxide nanomaterials for lithium-ion battery anodes. *Small, 12*, 2146–2172.

Zaghib, K., Guerfi, A., Hovington, P., Vijh, A., Trudeau, M., Mauger, A., et al., (2013). Review and analysis of nanostructured olivine-based lithium rechargeable batteries: Status and trends. *J. Power Sources, 232*, 357–369.

Zeng, X., Li, J., & Singh, N., (2014). Recycling of spent lithium-ion battery: A critical review. *Crit. Rev. Environ. Sci. Technol., 44*, 1129–1165.

Zhang, C. Q., Tu, J. P., Huang, X. H., Yuan, Y. F., Wang, S. F., & Mao, F., (2008). Preparation and electrochemical performances of nanoscale FeSn$_2$ as anode material for lithium-ion batteries. *J. Alloys Compd., 457*, 81–85.

Zhang, S. S., (2006). A review on electrolyte additives for lithium-ion batteries. *J. Power Sources*, *162*, 1379–1394.

Zhang, S. S., (2007). A review on the separators of liquid electrolyte Li-ion batteries. *J. Power Sources, 164*, 351–364.

Zhang, W. J., (2010). Comparison of the rate capacities of $LiFePO_4$ cathode materials. *J. Electrochem. Soc., 157*, A1040–A1046.

Zhang, W. J., (2011). Structure and performance of $LiFePO_4$ cathode materials: A review. *J. Power Sources, 196*, 2962–2970.

Zhang, Y., Huo, Q., Du, P., Wang, L., Zhang, A., Song, Y., Lv, Y., & Li, G., (2012). Advances in new cathode material $LiFePO_4$ for lithium-ion batteries. *Synth. Met., 162*, 1315–1326.

Zhang, Z., Hu, L., Wu, H., Weng, W., Koh, M., Redfern, P. C., Curtiss, L. A., & Amine, K., (2013). Fluorinated electrolytes for 5 V lithium-ion battery chemistry. *Energy Environ. Sci., 6*, 1806–1810.

CHAPTER 8

Effect of Morphology and Doping on the Photoelectrochemical Performance of Zinc Oxide

AKASH SHARMA,[1] POOJA SAHOO,[1] ALFA SHARMA,[2] and SASWAT MOHAPATRA[1]

[1]*Department of Physics, IIT(ISM), Dhanbad, Jharkhand, India, E-mail: akash.physics@gmail.com (A. Sharma)*

[2]*Discipline of Metallurgy Engineering and Materials Science, IIT, Indore, Madhya Pradesh, India, E-mail: alfasharma89@gmail.com*

ABSTRACT

As a clean and inexhaustible energy source, solar energy is the most important alternative to non-renewable fossil fuels to solve the problems of an energy shortage, global warming, and environmental pollution. The key issue is how to efficiently harvest and utilize solar energy. Photoelectrochemical (PEC) solar energy conversion is regarded as one of the most promising technologies for solar energy conversion and application. However, there are some challenges of employing this material for PEC water splitting application: (a) conduction band (CB) potential is not high enough to drive water reduction without bias, (b) a relatively low absorption coefficient because of indirect bandgap and (c) poor majority carrier conductivity and short hole diffusion length (2–4 nm). Metal oxides carved into hierarchical nanostructures are thought to be promising for improving photo-electrochemical performance by enhancing charge separation and transport. ZnO is studied as one of the most relevant optoelectronic materials, due to its unique optical (like large direct bandgap of 3.37 eV) and electrical properties (high exciton room temperature binding energy 60 meV, high electron mobility, etc.). However, the efficiency of the ZnO based photoanode has limited success because of its high recombination rate of e^-h^+ (electron-hole) pairs and poor catalytic activity. Over the years, several research efforts have been devoted to explore the effective methods to address those drawbacks. Such attempts have been employed through various approaches, such as doping, hetero-structure, dye sensitization, quantum dot sensitization, and co-catalysts modification. Out of these, nowadays, impurity-doped ZnO has been given more and more attention because the impurity element directly and simply enhances conductivity, carrier concentration, and optoelectronic performance of ZnO.

8.1   INTRODUCTION

The quest of achieving sustainable, clean, green energy has always led mankind to think off about alternate sources of energy. Several attempts are also done so far to achieve the same by scientists across the globe. Apart from all the available renewable resources; solar energy is considered to be the best among all with regard to its ability in generating energy at a significant level (Tee et al., 2017; Tsao, Lewis, and Crabtree, 2006). Several solar energy-driven applications are being tried nowadays to harvest sunlight at its fullest. Extraction of hydrogen from water is one of the possible solutions to meet the world's energy demands. When this evolved gas is further consumed it generates water as the only by-product, which proves its credibility as a clean green energy solution (Figure 8.1). In order to realize this, photoelectrochemical (PEC) water splitting has emerged as a broad area of research. Basically in this process hydrogen and oxygen are collected by splitting water in presence of photoelectrodes (semiconductors). But the lowered efficiency in these types of systems put hurdles in further commercialization. On one hand, we have stable oxide semiconductors but with considerably reduced efficiency while on the other hand, highly efficient (as compared with oxide semiconductors) semiconductors suffer with short life span. Henceforth to come up with highly efficient stable systems several approaches has been made.

After the first experiment by Fujishima and Honda (1972) to generate H_2 and O_2 from TiO_2 by means of water splitting several other semiconductors have been taken under trail for the same. In order to utilize hydrogen fuel, several research groups have developed many metal oxide nanostructures (Chen et al., 2017) such as ZnO (Djurišić et al., 2012; Kołodziejczak-Radzimska and Jesionowski, 2014), Cu_2O (Pan et al., 2018; Yang et al., 2016), TiO_2 (Chen et al., 2015; Guo et al., 2016; Schneider et al., 2014), SnO_2 (Outemzabet et al., 2015), Fe_2O_3 (Cha et al., 2011), WO_3 (Vidyarthi et al., 2011; Xu et al., 2015), $BiVO_4$ (Luo et al., 2008; Sun et al.,

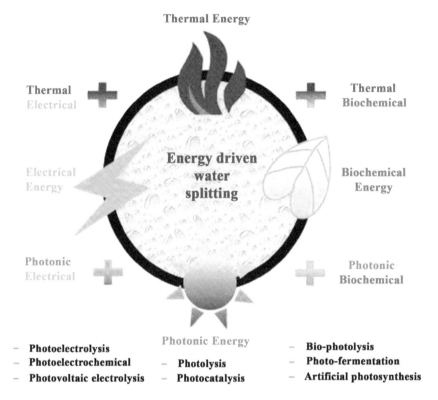

Thermal Energy

Thermal
Electrical

Thermal
Biochemical

Electrical
Energy

Energy driven
water
splitting

Biochemical
Energy

Photonic
Electrical

Photonic
Biochemical

Photonic Energy

- **Photoelectrolysis**
- **Photoelectrochemical**
- **Photovoltaic electrolysis**

- **Photolysis**
- **Photocatalysis**

- **Bio-photolysis**
- **Photo-fermentation**
- **Artificial photosynthesis**

FIGURE 8.1   Various energy-driven water splitting routes by using thermal, electrical, biochemical, and photonic energy or their combinations.
Source: Reprinted with permission from Tee et al. (2017). https://creativecommons.org/licenses/by/4.0/

2014), $CuWO_4$ (Salimi et al., 2019; Valenti et al., 2015), Ag_3PO_4 (Lin et al., 2012; Martin et al., 2015), V_2O_5 (Andrews et al., 2018), etc., through their potential application in PEC water splitting. Among all, the ZnO has been acknowledged as a promising photoanodic material due to its suitable band edges, high electron mobility and lower electrical resistance as well as its environmentally benign nature. As far as solar energy-driven water splitting application is concerned, the suitable band positions of ZnO, i.e., conduction band minimum (CBM) and valence band (VB) maximum (VBM) are –0.31 V and +2.89 V respectively (Here both of the values mentioned are in normal hydrogen electrode (NHE) scale). However, the large bandgap (3.37 eV) of ZnO nanomaterials substantially limits its photoconversion efficiency. Being a UV active material it always lacks with regard to its limited ability in further use of solar spectrum. This emphasizes more focus on increasing the visible region absorption ability of ZnO through various processes like sensitization with dyes or quantum dots, and doping with heteroatoms. Basically doping with heteroatoms or Impurity doping is widely investigated as an effective tool to increase or enhance electrical conductivity and narrowing optical band gap of semiconductors (Baruah and Dutta, 2009; Bharathi et al., 2014; Ilican, 2013; Panigrahy et al., 2010). Till now, various types of dopants have been introduced into ZnO in order to improve the photoconversion efficiency.

Another important aspect for ZnO is the facile tunable properties along with a wide range of morphologies. The morphology has also been known to have a substantial impact on the photoelectrochemical performance (Govatsi et al., 2018). Therefore, ZnO of different morphology like nanosheets (Hsu et al., 2011), nanorods (NRs) (Sharma et al., 2018a, b), nanowires (NWs) (Wang et al., 2019), nanopencils (Wang et al., 2015), nanotree (Ren et al., 2016), nanotriangles (Chandrasekaran et al., 2016), nanotertrapods (Qiu et al., 2012), nanocorals (Ahn et al., 2008; Shet, 2011), etc., were developed. The usually occurring fast surface recombination rate of photoinduced electron-hole pairs are well managed by 1D nanostructure. For a better photoactive semiconductor, a material should be capable of charge collection along with effective separation and transfer of the same to the surrounding medium. NRs possess large surface-area-to-volume ratio which provides large surface area to the surrounding medium; thus helping the minority carriers to get diffused easily at the semiconductor electrolyte interface, leading to an enhanced charge separation. Furthermore, the facile synthesis of 1D nanostructure along with tuneable diameter suffices with the carrier diffusion length leading to easy removal of the carriers from the surface. In view of these overall features, these NRs have emerged as a potential candidate for realizing efficient photoelectrochemical devices. Not only this but also ZnO has shown its immense fidelity in several other applications like transistors (Li et al., 2008), ultraviolet (UV) photodetectors (Boruah et al., 2017), UV nanolaser, Field emitter, Photovoltaics (Djurišić et al., 2014), Sensors (Wei et al., 2011), Gas sensors (Singh et al., 2012), etc. Materials which offer photoelectrochemical water splitting possess unique characteristics. Lin and group have reported the detailed physical and chemical processes involved in water splitting (Lin et al., 2011). The reaction involves the following steps:

1. Light Absorption and separation of charge carriers;
2. Charge transfer;
3. Charge transport;
4. Surface chemical reactions.

The material should possess a suitable bandgap (>1.23 eV) to permit light absorption. Also, the material should have proper band edge positions to split H_2O molecules. Since the number of incident photons received per unit area is fixed under certain standard condition (i.e., AM 1.5 with intensity of 100 mW/cm²); so to match the kinetics material should be catalytically active for both oxidation and reduction of H_2O. Furthermore, the semiconductor needs to be resistant to photo corrosion and should survive reactions in water under intense illumination (Lin et al., 2011).

In the case of defect-free ZnO, photogenerated carriers are vulnerable for surface-bulk recombination resulting in low photonic efficiency. The presence of defects such as oxygen vacancy, zinc interstitial, and oxygen interstitial can temporarily inhibit the charge carrier recombination process and magnificently improve the reaction rates (Kayaci et al., 2014; Pei et al., 2013). Impurity doping at the cationic/anionic sites is a promising approach to alter the structure-electronic properties of any semiconductors. The significant improvement in the photonic efficiency with all these dopants is successfully achieved. The following observations are largely encountered upon impurity doping: (1) change in the defect chemistry—substituting at lattice sites in the Zn-O-Zn (cationic doping) and Zn-O-O (anionic doping) framework leads to the formation of Zn-O-M. The change in the local atomic configuration modifies the electronic environment and simultaneously introduces structural defects, lattice disorder, and lattice stress/strain. The segregation of these dopants along the grain boundary region inhibits the growth of crystallite size; (2) mismatch in the ionic radius between the dopant and host ion is the basis for changes in the defect chemistry (Sushma and Girish Kumar, 2017).

In this chapter, we present a discussion regarding the effect of various morphology of ZnO on the photoelectrochemical performance of ZnO. We also tried to put an insight on the effect of various dopants on the water splitting performance of ZnO NRs (Figure 8.2). Also, emphasis has been given on the basics of photoelectrochemical water splitting and different synthesis methods of ZnO nanostructures.

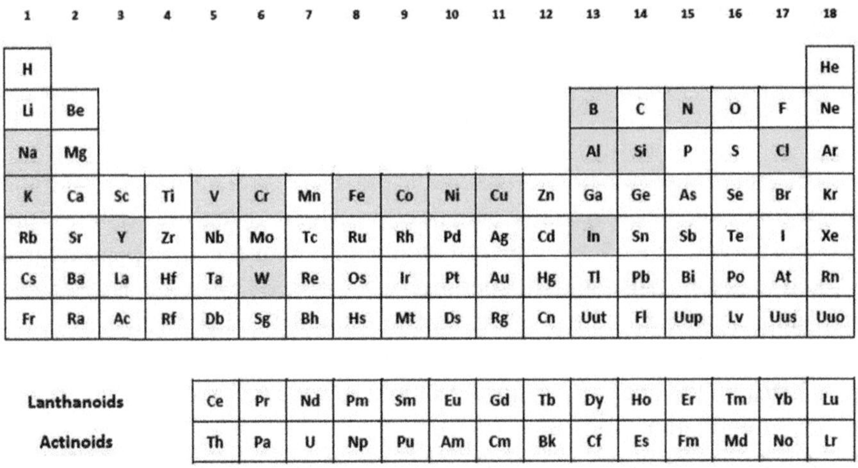

FIGURE 8.2 Periodic table highlighted with the elements doped in ZnO nanorods discussed in this chapter.

8.2 PHOTOELECTROCHEMICAL WATER SPLITTING

8.2.1 BASIC PHOTOELECTROCHEMICAL SET-UP

The photoelectrochemical set up comprises of a PEC cell, a light source, and a potentiostat. A photoelectrochemical cell basically consists of two electrodes immersed in an aqueous solution and either both or at least one are supposed to be photoactive for carrying out the processes by illumination of light. But for a 3-electrode configuration the system is accompanied with a counter electrode (CE) (e.g., Pt electrode), working electrode (WE) (e.g., thin-film) and a reference electrode (REs) (e.g., Ag/AgCl electrode) dipped in an electrolyte. The vessel containing the system should be transparent or contain an optical window in order to make maximum usage of the light energy. Pt coil or wire is used often as a CE. As per the purpose, the WEs are chosen p-type or n-type. The WEs are usually the photoactive semi-conductors deposited on transparent conducting oxide materials with high conductivity over glass substrates. The optically transparent nature allows the light to pass through it so as to reach the semiconductor layers, while the highly conducting nature provides pathways for carrier collection. Some of the commonly used transparent conducting substrates are Sn-doped In_2O_3 (ITO), F-doped SnO_2 (FTO); Al-doped ZnO (AZO), etc. In addition to this, the illumination from substrate to sample or vice versa also has a significant impact on the PEC performance. For example in the case of n-type semiconductor when light is illuminated from the sample side, maximum absorption occurs at the surface while a considerably lesser light reaches the substrate. Consequently, the charge carrier density becomes higher at the surface as compared with the electrode side. The holes easily gets diffused into the electrolyte and thus with a small minority (hole) diffusion length is enough. But on the other hand, the electrons diffuse through the entire thickness of the sample for reaching the substrate in order to take part in hydrogen evolution. Thus, it is very important that the thickness should suffice the electron diffusion length. On the contrary, upon back illumination, i.e., illumination direction from the substrate to sample side makes the electron density higher at the sample substrate interface. Thus a small electron diffusion length can be enough for this case, while the hole diffusion length must be higher to avoid recombination. Keeping into consideration this type of formalism, appropriate choices are made to meet the purposes. While conducting PEC measurements several type of REs are used to take care of the exact potential applied and the accurate value of current obtained. Silver (Ag/AgCl) and saturated calomel (Hg/Hg_2Cl_2) electrodes are the mostly used REs. When the system is illuminated with a light of wavelength greater than the bandgap of the material, electron-hole pairs are generated within the semiconductor (the essential criterion has been discussed later). In practice, two electrode configurations are used for fabricating PEC cells. Light sources mostly used in this type of experiments are considered to create an environment equivalent to sunlight. In order to do so, a solar simulator or xenon lamps equipped with appropriate filters are used.

8.2.2   PREPARATION OF PHOTOELECTRODES

In the entire photoelectrochemical setup, photoelectrode fabrication plays a key role for the success or failure of any experiment. Thus, it is advisable to take enough technical considerations during fabrication of photoelectrodes. The photoelectrode is usually prepared by depositing the semiconductor material (n-type or p-type) on the conducting substrates (a conducting substrate is a thin film of conducting material deposited on a glass substrate). A layer of some nm or μm (it is chosen in such a manner to avoid recombination, as well as the maximum portion of the spectrum, is used) is coated on the conducting side of the substrate leaving a small portion of it for electrical contacts. Copper wires are used along with silver paste on behalf of their highly conductive nature for making contact with the film. In the case of the semiconductor materials coated on the metal film, back contacts are taken for electrical purposes. Further, in order to protect the metal wire from getting reacted with the electrolyte, it is preferably kept within a glass tube. Epoxy resin is used to define the active area used for illumination of the sample.

8.2.3 CHOOSING AN ELECTROLYTE

The electrolyte selection is of utmost importance to the photoelectrochemical process. Not only this serve as a medium but also it is the factor responsible for photo effects (like photo corrosion) at the surface of the photoelectrodes. It consists of charged species as which take part in the redox process. They carry the photogenerated charge carriers to and forward between the electrodes, which finally ends up with the generation of electric current. Hence an electrolyte should possess the following properties for better performance: (i) a high charge transfer rate for effective transfer of carriers generated; (ii) least optical absorption to ensure the spectrum is mostly utilized by the photoelectrode; (iii) better photostability as well as temperature stability; (iv) ionic conductance to make up for the negligible ohmic loss; (v) non-reactive to the electrodes as well as the cell wall; and (vi) cost-effective as well as environmental friendly nature.

8.2.4 BASIC PRINCIPLES OF A PHOTOELECTROCHEMICAL PROCESS

The mechanisms governing the processes at the semiconductor, semiconductor/electrolyte interface are: (a) absorption of light of appropriate wavelength, (b) generation of photocarriers, (c) charge migration, (d) charge recombination, (e) redox reaction at the surface. It can be further elaborated as follows. The photoelectrochemical process begins with the immersion of semiconductor photoelectrodes within the electrolyte by applying a suitable bias under illumination. When the energy provided by the photons exceeds the bandgap energy of the semiconductor, electrons start moving towards the CB from the VB, thus leaving an empty hole behind it in the VB. Thus due to biasing, the photocarriers move through the photoelectrodes and reach the semiconductor electrolyte interface. At the interface redox reactions takes place and light energy is converted to chemical energy resulting in evolution of gases as applicable. For photoanodes (i.e., n-type photoelectrodes), holes easily diffuse to the electrolyte leaving electron to flow through the conducting substrate. Because of this, the holes react with water causing evolution of O_2. Meanwhile the electrons travels to the CE through the substrate and external circuit; resulting in the reduction (addition of electrons) of H^+ to H_2. On the other hand, the process gets reversed for photocathodes (i.e., p-type photoelectrodes) and causes evolution of H_2 gas at the semiconductor surface and O_2 gas at the CE respectively. Henceforth the basic reactions can be mentioned as:

$$Semiconductor + 2hv \rightarrow 2h^+ + 2e^-$$

$$2h^+ + H_2O \rightarrow \frac{1}{2}O_{2(gas)} + 2H^+$$

$$2H^+ + 2e^- \rightarrow H_{2(gas)}$$

Thus, as mentioned earlier the photoelectrodes are dipped in an electrolyte which provides a medium for the carriers to flow, and under illumination, the reactions are being carried out. Because of these reactions, water splitting occurs and gases are evolved at the corresponding electrodes. Although the process appears to be simple, the mechanism underlying beneath the process is very complicated. To elaborate further the semiconductor electrolyte interface can be divided into three regions as (i) space charge layer, (ii) Helmholtz layer, and (iii) Gouy–Chapman layer (Hagfeldt and Graetzel, 1995; Li et al., 2013; Xiao et al., 2015). The region formed within the semiconductor on behalf of the electric field developed due to the transfer of charge carriers is known as the space charge region. Also, the space charge region can be further divided into three separate layers (Li et al., 2013): (a) the accumulation layer (b) the depletion layer, and (c) the inversion layer. In a similar fashion, the electrostatic double layer is formed in the electrolyte side, which is known as the Helmholtz layer and Gouy–Chapman layer (Figure 8.3). The Helmholtz layer thickness is nearly of 0.3–0.5 nm irrespective of the photoelectrodes (Li et al., 2013). On the contrary, thickness of the space charge region and Gouy-Chapman layer is inversely related to the charge carrier concentration at the interface. The thickness of space charge layer thus can be estimated by using the formula:

$$L_{SC} = L_D \left[\frac{2q|\varphi_{sc}|}{kT} \right]^{1/2} = L_D \left[\frac{2q(V - V_{fb})}{kT} \right]^{1/2}$$

and

$$L_D = \left[\frac{\varepsilon_0 \varepsilon k T}{q^2 (n_0 + p_0)} \right]^{1/2}$$

Here L_D is Debye length and varies according to the concentration of charge carriers. The other symbols $\varepsilon_0, \varepsilon,\ q, k, T, \varphi_{sc}, V, V_{fb}, n_0, p_0$ stands for the vacuum dielectric constant, relative dielectric constant, charge of electron, temperature in °K, space charge width, applied potential, flat band potential, electron concentration and hole concentration.

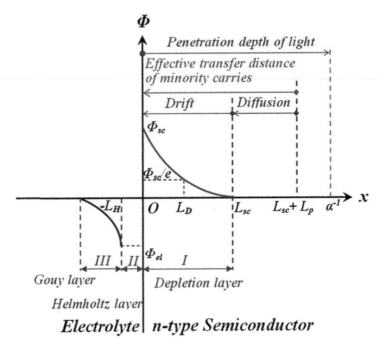

FIGURE 8.3   Structure of a semiconductor-electrolyte junction. $\emptyset_{sc}$ is the potential drop within the depleted layer; L_D is the Debye length that can be regarded as the characteristic length of potential attenuation $(\emptyset_{sc})$; fel is the potential drop within the Gouy layer; L_{sc} is the width of the space charge layer (the depletion layer is the most common and important type of space charge layer); L_p is the minority carrier diffusion length and a is the optical absorption coefficient of the photoelectrode. For the depletion layer, $L_{sc} > L_D$, while $L_{sc} < L_D$ for accumulation layer or inversion layer. The number e is a mathematical constant, approximately equal to 2.71828.

Source: Reprinted with permission from Li et al. (2013). © Royal Society of Chemistry.

8.2.5   *IMPORTANT CRITERION FOR AN EFFICIENT PHOTO ELECTRODE*

In order to behave as an efficient photoelectrode the materials should obey some basic criterion as mentioned below:

1. Materials are expected to have an absorption edge in the visible region and thus the bandgap of the material has a substantial impact on the solar-to-hydrogen efficiency of a PEC cell. So the bandgap should be small enough to absorb the maximum portion of the spectrum. Thermodynamically for water splitting reaction $\Delta G = 237 \, KJ \, / \, mol$ and for hydrogen and oxygen gas to evolve a minimum 1.23 eV is required (Bolton et al., 1985; Sivula and Van De Krol, 2016). Taking into account all the losses and overpotential required for the fast reaction kinetics, a bandgap value of ~1.8 eV is desirable. As per the reports of the US Department of Energy, a solar-to-hydrogen efficiency of 10% is achievable for a photoelectrode when it can have at least a photocurrent density of 8 mA/cm^2; which is related to a bandgap of 2.4 eV (Chen et al., 2013). On account of this, the semiconductor should have a bandgap value within the range of 1.8 to 2.4 eV to serve as a photoelectrode for water splitting reactions.
2. Not only bandgap but also the suitable band edges of materials should straddle the water redox potential. Only materials with small band gap values within the permissible limits also cannot help in dissociating the water molecules. To do so the position of CB and VB must be above and below of the oxygen and hydrogen generation potential levels respectively.
3. Also to achieve better performance the photogenerated charge carriers should be able to get transferred easily at the semiconductor electrolyte interface. Thus properties like mobility, diffusion length plays a critical role in deciding better performing electrodes.
4. Another important factor regarding the photoelectrodes is their stability in chemical environment. Generally, for a n-type semiconductor the oxidation potential of the material in the electrolyte should be lower than the oxygen evolution potential while for a material of p-type nature, the reduction potential should be higher than hydrogen evolution potential.
5. The cost-effectiveness as well as environmental friendly nature of the sample is also considered as essential criteria for photoelectrode selection.

Although this criterion are followed for any material still the appropriate-ness of any semiconductor photoelectrodes are relied upon by their efficiency in water splitting. In the next section several types of efficiency, which are commonly related to photoelectrochemical water splitting are described.

8.2.6   EFFICIENCY MEASUREMENT OF PHOTOELECTRODES

The efficiency measurement of photoelectrodes plays an important role in determining the light-sensitive ability of the photoelectrodes for future appli-cations. Several types of efficiency are often reported in articles, such as incident photon-to-current conversion efficiency (IPCE) or external quantum efficiency (EQE), absorbed photon-to-current conversion efficiency (APCE) or internal quantum efficiency (IQE), solar to hydrogen (STH) conversion efficiency, applied bias photon-to-current conversion efficiency (ABPE).

8.2.6.1   INCIDENT PHOTON-TO-CURRENT EFFICIENCY (IPCE)

It is defined as the amount of photocurrent generated per incident photon flux or in other words, it depicts the number of electrons generated with the number of photons incident on it. IPCE (also known as, EQE) of a photoelectrode depends on the optical absorption, width of the space charge layer, and on the minority carrier diffusion length (Xiao et al., 2015). It can be measured using a two or three-electrode configuration. For instance, if two-electrode set up is used for measurements then the photocurrent density has dependence over the type as well as the distance of the CEs. Thus, it can be erroneous to claim the appropriateness of the observed photocurrent density values. Thus to tackle with this a third electrode (REs) is used while carrying out the photoelectrochemical measurements. Thus, it stays indepen-dent of the type of CE, the distance between the electrodes, while keeping the photocurrent values unaffected. Hence, it characterizes the ability of the photoelectrodes to generate electrons/holes by incidenting photons of a particular wavelength and can be presented by the expression:

$$IPCE(\lambda) = EQE(\lambda) = \frac{\left[\left| J_{ph} \right| \left(\dfrac{mA}{cm^2} \right) \times hc \left(V\,nm \right) \right]}{\left[P_{\lambda} \left(\dfrac{mW}{cm^2} \right) \times \lambda \left(nm \right) \right]_{AM1.5G}}$$

8.2.6.2 ABSORBED PHOTON-TO-CURRENT EFFICIENCY (APCE)

While calculating IPCE the optical losses are not taken into account which ultimately makes the efficiency less apparent. To make the estimations accurate regarding the photoelectrochemical performance of the photoelectrodes the APCE or the IQE is estimated. It may be defined as the number of photogenerated charge carriers involved in the generation of photocurrent per absorbed photon and formulated mathematically by the equation:

$$APCE(\lambda) = \frac{IPCE(\lambda)}{Abs} = \frac{IPCE(\lambda)}{1-R-T}$$

where, A, R, T represents absorption, reflection, and transmission respectively.

8.2.6.3 SOLAR TO HYDROGEN CONVERSION (STH) EFFICIENCY

It is defined as the ratio of chemical energy produced to that of input solar energy under illumination and zero bias condition. Thus, STH efficiency is regarded as the water-splitting ability of a photoelectrode. The chemical energy stored is equal to the product of rate of hydrogen production and change in Gibbs free energy (ΔG) per mol of H_2 (237 kJ/mol), while the solar energy input is the product of incident power density (P, 100 mW/cm², AM 1.5G) with the sample area under illumination within the electrode. The formula thus can be stated as:

$$STH(\%) = \frac{\left[\left(mmol\frac{H_2}{s}\right)\times\left(237\frac{kJ}{mol}\right)\right]}{\left[P_{total}\left(\frac{mW}{cm^2}\right)\times Area\left(cm^2\right)\right]_{AM1.5G}}\times100(\%)$$

Most often, another modified equation is used for estimating the STH efficiency as follows:

$$STH(\%) = \frac{\left[|J_{ph}|\left(\frac{mA}{cm^2}\right)\times1.23(V)\times\eta_F\right]}{\left[P_{total}\left(\frac{mW}{cm^2}\right)\right]_{AM1.5G}}\times100(\%)$$

Here J_{ph} is the photocurrent density and η_F is the faradaic efficiency.

8.2.6.4 APPLIED BIAS PHOTON-TO-CURRENT EFFICIENCY (ABPE)

With the application of the external bias, the photocurrent density value increases and results in higher efficiency PEC processes. Several factors like cell configuration (two- or three-electrode), light source, CE, contact resistance, and the band structure of the semiconductor, etc., are known to influence the ABPE (Xiao et al., 2015; Zhou et al., 2013). This type of estimated efficiency is known as applied bias Photon-to-current efficiency (ABPE) and can be stated mathematically as:

$$ABPE\,(\%) = \frac{\left[\left|J_{ph}\right|\left(\frac{mA}{cm^2}\right)\times\left(1.23 - \left|V_{app}\right|\right)(V)\right]}{\left[P_{total}\left(\frac{mW}{cm^2}\right)\right]_{AM1.5G}}\times100\,(\%)$$

The symbols J_{ph}, V_{app}, P_{total} are meant for photocurrent density, potential applied and intensity of light used for illumination. Taking into consideration that faradaic efficiency is less than unity, the applied bias STH conversion efficiency (AB-STH) can be expressed as:

$$AB - STH\,(\%) = ABPE\times\eta_F\times100\,(\%)$$

And the equations are applicable for 2-electrode system. While, for a 3-electrode system, a REs is used, and the applied potential is thus converted into the reversible hydrogen electrode (RHE) by using the formula below:

$$V_{app} = V_{Ag/AgCl} + 0.059\,pH + E^0_{Ag/AgCl}\,;\,E^0_{Ag/AgCl} = +0.197V$$
$$V_{app} = V_{SCE} + 0.059\,pH + E^0_{SCE}\,;\,E^0_{SCE} = +0.241V$$

For Ag/AgCl and SCE are silver and saturated calomel electrodes (SCE) respectively. Thus, the half-cell STH conversion efficiency (HC-STH) is considered to be the ability of the photoelectrode in converting the solar energy into chemical energy.

So for a photoanode:

$$HC-STH\left(\%\right)=\frac{\left[\left|J_{ph}\right|\left(\frac{mA}{cm^2}\right)\times\left(1.23-E_{RHE}\right)(V)\right]}{\left[P_{total}\left(\frac{mW}{cm^2}\right)\right]_{AM1.5G}}\times100\left(\%\right)$$

While for a photocathode:

$$HC-STH\left(\%\right)=\frac{\left[\left|J_{ph}\right|\left(\frac{mA}{cm^2}\right)\times E_{RHE}\left(V\right)\right]}{\left[P_{total}\left(\frac{mW}{cm^2}\right)\right]_{AM1.5G}}\times100\left(\%\right)$$

The dependence of the efficiency of a PEC cell on optical absorption, the width of space charge layer, sluggish reaction kinetics, and minority carrier diffusion necessitates the optimization of all the parameters for achieving high efficiency in a cost-effective manner. The physical properties can thus be optimized by varying the morphology (for e.g., in case of 1D nanostructures). The high surface to volume ratio of 1D structure facilitates better pathways for carrier transport. Thus, it became inevitably a matter of interest for several researchers.

8.3   ZINC OXIDE NANOSTRUCTURES

The ZnO nanostructures are known to be prepared broadly via vacuum-based or non-vacuum based approaches. Several synthesis techniques, like chemical vapor deposition (CVD), magnetron sputtering, pulsed laser deposition, atomic layer deposition (ALD), electrochemical deposition, spray pyrolysis and hydrothermal have been reported till date for the fabrication of ZnO nanostructures (Baskoutas, 2014; Dhara and Giri, 2012). But the non-vacuum based hydrothermal (water is used as the solvent to prepare growth solution) approach caught greater attention for synthesis on behalf of its simplicity and cost-effectiveness nature. Mostly by using hydrothermal technique for growth of the 1D structure, it became convenient for fabrication of large area. Furthermore, the familiarity of the process caused due to the usage of aqueous solution in the growth process, which ultimately reduces the growth temperature to less than the boiling temperature of water.

Several researchers have reported the synthesis of ZnO by various solution-based approaches namely hydrothermal, solvothermal, spray pyrolysis, electrodeposition, etc. The major advantage of this type of procedures is their low cost, low synthesis temperature. This almost gives a substantial cost reduction in mass production as far as commercialization is concerned. In this technique, H_2O acts as a dopant source along with Zn precursor, and hexamethylenetetramine (HMTA) is used for growing 1D nanostructure. Prior to this, most reports suggest coating a seed layer of ZnO of few nanometers. As the purpose of dopant and Zn precursor is to provide the doped type of ZnO NRs only, it is important to discuss about the role of HMTA. It is a non-ionic tetra dentate cyclic tertiary amine usually preferred in hydrothermal technique with regards to its high solubility in water. On the application of temperature, thermal degradation occurs resulting in the formation of hydroxyl ions. They further react with the dopants as well as Zn present in the aqueous solution in order to grow doped ZnO nanostructures. In general, the sealed container containing the seed layer substrate immersed in the aqueous solution is kept at 90–95°C for 4–5 hours. Our group has prepared several doped ZnO NRs through a two-step procedure using a facile sol-gel, spin coating, and hydrothermal method. First is the deposition of the seed layer on cleaned substrates followed by the growth of NRs by dipping the seed layers in growth solution in the second step. All the growth conditions including the growth temperature were kept at 90°C for 5 hrs as per our earlier report.

Two main research approaches have emerged in the development of photocatalytic materials with improved photon-to-current conversion efficiency (PCE). The first approach is to tune the electronic structure of ZnO in order to extend its ability to harvest photons in the visible region of the light spectrum (bandgap engineering). The light-harvesting ability can be enhanced by controlling the morphology of the ZnO crystal (quantum confinement effect), modifying the carrier concentration within the ZnO crystal lattice (doping), or by functionalizing the surface of ZnO with photosensitizer molecules. The second approach is to promote effective photogenerated charge separation by controlling the defects in the crystal lattice or incorporating electron transfer (ET) agents. Alternatively, hybridizing ZnO with metallic or semiconductor co-catalysts has proven successful in accelerating water oxidation reactions.

8.3.1   EFFECT OF MORPHOLOGY ON PEC PERFORMANCE OF ZNO NANOSTRUCTURES

Various research groups have investigated different morphologies of ZnO nanostructures in order to enhance their photoelectrochemical performance. In the following section, we have discussed some useful reports available in the literature.

8.3.1.1   1D NANOSTRUCTURES

8.3.1.1.1   Nanorods (NRs)

Babu et al. (2015) reported the photoelectrochemical water splitting properties (under both UV and visible light illumination) of ZnO NRs by changing the diameters from 45 nm to 275 nm. ZnO NRs with 45 nm exhibited maximum photoconversion efficiency of 45.3% under 365 nm UV light illumination and 0.42% under Air Mass 1.5 Global simulated solar light illumination. Higher efficiency for smaller diameter NRs was attributed to the increased light absorption due to a decrease in diameter as verified by the theoretical simulation using the finite-difference time domain. Rokade et al. (2017) studied ZnO NRs and ZnO nanotubes as efficient photoelectrodes for photoelectrochemical water splitting. They reported an enhanced photocurrent density of 0.67 mA/cm^2 at 0.5V vs. SCE in the case of ZnO nanotubes in contrast with NRs (0.39 mA/cm^2). The difference in these values are due to the high surface area of ZnO nanotubes present in the vicinity of electrolyte which helped in harvesting a large number of photons that significantly increased the generated charge carriers and photocurrent density. Moreover, the ABPE for ZnO nanotubes was 0.50% as compared to NRs (0.29%) under visible light illumination (100 mW/cm^2 AM 1.5). In another report, Samad et al. investigated the photoelectrochemical performance of ZnO NRs by changing the applied potential in ZnCl$_2$ and KCl electrolyte. The optimized applied potential was 1V which exhibited the highest photocurrent density for UV (17.8 mA/cm^2) and visible illumination (12.94 mA/cm^2) in contrast with the applied potential of 2V (with UV illumination; 11.78 mA/cm^2 and visible illumination; 10.78 mA/cm^2). The reason behind the highest photocurrent density is that the applied potential of 1V produced ZnO NRs with the highest average aspect ratio. The other two applied potentials (2V and

3V) showed dense porous structures which prevented the charge transfer showing a decreased photocurrent density.

Co-catalyst is a noble metal which when loaded onto the surface of ZnO photocatalyst lowers the recombination rate of the charge carriers, enhances charge separation, increases the number of reactive sites, and reduces activation energy for gas evolution. Co-based catalyst deposited photochemically on the surface of ZnO NRs gives enhanced solar O_2 evolution (Steinmiller and Choi, 2009). This nanoparticulate morphology (10–30 nm of Co nanoparticles (NPs) uniformly distributed on the surface of ZnO NRs) increases the surface area of catalysts per unit mass while minimizing the blockage of the ZnO surface. The photocurrents measured at a constant bias of 0.0 V and 0.2 V against the Ag/AgCl REs show that the presence of Co-based catalyst enhanced the steady-state photocurrent by 2.6 and 1.5 times, respectively. ZnO NRs coated with a silver film on a flexible substrate like polyethylene terephthalate (PET) can be used as an efficient photoanode in PEC water splitting as reported by Wei et al. in 2012. They reported a maximum photocurrent density (0.616 mA/cm^2) and PCE (0.81%) achieved with an optimized Ag film thickness of 10 nm and a substrate bending radius of 6 mm. The improvement in PEC performance of ZnO NRs coated with Ag film may be attributed to the increase of light absorption capability of ZnO: the localized surface plasmonic effect of Ag islands deposited on the surface of ZnO NRs helps to increase the local field strength, resulting in higher absorption. It also helps to enhance light scattering and lengthen the light path thereby increasing the light trapping. Fabrication of platinum nanoparticles (PtNPs) on the surface of ZnO nanorod (ZnO NR) arrays can also act as an efficient photoelectrode in the process of water splitting as reported by Hsu et al. in 2014. They achieved a twofold enhancement in photocurrent density and greater PEC stability for Pt NPs/ZnO NR arrays as compared to pristine ZnO NR electrodes. Pt decorated ZnO NR arrays showed greater separation efficiency of photogenerated electron-hole pairs and faster charge transfer in comparison with pristine ZnO NRs (Hsu et al., 2014). The hybridization of two semiconductor materials also leads to the formation of heterojunction that proves to be more advantageous in charge transport (Moniz et al., 2015). AgSbS$_2$ catalyst on ZnO nanotube arrays plays a crucial rule in the enhancement of photoconversion efficiency and photocurrent density (Han et al., 2015). The three important factors influencing the PEC water splitting performance are photo response region, the stability of the semiconductor, and electrons transmission velocity. In order to broaden the photoresponse spectrum of ZnO/AgSbS$_2$, a suitable energy bandgap of this coupled nanostructure is obtained by using miargyrite $AgSbS_2$. As a result, the light absorption efficiency of $ZnO/AgSbS_2$ which ultimately leads to the enhancement of PEC performance of this nanotube arrays. A remarkably higher value of photocurrent density (5.08 mA/cm^2) and hydrogen production efficiency (5.76%) was reported for $ZnO/AgSbS_2$ nanotube arrays as compared to ZnO NRs (0.41 mA/cm^2 and 0.42%). 1D nanostructure (nanotube) without crystal boundary resistance have a very high-speed photo-induced charges transmission and lower recombination rate of photogenerated charge carriers. Thus, 1D $ZnO/AgSbS_2$ nanotubes help in providing a direct electrical path to ensure the rapid collection of the charge carriers.

8.3.1.1.2   Nano Rod @ Nano Platelet

Another strategy to enhance the photoelectrochemical conversion efficiency of ZnO nanostructures was reported by Zhang et al. in 2015, i.e., Au-sensitized ZnO nanorod@nanoplatelet (NR@NP) core-shell arrays. The introduction of Au NPs on ZnO NR@NP increases the absorption of light and promotes the charge transfer to the electrode/electrolyte interface resulting in the PEC performance. The maximum PCE for the Au-ZnO NR@NP arrays was found to be 0.69% (at 0.42 V vs. Hg/Hg_2Cl_2) which is 1.3 times higher than the ZnO NR@NP arrays (0.54% at 0.40 V). Furthermore, an enhanced value of photo-current density was observed for Au NPs sensitized ZnO NR@NP photoelec-trode (0.06 mA/cm^2) at 0.6 V as compared to the ZnO NR@NP having 0.04 mA/cm^2 at a potential of 0.6 V vs. Hg/Hg_2Cl_2. Heterostructures constructed using metal oxide co-catalyst or carbon-based ET agents like graphene are pervasively investigated materials to improve the transport of charge carriers in semiconductors. Development of photoanodes with enhanced light-harvesting efficiency is a key factor in improving PEC performance. Another approach to enhance the STH efficiency of ZnO photoanodes is the lowering of recombination rate of the photogenerated charge carriers in ZnO on semiconductor surface and improving the capability for instantaneous charge collection, separation, and transportation. ZnO NRs/RGO/$ZnIn_2S_4$ heterojunction gives a photoconversion efficiency of 0.46% as reported in previous results (Bai et al., 2015). In this heterojunction, ZnO NWs acts as core materials, RGO sheets behave as the charge transfer interlayer and $ZnIn_2S_4$ serves as a visible light sensitizer. Furthermore, RGO sheets also provide instantaneous pathways for photogenerated charge carriers resulting in a decrease of the recombination rate of the charge carriers.

8.3.1.1.3 Nanosheets

Bakranov et al. (2017) fabricated ZnO nanosheets and ZnO NRs by electrochemical deposition and further modified these morphologies with AgNPs in order to enhance the PEC performance. They observed approximately 1.4 times increase in photocurrent density of ZnO nanosheets after integration of plasmonic AgNPs on the surface. The photocurrent density was furthermore enhanced (~5 times) in case of ZnO NRs decorated by AgNPs. In the case of nanosheets, the charge carrier trapping zones could enter into force in ZnO/Ag sheet-like composites which was determined by the presence of structural defects, i.e., the difference observed in values of photocurrent density was due to presence of large number of defects in nanosheets structure. Zhang et al. (2016) also reported the fabrication of ZnO nanosheet films in a scalable manner. The photoelectrochemical properties observed for the solution-processed on ITO photoelectrode was optimized by varying $ZnSO_4$ concentration. As compared with all the prepared samples, the 7.5 mM sample performed better photoelectrochemical properties and showed a photocurrent density of 500 $\mu A/cm^2$. They explained the enhancement as a synergistic effect of high surface to volume ratio along with lowered resistance.

8.3.1.1.4 Nanotubes

Hsu et al. (2011) synthesized ZnO nanotubes and ZnO nanosheets as photoelectrodes in water splitting cells. They reported that the difference between the relative intensities of XRD (100) and (002) peaks of ZnO nanotubes and ZnO nanosheets is related to the different degrees of polarity in ZnO nanostructures which greatly affects the PEC performance of these photoanodes. A lower value of (100)/(002) ratio reveals the formation of nanotubes along the c-axis and a large fraction of non-polar facets. Conversely, a high (100)/(002) ratio indicates shortening along c-axis of ZnO nanosheets whose surface is dominated by polar facets. PEC measurements revealed that ZnO nanosheets show high photocurrent density than those of ZnO nanotubes. ZnO nanosheets with polar facets possessed a more negative flat band potential (–0.35V) than those of ZnO nanotubes (–0.23V). This difference in value of flat band potential was due to the fact that polar and nonpolar facets generated different surface states and atomic arrangements, affecting the absorption of reactant ions or molecules. Furthermore, a threefold enhancement was observed in PCE of ZnO nanosheets with polar facets (0.27%) as compared with ZnO nanotubes with non-polar facets (0.08%) because of the high surface energy, spontaneous polarization, and negative flat-band potential of a polar-oriented surface.

8.3.1.1.5   Nanowires (NWs)

Zhifeng Liu and group studied photoelectrochemical properties of three different morphology of ZnO NWs by varying their dipping time in the growth solution. They reported that dipping time of 5, 10, and 15 days yields NRs, nanotubes, and nanodisks respectively. Among these four morphologies, ZnO nanotubes exhibits the highest photocurrent density of 0.49 mA/cm^2 at 1.2V (vs. RHE) in comparison with ZnO NWs (0.38 mA/cm^2), ZnO NRs (0.38 mA/cm^2) and ZnO nanodisks (0.26 mA/cm^2) respectively. The tubular structure of ZnO nanotubes helps them to possess excellent light scattering ability for harvesting light through multiple reflections between the nanostructures. Thus, they could utilize sunlight fully to generate more electron-hole pairs for PEC water splitting (Liu et al., 2017). Similar efforts were carried out by Govatsi et al. (Govatsi et al., 2018) who explored the role of morphology of ZnO NWs on the photoelectrochemical performance. They classified the nanowire arrays into five different average diameters ranging from 40–260 nm. ZnO NWs with average diameter of 120 nm exhibited the maximum applied bias photoconversion efficiency of 6.3% upon illumination 11.5 mW at 365 nm. Moreover, the photoanodes displayed a stable performance up to 10 hours representing its stability for a long duration. Other attempts to enhance the PEC performance of ZnO NWs was reported by Harnandez et al. they deposited a shell of anatase TiO$_2$ on to the arrays of ZnO NWs. TiO$_2$@ZnO core-shell structure shows an enhancement by 15 times in photocurrent density and 5 times in photo conversion efficiency. Anatase shell of TiO$_2$ protects the ZnO NWs structure as a result of which long term lifetimes of the core-shell photoanodes are obtained for water splitting reaction. Furthermore, these core-shell structure NWs when annealed in air increases the PEC performances favoring charge separation and lowers the recombination rate of the photogenerated charge carriers. Similar results were reported by other researchers using TiO$_2$ shell (Liu et al., 2013). The ultrathin TiO_2 shell chemically protects the ZnO NWs and allows the PEC water splitting of ZnO in a strongly alkaline environment, which is very important for efficient mass transport near the photoanode and to achieve high PEC activity. Also, TiO_2 shell passivates ZnO arrays surface states through partially removal of deep hole traps, without affecting the minority carrier diffusion due to its almost negligible thickness, resulting in an increase of photocurrent density. Chen et al. demonstrated a photodevice based on photosensitization of ZnO NWs with CdTe quantum dots to enhance photocurrent and photoconversion efficiency. They achieved an enhancement of more than 200% in photoconversion efficiency (1.83%) for ZnO@CdTe as compared to bare ZnO NWs. CdTe sensitization improves visible light absorption, and also the amount of photogenerated charge carriers in CdTe quantum dot can be easily transferred to ZnO nanowire as a result of which efficiency increases.

Another type of enhancement strategy was reported by GuO and group which used Graphene quantum dots to enhance the PEC performance of ZnO NWs (Guo et al., 2013). Zheng et al. (2016) also reported a heterostructure where ZnO NWs are synthesized via MOCVD. They put carbon on the NWs and upon illumination with a 300 Watt Xe lamp obtained 8.6 μmol of H_2 (2.5 times higher than the pristine ZnO). Heejin Kim and group developed a stable and environment-friendly photoelectrode for PEC hydrogen generation by using carbon nanodots coupled with a 3D ZnO structure. This 3D ZnO structure comprises of 1D ZnO nanowire (core) and 2D ZnO nanosheets on the surface to enhance the effective surface area for C-dot sensitization. The result reveals a 4 fold increase in photocurrent density of C-dot modified ZnO nanostructure (0.72 mA/cm²) as compared to bare ZnO nanostructure (0.18 mA/cm²) at a potential of 1.23V vs. RHE. This increment in photocurrent density of C-dot modified ZnO nanostructure may be attributed to the cascade band structure that enables efficient charge transfer between C-dots and ZnO nanostructures through their interface due to C-dot sensitization. Furthermore, they achieved a significantly increased without the use of any sacrificial reagents for the overall water splitting system. The photocurrent density was stable for more than 8000 seconds under 1 sun illumination. The enhanced stability of the photoelectrode comes from the amine passivation of C-dots that removes the surface defect states and provides a chemical shield to protect undesirable oxidation to form strong amide bonding with the intrinsic surface carbonyl groups (Kim et al., 2015).

A novel approach was reported by Wang et al. (2019) who fabricated ZnO nanowires overlapping junction (ZnO NWs-OLJ) by a facile and low-cost method in order to enhance the PEC water splitting of ZnO nanowire photoanodes. The NWs overlap and touch each other during their growth process to form ZnO NWs-OLJ. The photocurrent density of these overlapping junction (OLJ) was found to be 57 $\mu A/cm^2$ which was almost double than that of the ZnO NR vertical arrays at a potential of 0V versus Ag/AgCl. The improved PEC performance of ZnO NWs-OLJ may be because of two reasons. The first reason for enhancement is the existence of inner electric field at the junctions. The OLJs can result in tunneling effects due to an infinitely small gap between ZnO NWs. This provides a potential barrier and an inner electric field is formed. So, the separation efficiency of the photo-induced charges can be improved with the help of inner electric field and further enhanced the PEC performance. Secondly, nanowire junction arrays can effectively increase the ratio of multiple reflections which proves to be beneficial for capturing more light. Further, to enhance the PEC performance even more, they decorated ZnO NWs-OLJ with AuNPs. The average photocurrent density of this heterostructure was calculated to be 87 $\mu A/cm^2$ at 0V vs. Ag/AgCl which was 1.5 times higher than the pure ZnO NWs-OLJ. They proposed a mechanism for electron-transfer in Au/ZnO NWs-OLJ heterostructure during water splitting. Upon UV illumination, electron-hole pairs are generated where electrons gets excited and jumps from VB to CB of ZnO equal amount of holes in VB. The newly formed Fermi energy level of the heterostructure will be lower than the bottom of the CB of ZnO. Due to the energy difference, the photo-induced electrons will be transferred from ZnO to Au and effectively reduce the recombination rate of the charge carriers. Moreover, due to surface plasmon resonance (SPR) excitation, AuNPs can absorb the resonant photons to generate hot electrons. These electrons will be injected into the platinum electrode via the CB to generate hydrogen. The holes present in VB of ZnO will be immediately consumed by producing oxygen.

8.3.1.1.6   Nanopencils

Photosensitization of semiconductors with noble metals helps in the increment of its PEC performance as reported by various researchers. One of the previous, reports reveal that Au nanoparticle sensitized ZnO nanopencil arrays yields a high photocurrent density of 1.5 mA/cm² at 1V versus Ag/AgCl (Wang et al., 2015). Gold does not undergo corrosion during photoreaction and has the capability to strongly interact with light in visible and infrared region because of its localized surface plasmon resonance (LSPR) properties. The enhancement in photocurrent density is due to the formation of long tips of the nanopencil arrays which provides semiconductor-electrolyte interface where the space charge layer is developed. Furthermore, it has been earlier reported that ZnO NRs having long tips have better electrical properties because of the presence of oxygen vacancies.

8.3.1.1.7 Nanotriangles

The facility to tune the physiochemical properties of graphene between a semiconductor and a semimetal makes this carbon sheet an important subject of investigation for increment in PEC performance of ZnO nanostructures. ZnO nanotriangles on graphene oxide (GO) at proper condition (pH = 9) yields high photocurrent density and photoconversion efficiency due to optimum amount of oxygen vacancies as investigated by the experimental results and various deformation models (Chandrasekaran et al., 2016).

8.3.1.1.8 Nanoflowers

Sohila et al. (2016) synthesized ZnO nanoflowers by hydrolysis of zinc acetate in the presence of DMF (dimethyl formamide) with traced amount of water and explored its usage as ZnO nanostructure based photoanode for PEC water splitting. They evaluated the PEC performance by measuring photocurrent density at the different applied voltage and achieved the maximum photocurrent density of 0.39 mA/cm^2 at 0.6V vs. Ag/AgCl. The improved PEC performance of ZnO nanoflowers based photoanode was attributed to its mesoporous nature constituted by well-connected ultra-small ZnO NPs with high surface to volume ratio that provides a larger area of electrode/electrolyte junction and more number of water oxidation sites at the electrolyte-ZnO interface. Also, the ultra-small ZnO NPs in the flower-like morphology promoted the efficient separation of photogenerated charge carriers.

8.3.1.1.9 Nanotetrapods

Apart from these 1D morphologies, branched ZnO nanotetrapods also facilitates better electron transport and network forming ability for efficient photoelectrochemical water splitting (Qiu et al., 2012). The PEC behavior of these photoanodes experiences even more dramatic enhancement in photocurrent density (0.99 mA/cm^2 at 0.31V vs. Ag/AgCl) by doping Nitrogen. There are basically three main reasons responsible for this significant enhancement. Firstly, the branched growth of nanotetrapods helps in increasing the light-harvesting capacity. Secondly, N-doping increases the roughness factor which boosts up the light-harvesting associated with the ZnO nanotetrapod branching. Furthermore, N-doping also shifts the absorption spectra of ZnO towards visible light region by narrowing its bandgap.

8.3.1.1.10   Nanofibers (NFs)

Qiang Li and group reported a facile synthesis protocol for caterpillar-like branched ZnO nanofibers (BZNs) and explored their use as photoanodes in photoelectrochemical water splitting cells. They observed a significantly enhanced photocurrent density (151% higher) in the case of caterpillar-like BZNs (0.524 mA/cm^2 at 1.2V vs. Ag/AgCl) as compared with ZnO NW arrays (0.348 mA/cm^2). Also, they achieved a PCE of 0.165% (at 0.89V vs. RHE) for caterpillar-like BZNs which was 147% higher than those of the vertically aligned ZnO NWs (0.112% at 0.85V vs. RHE). This improvement in PEC performance of caterpillar-like BZNs was attributed to three main reasons. Firstly, the ultra-dense secondary nanowire branches radially grew up on the parental ZnO NFs and offered a high spatial occupancy of the nanocomponents, which greatly increased the surface to volume ratio and roughness factor of the caterpillar like BZNs, and hence help capture more sunlight. Secondly, the nanowoven network consisting of caterpillar-like BZNs possesses an open micrometer porous structure which is beneficial to penetration of solar light and trapping the light via multireflections. Third, the fine branches with high length-to-diameter aspect ratios provided an advantage of efficient charge separation and hole diffusion at the electrode/electrolyte interface. With the elongated branches that generate joint sites among ZnO nanobranches to connect the parental ZnO NFs, the improved electron migration also reduces charge recombination to ensure efficient charge transport (Li et al., 2014).

8.3.1.1.11   Nanoforest

Sun et al. (2014) reported unique 3D ZnO nanoforests as a potential candidate to enhance the PEC water splitting performances of these ZnO nanobranches synthesized by hydrothermal method. ZnO nanoforests showed a high photocurrent density of 0.919 mA/cm^2 at 1.2 V vs. Ag/AgCl and PCE of 0.2999% at 0.89V vs. RHE. The large surface area and roughness factor of ZnO nanoforests resulted in increase of photocurrent density associated with efficient light and photon harvesting. Similar to nature trees, the upstanding nanotree arrays provided straight-forward light path and longer penetration depth, avoiding the thickness limitations of densely stacked NPs or thin films. Also, the nanosized branches of ZnO nanoforests helped to extend the light propagation and improved light trapping because of the multiple internal light reflections and scattering on the surface of branches. It was also worth noting that ZnO nanotrees have broader absorption in solar spectrum in comparison with ZnO NWs, because tree-like morphology can activate maximized excitonic band gaps of wurtzite ZnO. Due to the large surface area and high light trapping capacity, PCE was of ZnO nanoforests was greatly improved.

8.3.1.1.12   Nanocorals

Ahn et al. (2008) synthesized ZnO nanocorals and investigated their usage as photoanodes in PEC water splitting cells. ZnO nanocorals showed enhanced PEC performance (10 fold increment) as compared to ZnO compact and nanorod films because of their suitable electrical pathways for efficient charge collection as well as large surface area.

8.3.1.2   2D NANOSTRUCTURES

Wolcott et al. reported that the photoelectrochemical properties strongly depend on the porosity and morphology of ZnO photoelectrodes (2009). They prepared ZnO thin films by three different methods and compared the PEC performance of all the three electrodes. Pulsed layer deposition (PLD) with the substrate normal to vapor flux used to prepare ZnO thin film and the other two methods oblique-angle deposition (OAD) and glancing-angle deposition (GLAD) generated nanoplatelet. Similar efforts were put by Hamid et al. to enhance the applied bias photon to current efficiency (ABPE) of ZnO thin films by hybridization with reduced graphene oxide (RGO). Growing ZnO thin films on rGO increased the photocurrent density from 1.0 to 1.8 mA/cm^2 and ABPE from 0.55% (ZnO) to 0.95% (rGO/ZnO). Hybridization of ZnO with rGO increases the active surface area, increases light absorption range, decreases crystal strain, reduces defect based recombination, and increases ET at the semiconductor-electrolyte interface. These factors help in enhancing the PEC properties of ZnO thin films when hybridized with rGO.

8.3.2   EFFECT OF DOPING ON THE PHOTOELECTROCHEMICAL PERFORMANCE

8.3.2.1   ALKALI METALS

Although ZnO is an n-type material with excess electrons to donate, it can be modified with suitable cationic dopant to appear with shallow acceptors and thus behaving as a p-type. The p-type ZnO NRs has been known to achieve by doping with group I elements. Several research articles have been reported using dopants of group I elements such as Li, Na, K, and Cs. But among these Li is known to possess a smaller ionic radius as compared with Zn as well as it acts as a shallow acceptor/reactive donor. Thus, it is difficult for the Li doped ZnO to retain p-type nature. Henceforth in order to achieve a stable p-type ZnO semiconductor it is preferable to dope other group I elements with bigger cationic radius. Lee et al. (2016) has reported enhanced photoelectrochemical water oxidation for K and Na doped ZnO NRs synthesized by means of chemical bath deposition (CBD) technique. They showed a photocurrent density of 0.62 and 0.90 mA/cm^2 as compared with the undoped ZnO NRs.

8.3.2.2   THE SCANDIUM FAMILY

Qasim et al. (2018) reported the enhancement in photoconversion efficiency of Y doped ZnO NRs synthesized by a two-step process, i.e., sol-gel spin coating of seed layer followed by hydrothermal growth of nanostructures. Lowering of bandgap along with the creation of oxygen vacancies are the preferable reasons for choosing Y^{3+} as a dopant. They achieved a photocurrent density of 1.2 mA cm^{-2} for 1.2 mol% doping; which is nearly five times higher as compared with the pristine ZnO NRs (0.25 mA cm^{-2}) at a potential of 0.2V vs. Ag/AgCl.

8.3.2.3   THE VANADIUM FAMILY

The advantage of the ion implantation technique resides within the control of the doping percentage with variation of the ions dose. So the implantation causes modification of the electronic structure within the bulk of the semiconductor. Cai et al. (2015) have reported the doping of vanadium (V) into the ZnO matrix by using an ion implantation technique. Exposing ZnO nanostructures to V^{4+} ion allows forming an impurity level in the forbidden band and thus expands the optical absorption to visible range. Upon evaluation of their electrochemical performance under visible light illumination, the V ions implanted sample with 2.5×10^{15}. It shows a photocurrent density of 10.5 µA/cm^2 at 0.8V (vs. Ag/AgCl); which is nearly 4 times higher compared to that of the undoped ZnO NRs arrays. On further increase in the ions implantation dose, more number of defects is induced causing recombination of the photo carriers.

8.3.2.4   THE CHROMIUM FAMILY

Shen et al. (2013) engineered the intra-bandgap states by using Cr as a dopant in the ZnO nanosheet causing enhanced absorption in the visible range. At the same time, the usage of 1D nanostructure provides a facile pathway for carrier transport. Though the efficiency achieved for the isostructure is higher than the individual parent structure, still the overall efficiency is lower for this structure. The authors provide a concept for fabrication of further high efficient photoelectrochemical devices. Cai et al. (2017) doped tungsten (W) by ion implantation technique into hydrothermally grown ZnO NR arrays. The optical absorption gets extended to visible range due to the creation of defect and a photocurrent density of 15.2 µA/cm^2 was observed at 1V (vs. Ag/AgCl).

8.3.2.5   THE IRON FAMILY

Recently Selloum et al. (2019) electrochemically fabricated n-type Fe doped Zinc oxide NRs on ITO substrates at a temperature of 70°C. The photocurrent values obtained for all the samples under UV illumination showed the optimum Fe doping concentration is 2%. Furthermore, they have presented the change in morphology from NRs to nanoflakes with increase in Fe concentration afterwards.

8.3.2.6 THE COBALT FAMILY

Lee et al. (2016) also reported with performance enhancement of ZnO NRs by doping with Co. They obtained a photocurrent of 0.58 mAcm^{-2} as compared with the undoped ZnO nanostructures (0.42 mAcm^{-2}).

8.3.2.7 THE NICKEL FAMILY

Nickel has also been used by Lee et al. (2016) in an attempt to improve the photoconversion efficiency of ZnO NRs. It was also observed that the doping efficiency of Ni was higher which also suggests that Ni is an effective dopant for hydrothermally grown ZnO NRs. A slight increment, i.e., 0.48 mAcm^{-2} in the value of photocurrent density was observed as compared with pristine ZnO. The lowered value is reported because of the formation of lesser defects after doping (as evident from the onset potential of ZnO and Ni:ZnO as –0.35 and –0.39 V$_{Ag/AgCl}$). Reddy et al. (2019) also used Ni as a dopant for visible-light-driven hydrogen generation. They obtained a photocurrent density of ~3.28 mA/cm^2 for 1% Ni-doped sample in a 0.1M NaOH solution. The efficient extraction of photocarriers after doping is the main cause for this enhancement. Upon visible light illumination, numerous charge carriers are produced. The same group also reported that 1.5 mol% of Ni-doped ZnO NRs are capable of providing a photocurrent density of 4.6 mA/cm^2, in an electrolyte solution of 0.1 M KOH and thus emphasizes a substantial improvement as compared with undoped ZnO (1.4 mA/cm^2) (Neelakanta et al., 2018).

8.3.2.8 THE COPPER FAMILY

Incorporation of Cu^{2+} within the ZnO matrix has also been known to cause a remarkable impact on the photoelectrochemical performance of 1D ZnO nanostructures. On behalf of the very small difference in ionic radii of Cu and Zn, the Cu^{2+} ion can substitutionally replace Zn^{2+} in ZnO hexagonal wurtzite structure. Thus, it can be a suitable dopant for ZnO for photoelectrochemical hydrogen generation. Hsu et al. (Hsu and Lin, 2012) synthesized $Zn_{1-x}Cu_xO$ NRs by means of electrodeposition technique with varying concentration from 1% to 10%. Better photoresponse in the visible range was observed due to the bandgap narrowing effect. For 4% doping the maximum use of solar spectrum was observed resulting with an enhanced photoconversion efficiency of 0.21% as compared with undoped ZnO (~0.1%). Babu et al. (2018) also reported the fabrication of Cu doped ZnO NR arrays synthesized via thermal deposition technique. Enhanced photoelectrochemical performance was evident for the sample from photocurrent density of 0.92 mA/cm^2 and photon conversion efficiency of 0.349%. Wang et al. (2014) also doped Cu into the host matrix of ZnO by ion implantation technique. They achieved a significant enhancement (11 times) in photocurrent density of 18 μA/cm^2 at 0.8V (vs. SCE) after inclusion of Cu. The visible-light-driven photoelectrochemical process observed in case of Cu is expected to help in realizing efficient use of solar energy. Rasouli et al. (2019) reported the enhanced photoelectrochemical performance after doping of Cu in a graded manner upon ZnO NRs. The fabrication was carried out by means of electrodeposition technique. The doping causes reduction of bandgap, which allows the efficient charge transfer and enhanced photocurrent density as well. Khurshid et al. (2019) fabricated rGO coated Ag-doped ZnO NRs and a photocurrent density of 206 nA/cm^2 was obtained.

8.3.2.9   THE BORON FAMILY

In order to enhance the optical properties all the group III elements have proved beneficial as dopant for ZnO. Among all these B possess the highest electronegativity and lowest electronic radii. Our group reported the enhanced photoelectrochemical performance achieved via boron-doped ZnO NRs synthesized via facile hydrothermal technique. For a 6% B doped ZnO sample the almost five fold enhancement in photoconversion efficiency (i.e., 2.054%) was obtained as compared with undoped ZnO (0.491%). Wang et al. (2017) also fabricated B doped ZnO nanostructures on flexible PET substrates by means of hydrothermal method. Change morphology from sheet to sphere was observed in their case along with the increase in boron (B) dopant concentration. The B-doping causes efficient separation of charge carriers which is observed as an enhancement in photocurrent density of 0.055 mA/cm^2 from 0.016 mA/cm^2 (for undoped ZnO NRs) under UV illumination. Kant et al. (2018) reported the fabrication of AZO NRs from doped seed layer solution. For ZnO photoanode the photocurrent density was ~19 µAcm^{-2}, which becomes 182 µAcm^{-2} after Al inclusion. They presented that the enhancement in PEC performance is due to the extended absorption as well as improved morphology. Also, Al doping removes the trap centers from the ZnO host matrix and facilitates better charge transport. Indium (In) has been reported to be a dopant of interest in order to increase the electrochemical performance of ZnO NRs. Henni et al. (2016) synthesized the 1D nanostructure via electrodeposition. A photocurrent of 18 mA was obtained for ZnO while it increases up to 44 mA for 4% dopant concentration, showing the effect of doping In.

8.3.2.10   THE CARBON FAMILY

We have also studied photoelectrochemical performance of Si-doped ZnO (Sharma et al., 2018a). It was observed that doping with Si increases the stability of the ZnO NRs in a chemical environment. A PEC efficiency of ~0.87% was observed after doping.

8.3.2.11   THE PNICTOGENS FAMILY

Several reports have also suggested the importance of Nitrogen as a dopant to enhance the photoelectrochemical performance of ZnO. Yang et al. (2009) carried out the fabrication of N doped ZnO via facile hydrothermal technique. They also reported that the N substitution occurs at O sites up to ~ 4% doping. By means of these non-vacuum based photoanodes, they achieved a photo to hydrogen conversion efficiency of 0.15% at +0.5 V (vs. Ag/AgCl). Moreover, the IPCE measured for the samples demonstrated the appreciable enhancement at visible region. Wang et al. (2015) followed an ion implantation technique to fabricate N doped ZnO structure. They found a remarkable enhancement in visible light-driven PEC photocurrent density of 160 µA.cm^{-2} at +1.1V vs. SCE, for an ion dose of 10^{15} ions/cm^{2}. A gradient distribution of N dopants occurred throughout the structure which served the purpose of extending the optical band edge to visible range. Not only this but also a terraced structure was introduced due to this, which promotes efficient transfer of photo carriers finally resulting with an enhanced efficiency. Another important aspect is the post-growth doping strategy which can help in further designing solar energy harvesting devices.

8.3.2.12   THE HALOGEN FAMILY

Halogens are also known to contribute in the photoelectrochemical enhancement of ZnO NRs. Wang et al. (2014) designed Cl doped ZnO NRs along with TiO_2 in a core-shell manner and achieved a photon to current PEC efficiency of 1.2% at –0.61V versus SCE; which is almost 3 times higher than the figures obtained for pure ZnO.

8.4   CONCLUSION

In this chapter, we tried to put insights on the importance of morphology on the photoelectrochemical performance of ZnO. As seen in most of the branched structures the increase in high surface to volume ratio provides the necessary pathway for carrier separation. This leads to the efficient extraction of photogenerated charge carriers. Along with this, the core-shell or nanoparticle sensitization approach has been observed by several groups to develop a cost-effective, highly efficient protocol for water splitting. Not only is this doping also an efficient way to tune the optical as well as electrical properties. In this study, we found a significant number of elements doped with ZnO NRs for photoelectrochemical applications. These key factors can be used to make the maximum use of solar spectrum for increasing the efficiency in terms of commercial prospects.

KEYWORDS

- **conduction band minimum**
- **doping**
- **hydrogen production**
- **nanostructures**
- **photoelectrochemical**
- **solar water splitting**

REFERENCES

Ahn, K. S., Yan, Y., Shet, S., Jones, K., Deutsch, T., Turner, J., & Al-Jassim, M., (2008). ZnO nanocoral structures for photoelectrochemical cells. *Appl. Phys. Lett., 93*, 163117.

Andrews, J. L., Cho, J., Wangoh, L., Suwandaratne, N., Sheng, A., Chauhan, S., Nieto, K., et al., (2018). Hole extraction by design in photo catalytic architectures interfacing CdSe quantum dots with top chemically stabilized tin vanadium oxide. *J. Am. Chem. Soc., 140*, 17163–17174.

Babu, E. S., Hong, S. K., Vo, T. S., Jeong, J. R., & Cho, H. K., (2015). Photo electrochemical water splitting properties of hydrothermally-grown ZnO nanorods with controlled diameters. *Electron. Mater. Lett., 11*, 65–72.

Babu, E. S., Rani, B. J., Ravi, G., Yuvakkumar, R., Guduru, R. K., Ravichandran, S., Ameen, F., Kim, S., & Jeon, H. W., (2018). Vertically aligned Cu-ZnO nanorod arrays for water splitting applications. *Mater. Lett., 222*, 58–61.

Bai, Z., Yan, X., Kang, Z., Hu, Y., Zhang, X., & Zhang, Y., (2015). Photo electrochemical performance enhancement of ZnO photo anodes from $ZnIn_2S_4$ nanosheets coating. *Nano Energy, 14*, 392–400.

Bakranov, N., Aldabergenov, M., Ibrayev, N., Abdullin, K., & Kudaibergenov, S., (2017). The study of photoelectrochemical water splitting by ZnO nanostructures and ZnO/Ag nanocomposites. *Proc. 2017 IEEE 7th Int. Conf. Nanomater. Appl. Prop. N.*, 1–4.

Baruah, S., & Dutta, J., (2009). Hydrothermal growth of ZnO nanostructures. *Sci. Technol. Adv. Mater., 10*, 013001.

Baskoutas, S., (2014). *Zinc Oxide Nanostructures, Zinc Oxide Nanostructures.* https://doi.org/10.1201/b15661 (accessed on 16 May 2020).

Bharathi, V., Sivakumar, M., Udayabhaskar, R., Takebe, H., & Karthikeyan, B., (2014). Optical, structural, enhanced local vibrational and fluorescence properties in K-doped ZnO nanostructures. *Appl. Phys. A Mater. Sci. Process., 116*, 395–401.

Bolton, J. R., Strickler, S. J., & Connolly, J. S., (1985). Limiting and realizable efficiencies of solar photolysis of water. *Nature, 316*, 495–500.

Boruah, B. D., Majji, S. N., & Misra, A., (2017). Surface photo-charge effect in doped-ZnO nanorods for high-performance self-powered ultraviolet photodetectors. *Nanoscale, 9*, 4536–4543.

Cai, L., Ren, F., Wang, M., Cai, G., Chen, Y., Liu, Y., Shen, S., & Guo, L., (2015). V ions implanted ZnO nanorod arrays for photo electrochemical water splitting under visible light. *Int. J. Hydrogen Energy, 40*, 1394–1401.

Cai, L., Zhou, W., Ren, F., Chen, J., Cai, G., Liu, Y., et al., (2017). W ion implantation boosting visible-light photoelectrochemical water splitting over ZnO nanorod arrays. *J. Photonics Energy, 7*, 016501.

Cha, H. G., Song, J., Kim, H. S., Shin, W., Yoon, K. B., & Kang, Y. S., (2011). Facile preparation of Fe_2O_3 thin film with photoelectrochemical properties. *Chem. Commun., 47*, 2441–2443.

Chandrasekaran, S., Chung, J. S., Kim, E. J., & Hur, S. H., (2016). Exploring complex structural evolution of graphene oxide/ZnO triangles and its impact on photoelectrochemical water splitting. *Chem. Eng. J., 290*, 465–476.

Chen, S., Takata, T., & Domen, K., (2017). Particulate photocatalysts for overall water splitting. *Nat. Rev. Mater., 2*, 1–17.

Chen, X., Liu, L., & Huang, F., (2015). Black titanium dioxide (TiO_2) nanomaterials. *Chem. Soc. Rev., 44*, 1861–1885.

Chen, Z., Dinh, H. N., & Miller, E., (2013). *Photoelectrochemical Water Splitting.* Springer briefs in energy, Springer Briefs in Energy. Springer New York, New York, NY., doi: 10.1007/978-1-4614-8298-7.

Dhara, S., & Giri, K. P., (2012). ZnO nanorods arrays and heterostructures for the high sensitive UV photodetection. *Nanorods*, 1–32.

Djurišić, A. B., Chen, X., Leung, Y. H., & Man, C. N. A., (2012). ZnO nanostructures: Growth, properties and applications. *J. Mater. Chem., 22*, 6526.

Djurišić, A. B., Liu, X., & Leung, Y. H., (2014). Zinc oxide films and nanomaterials for photovoltaic applications. *Phys. Status Solidi-Rapid Res. Lett., 8*, 123–132.

Fujishima, A., & Honda, K., (1972). Electrochemical photolysis of water at a semiconductor electrode. *Nature, 238*, 37–38.

Govatsi, K., Seferlis, A., Neophytides, S. G., & Yannopoulos, S. N., (2018). Influence of the morphology of ZnO nanowires on the photo electrochemical water splitting efficiency. *Int. J. Hydrogen Energy, 43*, 4866–4879.

Guo, C. X., Dong, Y., Yang, H. B., & Li, C. M., (2013). Graphene quantum dots as a green sensitizer to functionalize ZnO nanowire arrays on F-doped SnO_2 glass for enhanced photoelectrochemical water splitting. *Adv. Energy Mater, 3*, 997–1003.

Guo, Q., Zhou, C., Ma, Z., Ren, Z., Fan, H., & Yang, X., (2016). Elementary photo catalytic chemistry on TiO_2 surfaces. *Chem. Soc. Rev., 45*, 3701–3730.

Hagfeldt, A., & Graetzel, M., (1995). Light-induced redox reactions in nanocrystalline systems. *Chem. Rev., 95*, 49–68.

Han, J., Liu, Z., Guo, K., Zhang, X., Hong, T., & Wang, B., (2015). $AgSbS_2$ modified ZnO nanotube arrays for photo electrochemical water splitting. *Appl. Catal. B Environ., 179*, 61–68.

Henni, A., Merrouche, A., Telli, L., Karar, A., Ezema, F. I., & Haffar, H., (2016). Optical, structural, and photoelectrochemical properties of nanostructured Ln-doped ZnO via electrodepositing method. *J. Solid State Electrochem., 20*, 2135–2142.

Hsu, Y. K., & Lin, C. M., (2012). Enhanced photo electrochemical properties of ternary $Zn_{1-x}Cu_xO$ nanorods with tunable band gaps for solar water splitting. *Electrochim. Acta., 74*, 73–77.

Hsu, Y. K., Fu, S. Y., Chen, M. H., Chen, Y. C., & Lin, Y. G., (2014). Facile synthesis of Pt nanoparticles/ZnO nanorod arrays for photoelectrochemical water splitting. *Electrochim. Acta, 120*, 1–5.

Hsu, Y. K., Lin, Y. G., & Chen, Y. C., (2011). Polarity-dependent photo electrochemical activity in ZnO nanostructures for solar water splitting. *Electrochem. Commun., 13*, 1383–1386.

Ilican, S., (2013). Effect of Na doping on the microstructures and optical properties of ZnO nanorods. *J. Alloys Compd., 553*, 225–232.

Kant, R., Dwivedi, C., Pathak, S., & Dutta, V., (2018). Fabrication of ZnO nanostructures using Al doped ZnO (AZO) templates for application in photo electrochemical water splitting. *Appl. Surf. Sci., 447*, 200–212.

Kayaci, F., Vempati, S., Donmez, I., Biyikli, N., & Uyar, T., (2014). Role of zinc interstitials and oxygen vacancies of ZnO in photo catalysis: A bottom-up approach to control defect density. *Nanoscale, 6*, 10224–10234.

Khurshid, F., Jeyavelan, M., Hudson, M. S. L., & Nagarajan, S., (2009). Ag-doped ZnO nanorods embedded reduced graphene oxide nanocomposite for photo-electrochemical applications. *R. Soc. Open Sci., 6*. doi: 10.1098/rsos.181764

Kim, H., Kwon, W., Choi, M., Rhee, S. W., & Yong, K., (2015). Photoelectrochemical hydrogen generation using C-dot/ZnO hierarchical nanostructure as an efficient photoanode. *J. Electrochem. Soc., 162*, H366–H370.

Kołodziejczak-Radzimska, A., & Jesionowski, T., (2014). Zinc oxide-from synthesis to application: A review. *Materials (Basel), 7*, 2833–2881.

Lee, W. C., Canciani, G. E., Alwhshe, B. O. S., & Chen, Q., (2016). Enhanced photo electrochemical water oxidation by Zn x M y O (M = Ni, Co, K, Na) nanorod arrays. *Int. J. Hydrogen Energy, 41,* 123–131.

Li, F. M., Hsieh, G. W., Dalal, S., Newton, M. C., Stott, J. E., Hiralal, P., Nathan, A., et al., (2008). Zinc oxide nanostructures and high electron mobility nanocomposite thin film transistors. *IEEE Trans. Electron Devices, 55*, 3001–3011.

Li, Q., Sun, X., Lozano, K., & Mao, Y., (2014). Facile and scalable synthesis of "caterpillar-like" ZnO nanostructures with enhanced photo electrochemical water-splitting effect. *J. Phys. Chem. C, 118*, 13467–13475.

Li, Z., Luo, W., Zhang, M., Feng, J., & Zou, Z., (2013). Photo electrochemical cells for solar hydrogen production: Current state of promising photo electrodes, methods to improve their properties, and outlook. *Energy Environ. Sci., 6*, 347–370.

Li, Z., Luo, W., Zhang, M., Feng, J., & Zou, Z., (2013). Photo electro chemical cells for solar hydrogen production: Current state of promising photo electrodes, methods to improve their properties, and outlook. *Energy Environ. Sci., 6*, 347–370.

Lin, Y. G., Hsu, Y. K., Chen, Y. C., Wang, S. B., Miller, J. T., Chen, L. C., & Chen, K. H., (2012). Plasmonic $Ag@Ag_3(PO_4)_{1-x}$ nanoparticle photosensitized ZnO nanorod-array photoanodes for water oxidation. *Energy Environ. Sci., 5*, 8917–8922.

Lin, Y., Yuan, G., Liu, R., Zhou, S., Sheehan, S. W., & Wang, D., (2011). Semiconductor nanostructure-based photo electrochemical water splitting: A brief review. *Chem. Phys. Lett., 507*, 209–215.

Liu, M., Nam, C. Y., Black, C. T., Kamcev, J., & Zhang, L., (2013). Enhancing Water splitting activity and chemical stability of zinc oxide nanowire photoanodes with ultrathin titania shells. *J. Phys. Chem. C, 117*, 13396–13402.

Liu, Z., Cai, Q., Ma, C., Zhang, J., & Liu, J., (2017). Photo electrochemical properties and growth mechanism of varied ZnO nanostructures. *New J. Chem., 41*, 7947–7952.

Luo, H., Mueller, A. H., McCleskey, T. M., Burrell, A. K., Bauer, E., & Jia, Q. X., (2008). Structural and photo electrochemical properties of $BiVO_4$ thin films. *J. Phys. Chem. C, 112*, 6099–6102.

Martin, D. J., Liu, G., Moniz, S. J. A., Bi, Y., Beale, A. M., Ye, J., & Tang, J., (2015). Efficient visible driven photo catalyst, silver phosphate: Performance, understanding and perspective. *Chem. Soc. Rev., 44*, 7808–7828.

Moniz, S. J. A., Shevlin, S. A., Martin, D. J., Guo, Z. X., & Tang, J., (2015). Visible-light driven heterojunction photo catalysts for water splitting-a critical review. *Energy Environ. Sci., 8*, 731–759.

Neelakanta, R., I., Venkata, R. C., Sreedhar, A., Shim, J., Cho, M., Yoo, K., & Kim, D., (2018). Structural, optical, and bifunctional applications: Super capacitor and photo electrochemical water splitting of Ni-doped ZnO nanostructures. *J. Electroanal. Chem., 828*, 124–136.

Outemzabet, R., Doulache, M., & Trari, M., (2015). Physical and photo electrochemical properties of Sb-doped SnO_2 thin films deposited by chemical vapor deposition: Application to chromate reduction under solar light. *Appl. Phys. A, 119*, 589–596.

Pan, L., Kim, J. H., Mayer, M. T., Son, M. K., Ummadisingu, A., Lee, J. S., Hagfeldt, A., Luo, J., & Grätzel, M., (2018). Boosting the performance of Cu_2O photocathodes for unassisted solar water splitting devices. *Nat. Catal., 1*, 412–420.

Panigrahy, B., Aslam, M., & Bahadur, D., (2010). Aqueous synthesis of Mn- and Co-doped ZnO nanorods. *J. Phys. Chem. C, 114*, 11758–11763.

Pei, Z., Ding, L., Hu, J., Weng, S., Zheng, Z., Huang, M., & Liu, P., (2013). Defect and its dominance in ZnO films: A new insight into the role of defect over photo catalytic activity. *Appl. Catal. B Environ., 142, 143*, 736–743.

Qasim, A. K., Jamil, L. A., & Chen, Q., (2018). Enhanced photo electrochemical water splitting of hydrothermally-grown ZnO and Yttrium-doped ZnO NR Arrays. *IOP Conf. Ser. Mater. Sci. Eng., 454*, 012033.

Qiu, Y., Yan, K., Deng, H., & Yang, S., (2012). Secondary branching and nitrogen doping of ZnO nanotetrapods: Building a highly active network for photo electrochemical water splitting. *Nano Lett., 12*, 407–413.

Rasouli, F., Rouhollahi, A., & Ghahramanifard, F., (2019). Gradient doping of copper in ZnO nanorod photoanode by electrodeposition for enhanced charge separation in photo electrochemical water splitting. *Superlattices Microstruct., 125*, 177–189.

Reddy, I. N., Sreedhar, A., Venkata, R. C., Cho, M., Kim, D., & Shim, J., (2019). Nickel-doped ZnO structures for efficient water splitting under visible light. *Mater. Res. Express, 6*, 055517.

Ren, X., Sangle, A., Zhang, S., Yuan, S., Zhao, Y., Shi, L., Hoye, R. L. Z., et al., (2016). Photo electrochemical water splitting strongly enhanced in fast-grown ZnO nanotree and nanocluster structures. *J. Mater. Chem. A, 4*, 10203–10211.

Rokade, A., Rondiya, S., Sharma, V., Prasad, M., Pathan, H., & Jadkar, S., (2017). Electrochemical synthesis of 1D ZnO nanoarchitectures and their role in efficient photo electrochemical splitting of water. *J. Solid State Electrochem., 21*, 2639–2648.

Salimi, R., Sabbagh, A. A. A., Mei, B. T., Naseri, N., Du, S. F., & Mul, G., (2019). Ag-functionalized $CuWO_4/WO_3$ nanocomposites for solar water splitting. *New J. Chem., 43*, 2196–2203.

Schneider, J., Matsuoka, M., Takeuchi, M., Zhang, J., Horiuchi, Y., Anpo, M., & Bahnemann, D. W., (2014). Understanding TiO_2 photo catalysis: Mechanisms and materials. *Chem. Rev., 114*, 9919–9986.

Selloum, D., Henni, A., Karar, A., Tabchouche, A., Harfouche, N., Bacha, O., et al., (2019). Effects of Fe concentration on properties of ZnO nanostructures and their application to photocurrent generation. *Solid State Sci., 92*, 76–80.

Sharma, A., Chakraborty, M., & Thangavel, R., (2018a). Enhanced photo electrochemical performance of hydrothermally grown tetravalent impurity (Si^{4+}) doped zinc oxide nanostructures for solar water splitting applications. *J. Mater. Sci. Mater. Electron., 29*, 14710–14722.

Sharma, A., Chakraborty, M., Thangavel, R., & Udayabhanu, G., (2018b). Hydrothermal growth of undoped and boron doped ZnO nanorods as a photoelectrode for solar water splitting applications. *J. Sol-Gel Sci. Technol., 85*, 1–11.

Shen, S., Kronawitter, C. X., Jiang, J., Guo, P., Guo, L., & Mao, S. S., (2013). A ZnO/ZnO:Cr isostructural nanojunction electrode for photo electrochemical water splitting. *Nano Energy, 2*, 958–965.

Shet, S., (2011). Zinc oxide (ZnO) nanostructures for photo electrochemical water splitting application. In: *ECS Transactions* (pp. 15–25). doi: 10.1149/1.3583510.

Singh, O., Kohli, N., Singh, M. P., Anand, K., & Singh, R. C., (2012). *Gas Sensing Properties of Zinc Oxide Thin Films Prepared by Spray Pyrolysis, 38*, 191–193. doi: 10.1063/1.4732411.

Sivula, K., & Van, D. K. R., (2016). Semiconducting materials for photoelectrochemical energy conversion. *Nat. Rev. Mater., 1*. doi: 10.1038/natrevmats.2015.10.

Sohila, S., Rajendran, R., Yaakob, Z., Teridi, M. A. M., & Sopian, K., (2016). Photoelectrochemical water splitting performance of flower like ZnO nanostructures synthesized by a novel chemical method. *J. Mater. Sci. Mater. Electron., 27*, 2846–2851.

Steinmiller, E. M. P., & Choi, K. S., (2009). Photochemical deposition of cobalt-based oxygen evolving catalyst on a semiconductor photo anode for solar oxygen production. *Proc. Natl. Acad. Sci., 106*, 20633–20636.

Sun, S., Wang, W., Li, D., Zhang, L., & Jiang, D., (2014). Solar light driven pure water splitting on quantum sized BiVO$_4$ without any cocatalyst. *ACS Catal., 4*, 3498–3503.

Sun, X., Li, Q., Jiang, J., & Mao, Y., (2014). Morphology-tunable synthesis of ZnO nanoforest and its photoelectrochemical performance. *Nanoscale, 6*, 8769–8780.

Sushma, C., & Girish, K. S., (2017). Advancements in the zinc oxide nanomaterials for efficient photocatalysis. *Chem. Pap., 71*, 2023–2042.

Tee, S. Y., Win, K. Y., Teo, W. S., Koh, L. D., Liu, S., Teng, C. P., & Han, M. Y., (2017). Recent progress in energy-driven water splitting. *Adv. Sci., 4*.

Tee, S. Y., Win, K. Y., Teo, W. S., Koh, L. D., Liu, S., Teng, C. P., & Han, M. Y., (2017). Recent progress in energy-driven water splitting. *Adv. Sci., 4*. doi: 10.1002/advs.201600337.

Tsao, J., Lewis, N., & Crabtree, G., (2006). Solar FAQs, US Department of Energy.

Valenti, M., Dolat, D., Biskos, G., Schmidt-Ott, A., & Smith, W. A., (2015). Enhancement of the photoelectrochemical performance of CuWO 4 Thin films for solar water splitting by plasmonic nanoparticle functionalization. *J. Phys. Chem. C, 119*, 2096–2104.

Vidyarthi, V. S., Hofmann, M., Savan, A., Sliozberg, K., König, D., Beranek, R., et al., (2011). Enhanced photo electrochemical properties of WO$_3$ thin films fabricated by reactive magnetron sputtering. *Int. J. Hydrogen Energy, 36*, 4724–4731.

Wang, B., Li, R., Zhang, Z., Xing, W., Wu, X., Cheng, G., & Zheng, R. T., (2019). An overlapping ZnO nanowire photoanode for photoelectrochemical water splitting. *Catal. Today, 321, 322*, 100–106.

Wang, F., Seo, J. H., Li, Z., Kvit, A. V., Ma, Z., & Wang, X., (2014). Cl-doped ZnO nanowires with metallic conductivity and their application for high-performance photoelectrochemical electrodes. *ACS Appl. Mater. Interfaces, 6*, 1288–1293.

Wang, M., Ren, F., Cai, G., Liu, Y., Shen, S., & Guo, L., (2014). Activating ZnO nanorod photoanodes in visible light by Cu ion implantation. *Nano Res., 7*, 353–364.

Wang, M., Ren, F., Zhou, J., Cai, G., Cai, L., Hu, Y., Wang, D., Liu, Y., Guo, L., & Shen, S. N., (2015). Doping to ZnO nanorods for photo electrochemical water splitting under visible light: Engineered impurity distribution and terraced band structure. *Sci. Rep., 5*, 1–13.

Wang, T., Lv, R., Zhang, P., Li, C., & Gong, J., (2015). Au nanoparticle sensitized ZnO nanopencil arrays for photoelectrochemical water splitting. *Nanoscale, 7*, 77–81.

Wang, W., Ai, T., & Yu, Q., (2017). Electrical and photocatalytic properties of boron-doped ZnO nanostructure grown on PET-ITO flexible substrates by hydrothermal method. *Sci. Rep., 7*, 1–11.

Wei, A., Pan, L., & Huang, W., (2011). Recent progress in the ZnO nanostructure-based sensors. *Mater. Sci. Eng. B Solid-State Mater. Adv. Technol., 176*, 1409–1421.

Wei, Y., Ke, L., Kong, J., Liu, H., Jiao, Z., Lu, X., et al., (2012). Enhanced photo electrochemical water-splitting effect with a bent ZnO nanorod photoanode decorated with Ag nanoparticles. *Nanotechnology*, p. 23. doi: 10.1088/0957-4484/23/23/235401

Wolcott, A., Smith, W. A., Kuykendall, T. R., Zhao, Y., & Zhang, J. Z., (2009). Photoelectrochemical study of nanostructured Zno thin films for hydrogen generation from water splitting. *Adv. Funct. Mater., 19*, 1849–1856.

Xiao, F. X., Miao, J., Tao, H. B., Hung, S. F., Wang, H. Y., Yang, H., et al., (2015). One-dimensional hybrid nanostructures for heterogeneous photocatalysis and photoelectrocatalysis. *Small, 11*, 2115–2131.

Xu, F., Yao, Y., Bai, D., Xu, R., Mei, J., Wu, D., et al., (2015). Au nanoparticle decorated WO$_3$ photoelectrode for enhanced photoelectrochemical properties. *RSC Adv., 5*, 60339–60344.

Yang, X., Wolcott, A., Wang, G., Sobo, A., Fitzmorris, R. C., Qian, F., et al., (2009). Nitrogen-doped ZnO nanowire arrays for photoelectrochemical water splitting. *Nano Lett., 9*, 2331–2336.

Yang, Y., Xu, D., Wu, Q., & Diao, P., (2016). Cu$_2$O/CuO bilayered composite as a high-efficiency photocathode for photoelectrochemical hydrogen evolution reaction. *Sci. Rep., 6*, 35158.

Zhang, B., Wang, F., Zhu, C., Li, Q., Song, J., Zheng, M., et al., (2016). A facile self-assembly synthesis of hexagonal ZnO Nano sheet films and their photo electro chemical properties. *Nano-Micro Lett., 8*, 137–142.

Zhang, C., Shao, M., Ning, F., Xu, S., Li, Z., Wei, M., Evans, D. G., & Duan, X., (2015). Au nanoparticles sensitized ZnO nanorod@nanoplatelet core-shell arrays for enhanced photoelectrochemical water splitting. *Nano Energy, 12*, 231–239.

Zhang, N., Shan, C. X., Tan, H. Q., Zhao, Q., Wang, S. P., Sun, Z. C., et al., (2016). Black-colored ZnO nanowires with enhanced photocatalytic hydrogen evolution. *Nanotechnology, 27*, 1–6.

Zhou, M., Lou, X. W., David, & Xie, Y., (2013). Two-dimensional nanosheets for photoelectrochemical water splitting: Possibilities and opportunities. *Nano Today, 8*, 598–618.

CHAPTER 9

Methanol and Formic Acid Oxidation: Selective Fuel Cell Processes

TAPAN KUMAR BEHERA,[1] PRAMOD KUMAR SATAPATHY,[1] and PRIYABRAT MOHAPATRA[2]

[1]*North Orissa University, Baripada – 757003, Odisha, India*

[2]*C. V. Raman College of Engineering, Bhubaneswar – 752054, Odisha, India*

ABSTRACT

This chapter describes how the methanol and formic acid (FA) in the fuel cell system aimed to develop novel ideal clean energy sources as suitable alternatives than other renewable energy sources. Based on the systematic study on various types of fuel cells, the concerned chapter will be exported to describe the importance of methanol, formic acid oxidation (FAO) than other else. Specifically, this chapter focuses on the role of different shapes of noble metal nanoparticles (NPs) for electrocatalytic activity towards methanol and FAO. NPs modified on the carbon supporting materials like graphene substrate possesses high electrocatalytic activity and durability in fuel cell than the individual ones. Among all carbon supporting materials the basic design of FA and methanol oxidation fuel cell with Pd, Pt, Ru, Au-Pd alloy, and AuNPs with graphene substrate finding an appropriate way to choose the best catalyst on basis of low cost, high activity, and high stability. We here also highlight a systematic approach in recent studies on methanol and FAO with their basic mechanistic principles for better understanding. The main objective of the study is on methanol and FA fuel cell by employing conventional and advanced electrochemical methods than other methods. A combined potential step, fast cyclic Voltammetry experiment, and electrochemical impedance spectroscopy (EIS) were employed to study the production of electric current by NPs catalysts in methanol and FAO. This chapter discusses facile and eco-friendly synthesis methods of metal

NPs, various techniques used for characterization, and their electrocatalytic applications towards methanol oxidation and FAO.

9.1   INTRODUCTION

In the fuel cell, electricity is generated by the exploitation of a chemical reaction. This is just like a compatible battery and having no chemical fuel itself. In battery no longer, electricity produced when chemicals are exhausted but the fuel cell can generate electric current as long as there is a source of the chemical fuel. In a battery, fuel-to-electricity efficiency is 50% and the fuel cell has efficiency more than 90% (Pellow et al., 2015). To fulfill the energy demands, the development of a clean and green environment energy economy built around by fuel cells rather than the fossil fuels that traditionally used combustion of fossil fuel with air at high temperature causing the greenhouse effect, acid rain, atmospheric pollution, etc. In the future, direct use of renewable sources like solar energy, production, and storage of energy carriers like hydrogen and energy conversion through eco-friendly ways is more important. Hydrogen possess the most unique property of renewable energy source and practically non-polluting compared to other energy carriers like gasoline and diesel and hence it is forecasted by many of researchers that hydrogen will be the superlative energy carrier in the future. Hence, the production of hydrogen is more important. But the low volumetric energy density of hydrogen causes board storage difficulty and directs to the introduction use of liquid hydrogen carriers. Many studies have been done on this subject and found that renewable biofuels include a whole range of biotechnology-derived fuels like methanol, ethanol, formic acid (FA), and ammonia (Nico et al., 1981). Methanol and FA have the simplest structure than any other organic compounds and so should also have the most straight forward reaction mechanisms. The methanol and formic acid oxidation (FAO) on noble metal nanoparticles (NPs) are well studied due to their promising character as fuel cell feeds (Hearth et al., 1996; Nicol et al., 1999). The electrocatalytic oxidation of organic fuel cell-like direct methanol molecule and FA with different NPs has received considerable attention over the years due to their promising effect as fuels in fuel cells. Methanol oxidation is suggested to proceed via several possible pathways. The overall conversion of methanol to carbon dioxide requires water, which can be written as:

$$CH_3OH + H_2O \rightarrow CO_2 + 2H^+ + 2e^- \tag{9.1}$$

The simple pathway for the net decomposition of FA, also namely, dehydrogenation (direct pathway), and dehydration (indirect pathway) as shown in the following:

$$HCOOH \rightarrow CO_2 + 2H^+ + 2e^-$$ (9.2)

It is well-known that for the oxidation of above mentioned organic molecules, different parameters are highly responsible such as metal NPs size and its distribution on a carbon-like graphene supported materials, the oxidation state of metal, and the metal support interactions (Weijiang et al., 2008).

Metal NPs (1–100 nm range) exhibit very unique and exciting electronic with fascinating structures, optical, and chemical properties. These NPs show excellent adsorption and dissociation of small organic molecules like methanol, FA, ethanol, hydrogen peroxide, and reduction of oxygen in low-temperature fuel cells (Li et al., 2012; Behera et al., 2016). But it has been concluded that due to decrease in active surface area of NPs, their catalytic power have a limited life span. In order to enhance their activities, these NPs are supported with conductive substrates such as carbon black, carbon nanoribbon, carbon nanotube, carbon nanohorn, graphene, and, etc. (Rouxoux et al., 2002; Jena et al., 2011). Among all supporting materials, graphene have most attracting advanced properties like high electronic current density and thermal/electrical conductivity in addition to their ballistic transport property, chemical inertness, optical transmittance, and the most amazing superhydrophobicity property at nanometer scale (Chen et al., 2008; Geim and Kim, 2008). Therefore, the composite of graphene and NPs act as an excellent transducer for methanol and FAO in the fuel cell. The synthesis of different shaped like Icosahedral, dendrite, flower, fractal structures and spherical controlled of different noble metal nanostructures (Au, Ag, Pt, Pd, Au-Pd) and carbon-based graphene supported metal nano-composites by the facile and environment friendly approaching methods are highly required for electrocatalytic methanol and FAO applications. Based on the type of electrolytes and fuels used, fuel Cells are broadly classified into five major categories such as: (1) proton exchange membrane fuel cell (PEMFC), (2) phosphoric acid fuel cell (PAFC) (3) alkaline fuel cell (AFC), (4) solid oxide fuel cell (SOFC), (5) molten carbonate fuel cell (MCFC), etc. Out of these fuel cells, PEM fuel cell has attention the most promising research interest due to most significant advantages like low bi-products emission, low operating temperature, a minimum mechanism for better understanding, quick igniting, high power density and easy to transport (Zhang et al., 2013). The principle of the fuel cell was known in the 19th century but in the 20th century that a practical fuel cell was built. AFC was supposed to be the first efficient fuel cell followed by PAFC in the second half of the 20th century. Before that, in the 19th century Friedrich Schonbein and the British scientist William Grove developed the concept of water splitting in 1883. According to them if, by an applying electrical current water could be split into its constituent elements then the process can be reversed to produced current. The mechanism process in the fuel cell is not meant for treatment to the Carnot limitations. In spite of more demand with continued investment, fuel cells could not able till date to put major impact on concerning to global power generation. At the end of 2014, global installed power generating capacity was of around 6,200,000 MW but fuel cell provides only 0.02% of this. Earlier NASA used to fuel cells for its space program. For this space application, the cost of constructing fuel cell was immaterial provided the device could function efficiently with zero error and reliably. Meanwhile in the second half of the 20th-century research programs led to the development of new range fuel cell technologies.

9.2   PERFORMANCES OF NANOPARTICLES (NPS) FOR METHANOL FUEL CELL

The electrocatalytic oxidation of small organic molecule, i.e., methanol molecules by different NPs has been received considerable research interest due to their promising act as fuels in fuel cells. The small organic molecule methanol is a unique stable liquid at atmospheric conditions, high energy-dense and easy in handling and transport. Now a day's noble metal NPs such as platinum (Pt), palladium (Pd), gold (Au), ruthenium (Ru) and their bimetallic complexes have made new promising materials in the field of methanol fuel cell owing to their unique magnetic, optical, and catalytic properties (Behera et al., 2016; Sahu et al., 2016; Jena et al., 2011). The electrocatalytic activity of NPs depends on the size, shape, and inters particle distances among them. Currently, various methods for synthesis of different shaped noble metal NPs and their composite with graphene and other carbon-based substrate have been attracting huge research interest. Further, various characterization techniques are employed to analyze their surface morphology, i.e., TEM (Transmission electron microscope), SEM (Scanning electron microscope), and AFM (atomic force microscope); functional group/

elemental composition, i.e., FTIR (Fourier transform infrared spectroscopy), UV-Vis (Ultraviolet-visible spectroscopy), XPS (x-ray photoelectron spectroscopy) and crystallinity, i.e., XRD (x-ray diffraction). Here we highlight the different noble metal nanostructures such as Pt, Pd, Au, and Ag and their graphene composites towards methanol oxidation. Also, different NPs have been discussed in detail by cyclic voltammetry, impedance, and Chronoamperometric study techniques. Here we discuss about the proton exchange membrane (PEM) fuel cell, in which methanol in liquid form is directly used as a fuel. By far pure fundamental studies on this PEM related to the direct methanol fuel cell grabbed the most attention. Here methanol is oxidized at anode and oxygen is reduced at the cathode and in meanwhile through the PEM, protons are transported to the cathode and caused in the reduction of oxygen gas to water.

Bagotzky et al. put forwarded a general mechanistic study from methanol to CO_2. This group suggested the possible eight intermediates of which CO, FA, and formaldehyde are the most important interesting ones (Bagotzky et al., 1977). Many intermediates are detected by many researchers through various methods for the oxidation of methanol. The exact structure of the adsorbed intermediates during the oxidation reaction was unknown except for the adsorbed carbon monoxide (CO) molecule (Kua and Goddard, 1999). After all we concluded that the tendency of methanol oxidation reactions (MOR) to self-poisoning make cleared that catalyst properties must be changed to avoid the poisoning condition. The general mechanism Eqn. (9.1) suggest for the importance of oxygen donating source at the surface in order to complete the oxidation of methanol to CO_2. Many of metal electrodes came forward to eradicate the difficulties but so far, Pt electrode plays the most important role for methanol oxidation (Sahu et al., 2015). The Pt electrode surface towards the methanol oxidation has been demonstrated several times. Vielstich et al. explained the oxidation of methanol on platinum electrode surfaces starts with the de-hydrogenation of the methanol molecule as shown by differential electrochemical mass spectrometry (DEMS) (Vielstich et al., 1995). They explained that the first dehydrogenation step was considered to be a rate-determining step on two Pt planes of Pt (111) and Pt (110) by measuring of the kinetic isotope effect and the Tafel slope of the reaction (Franaszczuk et al., 1992; Herrero et al., 1994). According to Munk et al. minimum of four adjacent adsorption sites are required for the complete remove of all three methyl hydrogen atoms at lower potentials, which occur on the flat terrace areas of the electrode surface (Munk et al., 1996). Shimazu and Kita reported that Pt oxide does not form below the potential of 0.75 V. According to them the oxygen donor is dissociated water (Eqn. (9.3)), or some form of activated water or oxide and hence they used an electrochemical quartz crystal microbalance, EQCM for the mass change in a cyclic voltammogram (CV) on a platinum electrode.

$$Pt(site) + H_2O \rightarrow Pt - OH + H^+ + e^- \tag{9.3}$$

This lower potential during a positive-going potential scan rate was due to water adsorption on the platinum surface (Shimazu and Kita, 1992). In another report, Iwasita and Xia employed FTIR reflection-absorption spectroscopy technique in perchloric acid electrolyte for the formation of adsorbed hydroxyl ions from water dissociation. The potential found at 0.5V vs. SHE, which was much lower than other ones (Iwasita and Xia, 1996). Similarly, Tilak and co-workers used a process of reversible electrosorption of hydroxyl species method for oxidation of the platinum surface and the potential was 0.75 V vs. RHE (Tilak et al., 1973). Li et al. suggested that for the higher oxidation states of Pt, e.g., Pt (OH)$_3$ and PtO$_2$, platinum surface oxides were formed at potentials above 1.24 V vs. RHE (Li et al., 1997). The most interesting fact came when Vander Geest et al. claimed the formation of adsorbed OH species in the initial stages of Pt oxide rather than OH (Vander et al., 1997). After all, considering all efforts the exact mechanism for methanol oxidation at platinum and platinum-based electrodes still remain unclear. Here the main issue is, whether the adsorbed CO intermediate poison formed during methanol electro-oxidation is a necessary intermediate in a serial pathway mechanism or an unwanted poison in a parallel reaction mechanism through short-lived intermediates. Early studies suggest the formation of the poison intermediate species; CO is an unwanted parallel reaction under certain reaction conditions. It was observed that an excessive charge needed for the formation of CO poison at potentials where adsorbed CO is not oxidized on Pt (111), Pt(110), and Pt(100) surfaces. This was indicated the possibility of a parallel pathway in the electro-oxidation of methanol. Chronoamperometric and DEMS studies confirmed the formation of CO$_2$ can only be detected on the same potentials at which adsorbed CO is also oxidized (Lu et al., 2000; Vielstich et al., 1995). Wang and co-workers also employed DEMS technique and found the presence of a parallel path mechanism on platinum in a sulfuric acid electrolyte by considering the current efficiencies of CO$_2$ during methanol oxidation (Wang et al., 2001). On Summarizing to all we got that in a parallel reaction, dissolved intermediates such as formaldehyde, CO, and FA formed whereas in a serial reaction CO$_2$ is formed via CO$_{ads}$ oxidation process.

Seland et al. reported a quantitative evaluation of methanol oxidation on platinum (111) and later on Pt (100) in sulfuric acid by using conventional voltammetry for the fast and slow time scales (Frode et al., 2012). Here they reported the formation of CO directly from methanol in solution and the potential region range prior to platinum oxide formation. They employed chronoamperometry and fast potential sweep techniques and concluded that both the adsorption time and potential rate of formation of CO from methanol oxidation increases gradually with potential at least until about 0.6 V. The saturation charge of CO_{ads} derived from methanol was 0.6 V and decreased at 0.35 V for FA. Here the parallel path was only active at short times and low potentials. Again, they did the most powerful electrochemical impedance spectroscopy (EIS) for examining the reaction mechanisms. The nucleation and growth behavior of Methanol oxidation on Pt electrodes was studied along with the rate-determining step. There was a competition between OH and CO adsorption for the released reaction sites. A variety of spectral lines were observed having regions of negative relaxation times (showing the inductive behavior) and second region of negative tau value (indicates growth of islands of Pt oxide) with methanol oxidation current by the oxide. Another work by Wieckowski and co-workers did the same potential-sweep sequence on Pt electrodes in sulfuric acid for methanol oxidation for the same purpose (Franaszczuk et al., 1992; Herrero et al., 1994).

The same NPs of different sizes possess unique properties for optical, electronic, catalytic, energy harvesting, storage, and electrocatalytic applications. In general, NPs are classified as nanospheres, quantum dot, nanopyramid, nanostars, nanoflower, nanodendrite, nanocube, nanoplates, nanoribbons, and nanosheets, etc. (Tiwari et al., 2012). According to Hao et al. anisotropic 1D, 2D, and 3D nanomaterials are the most running research work due to their reactive sites such as facets, kinks, steps, and edges (2004). The role of a capping agent is most vital for shape-controlled synthesis of nanocrystals. These capping agent such as polyvinylpyrrolidone (PVP), cetyl trimethyl ammonium bromide (CTAB), sodium polyacrylate (SPA), and sodium dodecyl sulfate (SDS) are selectively adsorbed on a specific facet of the NPs leading the growth and nucleation in other crystal sites. These actions of capping agents play the most important role for the directional growth of NPs with different shapes (Xiong, 2007; Xia et al., 2009). Different shapes of noble metal NPs such as spherical, triangular, hexagonal, pentagonal, squares, rectangular have been synthesized by some of the biological methods using an enzyme, microorganism, plant extract, and peptides. Additionally, other different techniques have been applied for the shape-controlled synthesis of nanocrystals such as sonoelectrochemical method (Zhu et al., 2000), electrodeposition method (Ying et al., 1997), electrochemical method (Song et al., 2009), Microwave-assisted method (Mallikarjuna et al., 2007), solvothermal (Yang et al., 2007), irradiation of UV light (Esumi et al., 1995), oxidative etching technique (Wiley et al., 2004), etc. Many researchers reported that metal nanocrystal with supporting materials possesses high electrocatalytic activity than the individual one. Among them, graphene is broadly used as an efficient support material (Chen et al., 2008; Geim and Kim, 2008). But the most important is in the tricks or intellectuals applied during the synthesis method of the graphene and nanocrystals composite. In general, two synthetic methods are applied for graphene supported metal nanocomposite such as in-situ and ex-situ approach. In the former case, both graphene oxide (GO) and metal have to be reduced to graphene and metal NPs simultaneously but in ex-situ, both will reduced separately and then mixed up for the composite preparation. Such reduction could be done by various methods such as chemical reduction (Zhang et al., 2011), electrochemical reduction (Yao et al., 2012), microwave reduction (Jasuja et al., 2010), hydrothermal reduction (Hu et al., 2012), chemical vapor deposition (CVD) method (Zhang et al., 2013), ultra-sonication technique (Park et al., 2011) and photochemical reduction (Zhang et al., 2012). Among these two methods of composite preparation, the ex-situ method is predominantly used (Mazumder et al., 2009).

MOR activity of platinum nanoparticles (PtNPs) can be modulated by changing the shape morphology and size of NPs. These nanostructured Pt possesses highly active, better resistance towards surface poisoning, and decreases the usage of Pt (Liu et al., 2006; Solla-Gullon et al., 2011). The branched Pt nanostructures such as nanostar, nanoflower, and nanodendrites have reported as highly active for electrocatalysis (Behera et al., 2016; Sahu et al., 2015). They explained branched Pt nanocrystal showing excellent electrocatalytic activity towards MOR. The mechanism of methanol oxidation by PtNPs are basically based on the adsorption of methanol molecule on Pt surface and formation of poisoning intermediate CO gas, the selectivity of Pt-CO and Pt-OH species and oxidation of H_2O molecule on Pt. Although Pt plays a significant role in electrocatalytic activity towards MOR, yet the cost-effectiveness of the Pt electrocatalyst has become the major burden for commercialization. Therefore, there was other nanocrystals comes to notice to overcome these obstacles. On extensively careful study of all NPs Pd have been the most valuable for electrooxidation of methanol. Pd is treated as the best and alternate material for MOR activity. Different shape of Pd such as nanocube, nanorod, nanowire, tripod/tetrapod, nanostars, nanodendrites, and nanoflower possesses the presence of active sites and has been utilized as promising electrocatalyst for methanol oxidation (Tang et al., 2012; Zhang et al., 2016). In addition to this Pd based bimetallic such as Pd-Au, Pd-Ru, Pd-Sn, Pd-Ni, and Pd-Ag have been reported as excellent electrocatalyst for MOR activity (Cameron et al., 2018; Gilbert et al., 2002; Balaji et al., 2017). Jena et al. reported an excellent dramatically shape-controlled structures of Pd nanocrystal and their variation electrocatalytic activity towards MOR (Jena et al., 2011). They synthesized Pd NPs by taking $PdCl_2$ as a precursor and plant extract rutin (commonly known as vitamin P) as an organic reducing/stabilizing agent. They synthesized different shape of Pd by varying the concentration of precursor ($PdCl_2$) and the concentration of stabilizing/reducing agent of rutin. In a sum-up discussion for synthesis process, they used 0.1 mM of $PdCl_2$ precursor and 15 mM of rutin, 15 mM were heated for 30 min in an aqueous medium, and solution cooled for at room temperature. The nanocrystals were characterized thoroughly by TEM and XRD techniques. It has been observed that, around 4–5 min on reaction processing, spherical, and anisotropic NPs size observed, around 12–15 min, one-dimensional structure and a complete growth of porous dendrite nanowires after 30 min. The growth of particles is highly with respect to the concentration of rutin and precursor. The different shape of Pd nanocrystal shows different electrocatalytic activity. Among different structures of Pd nanocrystals, porous Pd structure exhibits the higher current density (5.05 ± $0.22 A m^{-2}$) and lower oxidation potential (228 ± 15 mV) towards the methanol oxidation. The application experiment was performed by a three-electrode system in the electrochemistry method. This magnificent value may be due to edges atoms, corner atoms, and porous morphology of Pd NPs (Wang et al., 2010). Here we concluded that nanocrystal's architectures with highly surface roughness and surface steps exhibit more attractive for enhancing electrocatalytic methanol oxidation application. This work proposed the synthesis strategy to explore other alloy networks of porous NPs, which could able to possess a better candidate for the highly electrochemical catalyst of methanol oxidation. The electrochemical activity of nanoparticle can be enhanced by engineering the pores within alloy NPs. Recently Dinga et al. reported a different and innovative approach for MOR by the network-like structure of Pt-Co catalyst (2017). The Pt-Co catalyst displayed markedly improved activity for MOR and high storage stability and operational durability compared to the Pt_3Co networks made of solid NPs, Pt/C, and PtRu/C catalysts. From electrochemical data, it was observed that the composite of Pt-Co catalyst showed the MOR nearly 3.9, 2.0, and 2.1 times higher than that of commercially available Pt/C, Pt-Ru/C catalyst respectively. The mechanism behind this unique value was the porous NPs morphology of Pt-Co networks and exhibited high CO oxidation in the applied potential range of +0.756 V vs. RHE. Here they synthesized PtCo network-like porous NPs by using the Pt-Co as precursor via the electrochemical dealloying method.

The particle sizes of NPs have great effect on the MOR due to differences in Pt-OH and Pt-CO bond strengths. The coupling of NPs in the nanometer range of different sizes via soluble product imparts electro-oxidizing methanol to CO_2. In addition to this efficiency of methanol, electro-oxidation also optimized the CO_2 formation pathway otherwise mostly related to the partial oxidation of methanol to formaldehyde (Bergamaski et al., 2006). Bergamaski et al. discussed the size dependant distribution of platinum nanocrystals towards methanol oxidation. Again they confirmed that particle size below 10 nM perform high electro oxidizing methanol to CO_2 and more that 10 nM causes partial oxidation of methanol to formaldehyde. Here they used 60% Pt/C catalyst composite for study the enhanced activity for the MOR. Sahu et al. synthesized graphene supported dendrite-shaped Pd NPs and performed for methanol oxidation application. The well dispersed and formation of Pd dendrite shaped nanostructures (PdnDs) on the graphene was confirmed by using transmission electron microscopy technique. On comparing electrochemical (EC), activity of graphene supported PdnDs with PdnDs alone, and a commercial available Pd/C towards methanol oxidation, it was observed that RG-PdnDs possess high electrocatalytic activity than all. From this unique result, we concluded that graphene support plays a vital role in directing towards the shaping of the nanostructures on its surface and caused for development of high electrocatalytic activity in MOR. Bi-metallic NPs also exhibit excellent electrocatalytic activity towards MOR. Sahu et al. synthesized a unique flower-like Au/Pd alloy nanostructures and their composite on graphene support in an aqueous medium. This bimetallic shows superior electrocatalytic activity for MOR compared to their individual ones (Sahu et al., 2015). This extraordinary activity was due to the electrocatalytic performance of Au/Pd alloy nanoflower. Ongoing through a recent study on bimetallic for the MOR fuel cell, Vilian et al. studied on PtNPs, highly dispersed on reduced graphene oxide (RGO) substrate with AuNPs. Here they synthesized a facile and environmentally method for the decoration of PtNPs on Au-RGO substrate sheets. This bimetallic nanostructure possesses a higher electrochemically active surface area (ECSA) than the Pt-RGO electrocatalyst and also than a commercial Pt/C catalyst. On voltammetric study, the ratio of the forward peak current to the reverse peak current ($I^f/I^b = 2.33$) for the Pt-Au-RGO electrocatalyst is much higher than that of the Pt-RGO electrocatalyst ($I^f/I^b = 1.16$). It clearly suggests the synergistic effects of the Au and RGO substrate in enhancing the electrochemical activity of PtNPs for methanol oxidation and CO_2 poisoning resistance. They explained the presence of Au nanoparticle on the RGO sheets helps in the formation of small, highly concentrated, and uniformly decorated PtNPs. This attributed to the increasing conductivity of the bimetallic with the removal of carbonaceous. This composite possesses an excellent methanol oxidation in fuel cell applications (Vilian et al., 2016). In addition, this Pt nanocrystal also possesses higher MOR on forming composite with another metal Ni with graphene substrate (Xiu et al., 2015). In this work, platinum-nickel (PtNi) bimetallic NPs with a three-dimensional graphene (3DGN) were synthesized by using the electrodeposition method. From the SEM and other techniques resulted the Pt/Ni NPs were uniformly distributed on the porous 3DGN. This bimetallic with graphene substrate possess higher ECSA (87.4 M^2/g_{pt}) than the Pt/3DGN (53.8 M^2/g_{pt}). Here the PtNi/3DGN composites exhibited the forward anodic peak current density of 822.1 mA/mgPt, only Pt 251.2 mA/mgPt and Pt/3DGN (416.9 mA/mgPt) towards methanol oxidation and also excellent electro-catalyst for CO tolerance. The Pt-Ni/3DGN composites exhibited a significantly improved, high stability reaction rate for methanol oxidation. Nickel is considered an excellent alternative catalyst in place of Pt nanocrystal, although Ni has a relatively lower stability under operational conditions, even in alkaline media (Li et al., 2018). Such NPs composite with carbon black applied for MOR in alkaline medium. The composite exhibited the highest electrocatalytic activity towards MOR with a mass current density of 820 $m^{-2}A$ mg^{-1} at 0.70 V vs. Hg/HgO, stabilities over periods of 10000 s at same potential in alkaline medium. That is significantly improved stability of Ni-Sn catalysts compared with that of Ni although having slightly increase of composite towards MOR than only Ni. The electrocatalytic MOR performance of the nanostructured bimetallic with graphene possesses the highest excellent transducer for MOR than the individuals. The shape of NPs play a more important role in catalytic activity for fuel cell MOR applications.

9.3 PERFORMANCES OF NANOPARTICLES (NPS) FOR FORMIC ACID (FA) FUEL CELL

As like methanol, FA also one type of PEMFC in which FA is directly used as fuel. FA is a small organic molecule and reasonably stable liquid at ambient condition of temperature and pressure. It is a natural biomass and a CO_2 reduction product. The storage of FA is easy and safe than hydrogen. Nearly about five decades ago, the FAO fuel cell investigation started with the advent of modern potentiodynamic techniques. The oxidation of FA on nanocrystals is considered as a model reaction in electrocatalysis due its simple structure where only two electrons involved in the total FAO reaction process to yield carbon dioxide. In 1960s, electrochemistry fuel cell FA experiment performed by many pioneers and resulted (i) Clarifying the chemical nature of the active intermediate in the direct pathway, (ii) Diminishing CO adsorption on nanocrystal electrodes (iii) Phenomenological interpretation of anodic polarization curves for FAO (iv) Discover of the catalytic effect with respect to NPs for FAO (v) Concentration, temperature effects in a fuel cell. In direct formic acid fuel cell (DFAFC), it is oxidized to CO_2 at anode, and oxygen reduction takes place at the cathode. The typical reactions in PEMFC are the proton passes to the cathode via PEM and reacts with oxygen on the catalyst layer at the cathode.

The cell reaction:

$$HCOOH + \frac{1}{2}O_2 \rightarrow CO_2 + H_2O \tag{9.4}$$

Jiang et al. explained the chemical reaction of FAO as a model reaction in the study of electrocatalysis of C1 molecules. They brought about deeper mechanisms in fundamental investigations of FAO on Pt and Pd electrode surfaces and put the concept "synthesis-by-design." According to them through the different synthesis methods, there was tailoring in geometric of nanocatalysts. They synthesized Pd- and Pt nanocatalysts and put their application for FAO fuel cell. There are two simple pathways to describe the net decomposition of FA (1) Dehydrogenation Eqns. (9.5) and (9.2) Dehydration (Eqn. (9.6)). The equations are as follows:

$$HCOOH - active\,intermediate \rightarrow CO_2 + 2H^+ + 2e^- \tag{9.5}$$

$$HCOOH \rightarrow CO_{ads} + H_2O \tag{9.6}$$

This team controlled the electronic and geometric structures of metal NPs as well as the catalyst supports and made Progresses in the development of miniature DFAFCs and the corresponding operation conditions (Jiang et al., 2014). The oxidation of FA on platinum electrodes has received considerable research attention in electrocatalysis research (Wilkinson et al., 2014;

Mosdale et al., 1995). They studied electrooxidation of FA on polycrystalline platinum electrodes in sulfuric acid as electrolyte. The electrode oxidation system carried out by AC voltammetry and corresponding impedance spectra was in the frequency range between 0.7 Hz to 30 kHz. Again they found potential regions of negative differential resistance (NDR) and hidden negative differential resistances (HNDR) where two potential regions in the positive-going scan and one in the negative-going scan. More interestingly, they found zero impedance which shows the possibility of oscillations arising from purely chemical reasons. The poisoning intermediate CO deactivates the electro-oxidation of FA at anodes. Hence, the oxidation of simple organic molecule becomes a changeling reaction system. Mohamed et al. studied to electrooxidation of FA at manganese oxide nanorods (NRs)-modified Pt planar and nanohole-arrays. Here they used Pt electrodes of two geometries such as a planar Pt disk electrode sealed in a Teflon jacket (2.0 mm in diameter and having an exposed geometric surface area of 0.031 cm^2) and a Pt nanohole-array supported on a glass substrate (4.0 mm in width, 5.0 mm in length and 0.5 mm in thickness). They electrodeposited MnO_x on the surface of Pt (planar and nanohole-arrays) by cycling the potential at 20 mV s^{-1} between −0.05 and 0.35 V vs. SCE in 0.1M Na_2SO_4 containing 0.1M $Mn(CH_3COO)_2$. The direct oxidation of FA to CO_2 was observed at the modified electrodes, while the formation of the poisoning intermediate CO was suppressed. Two oxidation peaks in pH 3.45 of 0.3M FA at 0.2 and 0.55 V are observed which are corresponds to direct and indirect pathways of oxidation respectively. The composite of Pt anodes with manganese oxide NRs enhances electrocatalytic activity of FAO along with high tolerance to CO. Modification of two geometry of Pt electrodes with nano-MnO_x composite resulted in a significant enhancement of the direct oxidation pathway for FA up to 14-times higher than the indirect pathway. Grozovski et al. studied FAO on shape-controlled PtNPs by pulsed voltammetry. They resulted out that surface structure of the nanocrystal electrodes, particular on the presence of domains with (100), (100–111), and (111) symmetry plays an important role on the reactivity and kinetics of FAO. Electrocatalytic activity depends on the relative amount of the surface sites of different NPs. Kinetic data obtained from the pulsed voltammetry experiment for these different shapes and concluded that the planes (111) PtNPs possesses lowest CO formation constant with moderate reaction rate and (100) exhibits most active toward FA electrooxidation via the active intermediate reaction path with more poisoning rate (Grozovski et al., 2010). These works open the research activity of FAO kinetics for decorating new better electrocatalytic materials.

Besides PtNPs, Pd nanocrystals exhibit excellent catalytic activity towards FAO. Fabrication of active electrocatalysts having low noble metal loading have been focused on the development of techniques and new materials to achieve high catalytic activity toward FAO. In recent investigations, low cost and higher catalysis toward FA electro-oxidation as compared to those of traditional Pt/C, Pd/C becomes the more changeling factor. To increase the FAO activity and further reduce noble-metal loading Wang et al. decorated palladium NPs on nickel supported on carbon and applied for FAO. The bi-functional palladium active sites are highly responsible for FAO to form adsorbed CO, which poisons the catalyst surface for further fuel oxidation in meanwhile the nickel sites provide adsorbed hydroxyl groups (OH_{ads}), which is favor the oxidant for the removal of CO_{ads} (Du et al., 2010; Li et al., 2011). Concerning the electronic effect, the presence of nickel changes the electronic structure of palladium in such a way that it lowers the CO adsorption energy in the reaction process. The Pd@Ni/C structure has been proven by various techniques such as SEM, TEM, and the presence of two NPs are confirmed by XRD and electrochemical techniques. The electrocatalytic activity of Pd on Ni has a higher catalytic activity and can be used as improved noble-metal utilization. The work further proves the electrocatalytic activity of Pd catalyst towards FAO can be controlled by varying the Pd size and shape. In the same year, Dai et al. synthesized the bimetallic NPs and performed the enhanced FAO on Cu-Pd NPs. They synthesized copper-palladium NPs through a galvanic replacement process in a stepwise manner. At first, selective dissolution of surface Cu carried out which followed by Pd-enriched deposited through galvanic redox reactions method to form Cu-Pd NPs (Wu et al., 2006). Then through potential cycling, the Cu has been selectively removed from the surface forming Pd enriched Cu-Pd NPs. These particles exhibit much higher FAO activities than that on pure Pd NPs, and they are much more resistant to the surface poisoning. These experiments at first used Cu NPs instead of Pt, Au, and other particles for FAO. This unique Cu-Pd composite are promising anode catalysts for DFAFCs.

The Pd NPs have initial activity superior to that of even platinum. However, initial performance of Pd cannot be sustained over long periods of time, as the highly active Pd surface is vulnerable to the process of catalytic oxidation. Although the Pd and its bimetallic alloy exhibits as an excellent transducer for FAO still it is difficult for palladium NPs to act as true catalyst. This is due to its highly active surface and which caused poisoning. To overcome these difficulties Zhang et al. have prepared high activity PdO/Pd-CeO$_2$ hollow spheres through an *in situ* electrochemical reduction process for creating available fresh surface via the *in situ* electrochemical reduction of PdO (2014). That prepared catalyst exhibits an enhanced mass activity and durability for FAO compared to the Pd-CeO$_2$ hollow spheres. Here, the reactive oxygen species of ceria promote the oxidation of the intermediates adsorbed on the Pd surfaces, which caused for improved resistance to poisoning for the PdO/Pd-CeO$_2$ hollow spheres. The excellent electrocatalytic performance of the PdO/Pd-CeO$_2$ hollow sphere provides a new approach to design the Pd-based electrocatalysts with high activity and durability for FAO (Zhang et al., 2014).

FAO also been studied by bimetallic core cells with their support to carbon-based materials. Tripkovi et al. explained the temperature-dependent on FAO by Pt/Ru NPs in acidic solution. They optimized the cuboctahedral Pt/Ru particle size of 4.3±0.3 nm at elevated temperatures of 313 K, 333 K, and at this particular temperature, FAO is more activated. The FAO on Pt/Ru alloys follows dual pathways with opening the dehydration channel at steady-state to lower the CO adsorption. The oxidation of FA on the Pt/Ru alloys NPs is the oxidative removal of CO in the Langmuir-Hinshelwood (L-H) type reaction (Tripkovi et al., 2005). Discussing another recent article based on a bimetallic alloy of the same PtNPs with other metals showed excellent active catalytic activity towards FA. Iyyamperumal et al. have explained efficient electrocatalytic oxidation of FA using monometallic and bimetallic Au@Pt dendrimer-encapsulated nanoparticles (DENs) (2013). The synthesis of NPs is more important. The synthesized NPs have interestingly countable 147-atom Au core (2-nm-diameter) and a Pt shell of ratio Au147: 1Pt. This core-shell mechanism with CO adsorbents can be clearly understood in Figure 9.1.

The optimized structures for Au/Pt ratio are illustrated in which (a) the Pt-only DEN mode (Pt$_{147}$), (b) the complete shell model (Au@Pt$_{cs}$), which is equivalent to Au$_{55}$@ Pt$_{92}$, and (c) the deformed partial-shell model (Au@Pt$_{dps}$), which is equivalent to Au$_{147}$@Pt$_{102}$. The marked numerical numbers represent possible CO binding sites for each model. This core bimetallic facilitates the direct oxidation of FA to carbon dioxide. These structural distortions suppress the formation of adsorbed CO (CO$_{ads}$) and resulted out the dramatic improvement in catalytic performance. The Au$_{147}$@Pt DENs exhibit better electrocatalytic activity for HCOOH oxidation reaction than Pt$_{147}$ DENs. The smaller particle size in the range of 1–2 nm of DENs is more studied likely due to the lower stability of the smaller particles, and thus their enhanced ability to deform in the presence of adsorbates and during catalysis. The Pt nanoparticle also studied in core-shell with Ru nanoparticle, which exhibits excellent catalytic activity towards FAO reaction. Ehab et al. studied at Ru@Pt core-shell NPs ranging from 0.4 to 1.9 monolayers (MLs) to determine the electronic effect of the Ru core and compression of the Pt lattice influence activity. From voltammetry study, it has been observed that the electronic structural effect of the Ru core on CO oxidation is the responsible factor for influencing FAO. That causes the adsorption of CO starts at lowest potential in ML Pt coverage and increases when Pt coverage is increased towards one ML. Again, they demonstrate a beautiful conclusion that electronic effect of the Ru becomes lower when second Pt layer is added and CO oxidation is shifted to higher potentials and FAO activity drops. So here the optimum Pt (0.85 ML) coverage causes the CO oxidation and also FAO (0.5 M), while 1.3 ML of Pt possesses particular high for 2 M FA at lower potential activity (Ehab et al., 2016).

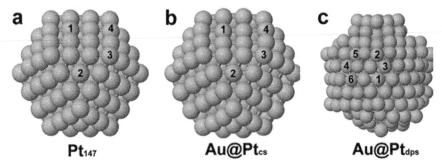

FIGURE 9.1   (a) Pt-only DEN model (Pt$_{147}$), (b) (Au@Pt$_{cs}$), (c) Deformed partial-shell model (Au@Pt$_{dps}$).
Source: Reprinted with permission from Iyyamperumal et al. (2013). © Royal Chemical Society.

To increase the catalytic activity of nanoparticle or carbon supports, it must be modified either by functionalization or doping. Again higher, the surface area of the supporting material better is the dispersion of the catalyst on a high surface area material (Chakrabarty, 1990). The surface area of the supported material can be increased by developing cracks and pores on it. The physical and chemical properties of the supported structures can be created to new states by modifying their electronic structure. Among all elements in periodic table, boron, and nitrogen are the most effective dopants due to their small atomic size, similar to that of the carbon atom. Hence, they can easily enter the carbon support at any materials lattice (Iijima, 1991; Villalpando-Paez et al., 2006). Between these two, nitrogen have made strong attention in doping on carbon-based supported materials due to its significant changes in hardness, electrical conductivity and chemical reactivity (Maldonado et al., 2006). Nguyen et al. improved the catalytic performance for FAO reactions by N-doped Cdot/PtPd nano network hybrid materials. The synthesized nano-network hybrid material that exhibits higher electrocatalytic activity for FA with current densities of 1919.5 mA/mg than with the commercial ones (of 806.02 mA/mg). Furthermore, the hybrid material exhibit unique architecture of abundant accessible sites which promotes excellent stability and hydrophilic dispersibility at room temperature (Nguyen et al., 2018). Jia et al. designed a simple and controllable approach for monodispersed Cu_3PdN NPs through the thermal decomposition of metal precursors and show improved activity and enhanced stability for FAO compared with the corresponding Pd and Cu_3Pd NPs. Here the doping of a heteroatom, i.e., N, into Cu_3Pd, changes the physical and chemical state of materials with the d-band center shifting away from the Fermi energy level.

That leads to for better catalyzing of FA. From CV curve, the current density at 0.1–0.3 V is much higher than that of Cu_3Pd and Pd electrodes with mass current density of 870 mA/mg for FAO (Figure 9.2 a–c). The sample shows higher storage and operational stability at a fixed potential of 0.15 V (Figure 9.2d) demonstrating a higher tolerance to the intermediate poisoning species. In the current year, revolution of fuel cell in field of nanoscience have attracted huge research demands and motivated towards FA fuel cell due to their enhanced efficiency, moving flexibility, and reduced contamination. Al-Akraa et al. have synthesized and tailored the design of Au/Pt nanoanode with platinum (PtNPs) and gold nanoparticles (AuNPs) to fabricate a nano sized anode for FA reaction. For the electrooxidation of FA, fabrication of working electrode (WE) is more important. Here they fabricated by a stepwise manner. At first Pt NPs was drop casted on glassy carbon electrode followed by AuNPs (surface coverage ≈ 32%). The Au/Pt/GC electrode exhibited the best catalytic performance. AuNPs play most important role, as it immunized the first layer of PtNPs against CO poisoning (an intermediate during the oxidation). This is the most simple preparation method of sample and having great performance to resist the critical CO poisoning of nanoparticles (PtNPs) by AuNPs. The surface modification of PtNPs with AuNPs stands as a major role in fuel cell application. Till now, the intellectuals used here have attracted most research scholar for fabricating nanoparticles electrodes to endure CO poisoning.

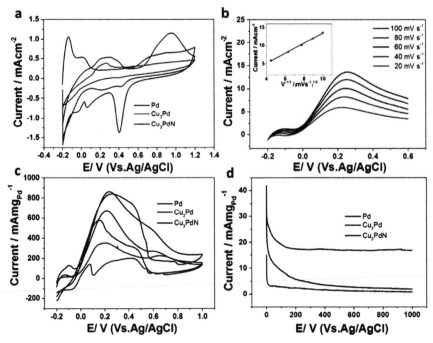

FIGURE 9.2   (a) Cyclic voltammetry of Pd, Cu₃Pd and Cu₃PdN NPs in 0.5 M H₂SO₄ at a scan rate of 50 mV/s; (b) linear sweep voltammetry of Cu₃PdN at different scan rates; (c) cyclic voltammetry and (d) chronoamperometry of Pd, Cu₃Pd and Cu₃PdN NPs in 0.5 M H₂SO₄ with 0.5 M formic acid.
Source: Reprinted with permission from Jia et al. (2016). © Elsevier.

9.4   CONCLUSION

The electrocatalytic properties of noble metal nanoparticle highly depend on the shape, size, morphology, and supporting substrate like carbon black, CNT, or graphene to them. The electrocatalytic activity of bare nanoparticles has the lowest catalytic power than the supporting materials of the same particle. Further by forming bimetallic NPs (with symmetry allowed metals), the activity increases tremendously. The porous and anisotropic nature particles have excellent catalytic activity than their corresponding nanocrystal. The synthesis method (in-situ or ex-situ synthetic route) and the role of stabilizing/reducing agent play the most important role for the shape-controlled morphology development of NPs. The main key mechanism of this simple organic molecule methanol and FAO fuel cell is to counter the critical CO intermediate poisoning for improvement of the catalytic performance of NPs.

This problem acts as the key issue for the development of a new generation methanol and FA fuel cells.

ACKNOWLEDGMENTS

One of the authors, Priyabrat Mohapatra is thankful to Science and Engineering Research Board (SERB), Department of Science and Technology (DST), Government of India, New Delhi for the financial assistance vide the sanction order no EMR/2016/003370 Dated 13/10/2017 under Extra Mural Research Funding Scheme.

KEYWORDS

- cyclic voltammetry
- electrochemical impedance spectroscopy
- formic acid oxidation
- methanol oxidation
- nanoparticles
- transmission electron microscope

REFERENCES

Bagotzky, V. S., Vassiliev, Y. U. B., & Khazova, O. A., (1977). Organic compounds, generalized scheme of chemisorption, electro oxidation and electro reduction of simple on platinum group metals. *J. Electroanal. Chem., 81*, 229–238.

Behera, T. K., Sahu, S. C., Satpati, B., Bag, B. P., Sanjay, K., & Jena, B. K., (2016). Branched platinum nanostructures on reduced graphene: An excellent transducer for nonenzymatic sensing of hydrogen peroxide and biosensing of xanthine. *Electrochimica Acta, 206*, 238–245.

Bergamaski, K., Alexei, L., Pinheiro, N., Erico, T. N., & Francisco, C. N., (2006). Nano particle size effects on methanol electrochemical oxidation on carbon supported platinum catalysts. *J. Phys. Chem. B, 110*, 19271–19279.

Cameron, H., Kelly, W., Tania, M., Benedetti, A. A., Suhmann, W., Justin, J., & Richard, D., (2018). Understanding the effect of Au in Au-Pd bimetallic nanocrystals on the electrocatalysis of the methanol oxidation reaction. *J. Phys. Chem. C, 122*, 21718–21723.

Chakrabarty, D. K., (1990). *Adsorption and Catalysis by Solids*. Wiley Eastern Ltd., New Delhi.

Chen, J. H., Jang, C., Xiao, S., Ishigami, M., & Fuhrer, M. S., (2008). Intrinsic and extrinsic performance limits of graphene devices on SiO_2. *Nature Nanotech, 3*, 206–209.

Dai, L., & Zou, S., (2011). Enhanced formic acid oxidation on Cu-Pd nanoparticles. *Journal of Power Sources, 196*, 9369–9372.

Dinga, J., Shan, J., Wang, H., Bruno, G., & Wang, R., (2017). Tailoring nanopores within nanoparticles of PtCo networks as catalysts for methanol oxidation reaction. *Electrochimica Acta, 255*, 55–62.

Du, C., Chen, M., Wang, W., & Yin, G. P., (2010). Electrodeposited $PdNi_2$ alloy with novelly enhanced catalytic activity for electrooxidation of formic acid. *Electrochem. Commun., 12*, 843–846.

Ehab, N., Sawy, E., & Peter, G., (2016). Formic acid oxidation at Ru@Pt core-shell nano particles. *Electrocatalysis, 7*, 477–485.

Esumi, K., Matsuhisa, K., & Torigoe, K., (1995). Preparation of rodlike gold particles by UV irradiation using cationic micelles as a template. *Langmuir, 11*, 3285–3287.

Franaszczuk, K., Herrero, E., Zelenay, P., Wieckowski, A., Wang, J., & Masel, R. I., (1992). A comparison of electrochemical and gas-phase decomposition of methanol on platinum surfaces. *J. Phys. Chem., 96*, 8509–8516.

Franaszczuk, K., Zelenay, E. P., Wieckowski, A., Wang, J., & Masel, R. I., (1992). A comparison of electrochemical and gas-phase decomposition of methanol on platinum surfaces. *J. Phys. Chem., 96*, 8509–8516

Gilbert, J., Matare, M. E., Tess, Y. Y., Khalil, A., & White, E., (2002). Electrochemical oxidation of methanol with Ru/Pd, Ru/Pt, and Ru/Au heterobimetallic complexes. *Organometallics, 21*, 711–716.

Grozovski, V., Solla, J., Climent, V., Herrero, E., & Feliu, J. M., (2010). Formic acid oxidation on shape-controlled pt nanoparticles studied by pulsed voltammetry. *J. Phys. Chem. C, 114*, 13802–13812.

Hao, E., Schatz, G. C., & Hupp, J. T., (2004). Synthesis and optical properties of anisotropic metal nano particles. *Journal of Fluorescence, 14*, 331–441.

Hearth, M. P., & Hards, G., (1996). Catalysis for low temperature fuel cells. *A Platinum Met. Rev., 40*, 150–159.

Herrero, E., Franaszczuk, K., & Wieckowski, A., (1994). Electrochemistry of methanol at low index crystal planes of platinum: An integrated voltammetric and chronoamperometric study. *J. Phys. Chem., 98*, 5074–5083.

Herrero, E., Franaszczuk, K., & Wieckowski, A., (1994). Electrochemistry of methanol at low index crystal planes of platinum: An integrated voltammetric and chronoamperometric study. *J. Phys. Chem., 98*, 5074–5083.

Hu, C., Cheng, H., Zhao, Y., Hu, Y., Liu, Y., Dai, L., & Qu, L., (2012). Newly-designed complex ternary Pt/PdCu nano boxes anchored on three-dimensional graphene framework for highly efficient ethanol oxidation. *Adv. Mater., 24*, 5493–5498

Iijima, S., (1991). Helical microtubules of graphitic carbon. *Nature, 354*, 56–58.

Islam, M., Akraa, A., Yaser, M., & Mohammad, A., (2019). Journal of nanomaterials, facile synthesis of a tailored-designed Au/Pt nanoanode for enhanced formic acid, methanol, and ethylene glycol electrooxidation. *Journal of Nanomaterials, 1–9.* doi: 10.1155/2019/2784708.

Iwasita, T., & Xia, X., (1996). Adsorption of water at Pt(111) electrode in $HClO_4$ solutions. The potential of zero charge. *J. Electroanal. Chem., 411*, 95–102.

Iyyamperumal, R., Zhang, L., Henkelman, G., & Richard, M. C., (2013). Efficient electrocatalytic oxidation of formic acid using Au@Pt dendrimer-encapsulated nano particles. *J. Am. Chem. Soc., 135*, 5521–5524.

Jasuja, K., Linn, J., Melton, S., & Berry, V., (2010). Microwave-reduced uncapped metal nanoparticles on graphene: Tuning catalytic, electrical, and Raman properties. *J. Phys. Chem. Lett., 1*, 1853–1860.

Jena, B. K., Sahu, S. C., Satpati, S., Sahu, R. K., Behera, D., & Mohanty, S., (2011). A facile approach for morphosynthesis of Pd nano electrocatalysts. *Chem. Commun., 47*, 3796–3798.

Jia, J., Shao, M., Wang, G., Deng, W., & Zhenhai, W., (2016). Cu$_3$PdN nanocrystals electrocatalyst for formic acid oxidation. *Electrochemistry Communications, 71*, 61–64.

Jiang, K., Zhang, H. X., Zou, S., & Cai, W. B., (2014). Electrocatalysis of formic acid on palladium and platinum surfaces: From fundamental mechanisms to fuel cell applications. *Chem. Phys., 16*, 20360–20376.

Kua, J., & Goddard, W. A., (1999). Oxidation of methanol on 2nd and 3rd row group VIII transition metals (Pt, Ir, Os, Pd, Rh, and Ru): Application to direct methanol fuel cells. *J. Am. Chem. Soc., 121*, 10928–10941.

Li, J., Luo, Z., Zuo, Y., Liu, J., Zhang, T., Tang, P., Arbiol, J., Llorca, J., & Cabot, A., (2018). NiSn bimetallic nano particles as stable electrocatalysts for methanol oxidation reaction. *Applied Catalysis B: Environmental, 234*, 10–18.

Li, N. H., Sun, S. G., & Chen, S. P., (1997). Studies on the role of oxidation states of the platinum surface in electrocatalytic oxidation of small primary alcohols. *J. Electroanal. Chem., 430*, 57–67.

Li, R., Wei, Z., Huang, T., & Yu, A., (2011). Ultrasonic-assisted synthesis of Pd–Ni alloy catalysts supported on multi-walled carbon nano tubes for formic acid electro oxidation. *Electrochim. Acta, 56*, 6860–6865.

Liu, H., Song, C., Zhang, L., Zhang, J., Wang, H., & Wilkinson, D. P., (2006). A review of anode catalysis in the direct methanol fuel cell. *J. Power. Sources, 155*, 95–110.

Lu, G. Q., Chrzanowski, W., & Wieckowski, A., (2000). Catalytic methanol decomposition pathways on a platinum electrode. *J. Phys. Chem. B, 104*, 5566–5572.

Maldonado, S., Morin, S., & Stevenso, K. J., (2006). Structure, structure, composition, and chemical reactivity of carbon nanotubes by selective nitrogen doping. *Carbon, 44*, 1429–1437

Mallikarjuna, N. N., & Varma, R. S., (2007). Microwave-assisted shape-controlled bulk synthesis of noble nanocrystals and their catalytic properties. *Cryst. Growth Des., 7*, 686–690.

Matthew, A. P., Christopher, J. M., Emmott, C. J., & Sally, M. B., (2015). A membrane-less electrolyzer for hydrogen production across the pH scale. *Energy and Environmental Science, 8*, 1–45.

Mazumder, V., & Sun, S., (2009). Oleylamine-mediated synthesis of Pd nano particles for catalytic formic acid oxidation. *J. Am. Chem. Soc., 131*, 4588–4589.

McNicol, B. D., (1981). An electrochemical and spectroscopic investigation into carbon monoxide surface poisoning. *J. Electoanal. Chem., 118*, 71–87.

McNicol, B. D., Dand, A. J., & Wiliams, K. R., (1999). Direct methanol-air fuel cells for road transportation. *J. Power Sources, 83*, 15–31.

Mohamed, S., & Deab, E., (2010). Electrocatalysis by nano particles: Oxidation of formic acid at manganese oxide nanorods-modified Pt planar and nano hole-arrays. *Journal of Advanced Research, 1*, 87–93.

Mosdale, R., & Srinivasan, S., (1995). Analysis of performance and of water and thermal management in proton exchange membrane fuel cells. *Electrochim. Acta, 40,* 413.

Munk, J., Christensen, P. A., Hamnett, A., & Skou, E., (1996). The electrochemical oxidation of methanol on platinum and platinum + ruthenium particulate electrodes studied by *in-situ* FTIR spectroscopy and electrochemical mass spectrometry. *J. Electroanal. Chem., 401,* 215–222.

Muthukumaran, B., & Balaji, S., (2017). Electro catalyzed oxidation of methanol by Pd-Ni nano particles supported on Vulcan carbon/nafion composite in alkaline medium. *Journal of Chemistry and Chemical Sciences, 10,* 863–870.

Nguyen, V. T., Tran, Q. C., Quang, N. D., Nguyen, N. A., Bui, V. T., Dao, V. D., & Choi, H. S., (2018). N-doped C dot/PtPd nanonetwork hybrid materials as highly efficient electrocatalysts for methanol oxidation and formic acid oxidation reactions. *Journal of Alloys and Compounds, 766,* 979–986.

Park, G., Lee, K. G., Lee, S. J., Park, T. J., Wi, R., & Kimdo, H., (2011). Synthesis of graphene-gold nano composites via sonochemical reduction. *J. Nanosci. Nanotechnol., 11,* 6095–6101.

Sahu, S. C., Behera, T. K., Dash, A., Jena, B., Ghosh, A., & Jena, B. K., (2016). Highly porous Pd nanostructures and reduced graphene hybrids: Excellent electrocatalytic activity towards hydrogen peroxide. *New J. Chem., 40,* 1096–1099.

Sahu, S. C., Samantara, A. K., Dash, A., Juluri, R. R., Sahu, R. K., Mishra, B. K., & Jena, B. K., (2013). Grapheneinduced Pd nanodendrites: A high performance hybrid nano electrocatalyst. *Nano Research, 6,* 635–643.

Sahu, S. C., Satpati, B., Besra, L., & Jena, B. K., (2015). A bifunctional nano-electrocatalyst based on a flowerlike gold/palladium bimetallic alloy nanostructure and its graphene hybrid. *Chem. Cat. Chem., 24,* 4042–4049.

Seland, F., David, A., & Harrington, R. T., (2012). Electrohemically fabricated nickel-tungsten nanowires. *ECS Transactions, 41,* 35–47.

Shimazu, K., & Kita, H., (1992). *In situ* measurements of water adsorption on a platinum electrode by an electrochemical quartz crystal microbalance. *J. Electroanal. Chem., 341,* 361–367.

Solla, G, J., Vidal, F. J., & Feliu, J. M., (2011). Shape dependent electrocatalysis. *Annu. Rep. Prog. Chem., Sect. C: Phys. Chem., 107,* 263–297

Song, Y., Kim, J., & Park, K. W., (2009). Synthesis of Pd dendritic nano wires by electrochemical deposition. *Cryst. Growth Des., 9,* 505–507.

Tang, S., Vongehr, S., Zheng, Z., Ren, H., & Meng, X., (2012). Facile and rapid synthesis of spherical porous palladium nanostructures with high catalytic activity for formic acid electro-oxidation. *Nanotechnology, 23,* 255606–255608.

Tilak, B. V., & Conway, B. E., (1973). The real condition of oxidized pt electrodes: Part III. Kinetic theory of formation and reduction of surface oxides. *J. Electroanal. Chem., 48,* 1–23.

Tiwari, J. N., Tiwari, R. N., & Kim, K. S., (2012). Zero-dimensional, one-dimensional, two-dimensional and three-dimensional nano structured materials for advanced electrochemical energy devices. *Progress in Materials Science, 57,* 724–803.

Tripkovi, A. V., Popovi, K. D., Lovi, J. D., Markovi, N. M., & Radmilovi, V., (2005). Formic acid oxidation on Pt/Ru nano particles: Temperature effects. *Materials Science Forum, 494,* 223–228.

Van, D. G. M. E., Dangerfield, M. J., & Harrington, D. A., (1997). *J. Electroanal. Chem., 420*, 89.

Vielstich, W., & Xia, X. H., (1995). Comments on electrochemistry of methanol at low index crystal planes of platinum: An integrated voltammetric and chronoamperometric study. *J. Phys. Chem., 99*, 10421–10423.

Vielstich, W., & Xia, X. H., (1995). Comments on electrochemistry of methanol at low index crystal planes of platinum: An integrated voltammetric and chronoamperometric study. *J. Phys. Chem., 99*, 10421–10422.

Vilian, A. T., Hwang, S. K., Kwak, C. H., Yeong, S., Kim, C., Lee, G., Lee, J. B., Huh, Y. S., & Han, Y. K., (2016). Synthesis of Au/Pd alloy nano flower for methanol oxidation. *Synthetic Metals, 21*, 952–959.

Villalpando, F., Zamudio, A., Elias, A., Son, H., Barros, E. B., Chou, S. G., Kim, Y. A., et al., (2006). Synthesis and characterization of long strands of nitrogen-doped single-walled carbon nano tubes. *Chem. Phys. Lett., 424*, 345–352.

Wang, H., Wingender, C., Baltruschat, H., Lopez, M., & Reetz, M. T., (2001). Methanol oxidation on Pt, PtRu, and colloidal Pt electrocatalysts: A DEMS study of product formation. *J. Electroanal. Chem., 509*, 163–169.

Wang, L., Wang, H., Nemoto, Y., & Yamauchi, Y., (2010). Rapid and efficient synthesis of platinum nano dendrites with high surface area by chemical reduction with formic acid. *Chem. Mater., 22*, 2835–2841.

Wang, R., Wang, H., Feng, H., & Shan, J., (2013). Palladium decorated nickel nano particles supported on carbon for formic acid oxidation. *Int. J. Electrochem. Sci., 8*, 6068–6076.

Weijiang, Z., & Jim, Y. L., (2008). Particle size effects in Pd-catalyzed electro-oxidation of formic acid. *J. Phys. Chem. C, 112*, 3789–3793.

Wiley, B., Herricks, T., Sun, Y., & Xia, Y., (2004). Polyol synthesis of silver nano particles: Use of chloride and oxygen to promote the formation of single-crystal, truncated cubes and tetrahedrons. *Nano Lett., 4*, 1733–1739.

Wilkinson, D. P., Voss, H. H., & Prater, K., (1994). Water management and stack design for solid polymer fuel cells. *J. Power Sources, 49*, 117.

Williams, K. R., & Burstein, G. T., (1997). Low temperature fuel cells: Interactions between catalysts and engineering design. *Catal. Today, 38*, 401–410

Wu, C., Mosher, B. P., & Zeng, T., (2006). One-step green route to narrowly dispersed copper nanocrystals. *J. Nanoparticle Res., 8*, 965–969.

Xia, Y., Xiong, Y., Lim, B., & Skrabalak, S. E., (2009). Shape-controlled synthesis of metal nanocrystals: Simple chemistry meets complex physics? *Angew. Chem. Int. Ed., 48*, 60–103.

Xiong, Y., & Xia, Y., (2007). Shape-controlled synthesis of metal nanostructures: The case of palladium. *Adv. Mater., 19*, 3385–3391.

Xiu, R., Zhang, F., Wang, Z., Yang, M., Xia, J., Gui, R., & Xia, Y., (2015). *RSC Adv.*, Electrodeposition of PtNi bimetallic nano particles on three dimensional graphene for highly efficient methanol oxidation. *RSC Advances, 5*, 86578–86583.

Yang, Y., Matsubara, S., Xiong, L., Hayakawa, T., & Nogami, M., (2007). Solvothermal synthesis of multiple shapes of silver nano particles and their SERS properties. *J. Phys. Chem. C, 111*, 9095–9104

Yanyan, L., Jiang, Y., Chen, M., Liao, H., Huang, R., Zhou, Z., Tian, N., Chen, S., & Sun, S., (2012). Electrochemically shape-controlled synthesis of trapezohedral platinum nanocrystals with high electrocatalytic activity. *Chem. Commun., 48*, 9531–9533.

Yao, Z., Zhu, M., Jiang, F., Du, Y., Wang, C., & Yang, P., (2012). Highly efficient electrocatalytic performance based on Pt nanoflowers modified reduced graphene oxide/carbon cloth electrode. *J. Mater. Chem., 22*, 13707–13713

Ying, Y., Chang, S. S., Lee, C. L., & Wang, C. R., (1997). Gold nanorods: Electrochemical synthesis and optical properties. *J. Phys. Chem. B, 101*, 6661–6664.

Zhang, K., Wang, C., Bin, D., Wang, J., Yan, B., Shiraishi, Y., & Du, Y. (2016). Fabrication of Pd/P nano particle networks with high activity for methanol oxidation. *Catal. Sci. Technol., 6*, 6441–6447.

Zhang, L., Ding, L. X., Luo, Y., Zeng, Y., Wang, S., & Wang, H., (2014). PdO/Pd-CeO$_2$ hollow spheres with fresh Pd surface for enhancing formic acid oxidation. *Chemical Engineering Journal, 347*, 193–201.

Zhang, S. Y., Shao, G. Y., & Lin, Y., (2013). Recent progress in nanostructured electrocatalysts for PEM fuel cells. *J. Mater. Chem. A, 1*, 4631–4633.

Zhang, Y., Yuan, X., Wang, Y., & Chen, Y., (2012). One-pot photochemical synthesis of graphenecomposites uniformly deposited with silver nano particles and their high catalytic activity towards the reduction of 2-nitroaniline. *J. Mater. Chem., 22*, 7245–7251.

Zhang, Y., Zhang, L., & Zhou, C., (2013). Review of chemical vapor deposition of graphene and related applications. *Acc. Chem. Res., 46*, 2329–2339.

Zhang, Z., Xu, F., Yang, W., Guo, M., Wang, X., Zhang, B., & Tang, J., (2011). A facile one-pot method to high-quality Ag-graphene composite nano sheets for efficient surface-enhanced Raman scattering. *Chem. Commun., 47*, 6440–6442.

Zhu, J., Liu, S., Palchik, O., Koltypin, Y., & Gedanken, A., (2000). Shape-controlled synthesis of silver nano particles by pulse sonoelectrochemical methods. *Langmuir, 16*, 6396–6399.

Index

Printed and bound by CPI Group (UK) Ltd, Croydon, CR0 4YY

23/10/2024

01777702-0014